AF869526

Yong Huang
Jinlong Yang

Novel Colloidal Forming of Ceramics

Yong Huang
Jinlong Yang

Novel Colloidal Forming of Ceramics

With 324 figures

TSINGHUA UNIVERSITY PRESS

Springer

Authors
Prof. Yong Huang
Department of Material Science and Engineering, Tsinghua University
100084, Beijing, China
E-mail: hy-dms@tsinghua.edu.cn

Prof. Jinlong Yang
Department of Material Science and Engineering, Tsinghua University
100084, Beijing, China
E-mail: yangjinlong999@126.com

ISBN 978-7-302-21324-6
Tsinghua University Press, Beijing

ISBN 978-3-642-12280-4
e-ISBN 978-3-642-12281-1
Springer Heidelberg Dordrecht London New York

Library of Congress Control Number: 2010924747

© Tsinghua University Press, Beijing and Springer-Verlag Berlin Heidelberg 2010
This work is subject to copyright. All rights are reserved, whether the whole or part of the material is concerned, specifically the rights of translation, reprinting, reuse of illustrations, recitation, broadcasting, reproduction on microfilm or in any other way, and storage in data banks. Duplication of this publication or parts thereof is permitted only under the provisions of the German Copyright Law of September 9, 1965, in its current version, and permission for use must always be obtained from Springer-Verlag. Violations are liable to prosecution under the German Copyright Law.
The use of general descriptive names, registered names, trademarks, etc. in this publication does not imply, even in the absence of a specific statement, that such names are exempt from the relevant protective laws and regulations and therefore free for general use.

Cover design: Frido Steinen-Broo, EStudio Calamar, Spain

Printed on acid-free paper

Springer is a part of Springer Science+Business Media (www.springer.com)

Yong Huang
Jinlong Yang

Novel Colloidal Forming of Ceramics

Yong Huang
Jinlong Yang

Novel Colloidal Forming of Ceramics

TSINGHUA UNIVERSITY PRESS

Springer

About Authors

(1) Yong Huang

Born in 1937, graduated from Tsinghua University in 1962. He has been engaging in inorganic nonmetallic material teaching and research for many years. He was the visiting scholar at the Department of Materials Science and Engineering in the University of Michigan in U.S.A., the research scientist at the material center of Massachusetts Institute of Technology in America, the visiting professor at the Department of Materials in Swiss Federal Institute of Technology (ETH Zurich), and also carried out a cooperative research with the University of Monash in Australia. His research interests include the composing, structure and performance of the advanced ceramic, as well as the preparation of high performance ceramic and the strengthening mechanism. He has won many top honors, including the second class National Technology Invention Award for a research program by State Council of China and "Science and Technology Advancement Prize" for other nine programs awarded by the Ministry of Education and the government of Beijing, Fujian Province respectively. He has delivered more than 400 papers, got 25 state patents for invention, published 14 teaching materials, text book, handbook etc. He was rewarded a national excellent text book and the first class award of excellent text book for building materials. He also was rewarded for the excellent teacher of Beijing, advanced person of the State Plan for Development of Basic Research and the winner of special government allowance.

(2) Jinlong Yang

Born in 1966, graduated from Beijing Institute of Technology in 1987, graduated and got the master degree from North University of China (NUC) in 1990, got the doctor degree from Tsinghua University in 1996. He had the post-doctoral experience at Department of Materials in Swiss Federal Institute of Technology (ETH-Zurich) from 1999 to 2000. He became a full professor in Tsinghua University in 2004, and had the position of special-engaged professor and director of laboratory of advanced ceramics in NUC since 2006.

Research fields include: structured ceramics, ceramic matrix composites, colloidal forming technology of ceramics and laser green machining of ceramics. He had won some top honors, including the second class prize of National Technology Invention for a research program by State Council of China, and "Science and Technology Advancement Prize" for other three programs awarded by the Ministry of Education and the government of Beijing respectively. He has delivered more than 100 papers, and got 30 state patents for invention.

Introduction

In ancient times, ceramic vessels or crafts were usually manufactured using clay-based natural raw materials. It is well known that the mud mixed clay with water had a good plasticity, and can be easily processed into various products with different shapes. However, their forming techniques used earlier were mainly manual procedures. So, to a large extent, the forming of ceramic wares was just as a kind of skill or workmanship. By the 1960s, new ceramic materials had already developed into an independent scientific system. At the same time, the raw materials that were used to manufacture ceramics began to transition from the clay-based system to the one with accurate chemical composition. In particular, for preparing high-performance ceramics, the synthetic chemical raw materials, such as Al_2O_3, ZrO_2, Si_3N_4, SiC and so on, were mainly used. These ceramics had excellent properties because of their structure characteristics of covalent bond and ionic bond, and were widely considered as candidate materials in many fields requiring high-temperature resistant, wear-resistant, and corrosion resistant substances. It had been predicted that such materials would be developed rapidly, and various new types of materials with excellent properties would be explored.

By the late 1970s, the emergence of the worldwide oil crisis caused many developed countries led by America and Japan to draft national development plans for high-temperature structural ceramic materials used in the field of internal combustion engine especially the automobile engines. With excellent properties such as resistance to high temperature, wear, and corrosion, the high-performance ceramics were acted as the optimal material for non-water cooling and adiabatic ceramic engine parts. The forming technique of ceramic was also been ranked the important research topics.

During the period of the "7th Five-Year Plan" (1986 – 1990) and "8th Five-Year Plan" (1991 – 1995) in China, around the key components of ceramic insulation engines, an in-depth research on the ceramic injection molding, extrusion molding, slip casting, and pressure filtration was done. Moreover, a few engine parts samples with high performance were prepared. However, because of the high cost, poor performance repeatability, and low yield, the process of industrialization of high-performance ceramics was greatly restricted. After years of research and development, there was a growing recognition as the key technology for forming high-performance ceramic materials and parts. The forming technique is not only the precondition for materials design and formulation, but also the important factor in reducing the manufacturing costs, improving the yield and the performance repeatability of the products. Simultaneously, several research booms were set off around the new forming technique of ceramics at home and abroad. During the period of the "9th Five-Year Plan" (1996 – 2000), in order to achieve high-tech ceramic industrialization as soon as possible, the ceramic forming technique with high performance and low cost was granted special funds from the "863 Program" (the National High-Tech Research and Development Program of China).

Along with the research upsurge in the field of ceramic engines, there was also considerable interest in the field of the injection molding of ceramics. On the basis of the theory of plastic injection molding, thermoplastic, thermosetting, and water-soluble organic compounds were used as binders, and then mixed with ceramic powder to prepare a suspension with high solid loading. In addition, ceramic parts with high size precision and complex shape can be prepared by injection molding. It is suitable for automotive and large-scale production. After several decades, owing to in-depth studies on ceramic injection molding, it developed into an integrated science and technology involving rheology, the dynamic molding process of injection suspension, and thermal degradation of organic compounds as well as other interdisciplinary technology. However, some problems which were caused by organic enrichment or particle rearrangement were exposed during the time and energy consuming process of de-bindering, such as poor uniformity and easy cracking. Therefore, the de-bindering was considered as the key issue to be solved, and the solution to lower the organic content gradually became an important research topic. In order to simplify the process of de-bindering, the low pressure injection molding with some small molecular organics was paid more attention.

After the 1990s, quickset injection molding was invented by B. E. Novich of the U.S.A. The pore fluid was used as a carrier in the process, its volume did not change with temperature. After the suspension was injected into the container, the carrier was sublimated, and then the green body was solidified by controlling the temperature and pressure. Because of avoiding the polymer organic carrier with large molecule, the problem of organic de-bindering was solved ingeniously. At present, due to those significant advantages such as high automation and good size precision, the injection molding continues to be highly used and is considered to be a highly competitive forming process.

In the mid-1980s, in order to avoid the difficulty of de-bindering in the injection molding, which was caused by the large number of organic binders, the traditional slip casting attracted more interests, as it involved less organic content and low cost. Moreover, the operation and control were easy in this method. However, because of the green body with low green density and poor strength, it was not suitable for high reliability ceramics. On the basis of traditional slip casting, the pressure filtration and centrifugal casting techniques were developed thereafter. The green body's density and strength were improved by applied pressure and centrifugal force, and at the same time the complicated de-bindering process was avoided. However, such processes were also unable to meet the production of the green bodies with high reliability and high performance due to poor uniformity of the green bodies.

After the 1990s, in order to improve the uniformity and reliability of ceramic bodies, the forming *in-situ* techniques, such as gel-casting, temperature induced flocculation, colloidal vibration casting and direct coagulation casting were developed. The *in-situ* solidification process is highly regarded because it is the precondition to ensure the uniformity and is an important way to improve the reliability of ceramics.

Gel-casting, a novel colloidal forming technology of ceramics, was first invented by Oak Ridge National Laboratory (ORNL), U.S.A., in 1990. In the gel-casting process, about 2 – 4 wt% acrylamide monomer is added into the ceramic suspension, and then it polymerizes *in-situ* by the interaction of catalysts and initiators. Furthermore, the drying process should take place at room temperature and high humidity for a long time, otherwise the green bodies would easily crack. In addition, the degree of automation and industrialization of the gel-casting process is poor when compared to the injection molding. The ceramic bodies prepared by gel-casting have the obvious advantages of high strength and excellent machinability. Thus, some ceramic parts with complex shape, or that are difficult to be demoulded, just like inside screw, can be passed through green body machining after drying to achieve the required shape and precision. As a kind of brittle and difficult-to-machine material, it is very important, even necessary for the bodies to be machined partly, which also provides people a very good idea.

Temperature-induced flocculation was developed by L. Bergstrom at the Institute of Surface Chemistry of Stockholm University, Sweden, in 1993. A special amphoteric polyester surfactant or dispersant changing with temperature is introduced into the concentrated suspension to make particles to disperse, one end of the dispersant is adsorbed on the particle surface, the other end goes inside the solvent. As the temperature is reduced, the solubility of dispersant declines and the function of dispersion fails and then the suspension is flocculated *in-situ*. The outstanding advantage of the method is the recycling of the unqualified green bodies, but it is restricted to use such dispersant for the different ceramic system.

Colloidal vibration casting was first developed by F. F. Lange at the California University in Santa Babara, U.S.A., in 1993. The prepared dilute suspension (20 – 30 vol%) with high ionic concentration will be pressure-filtered or centrifuged to obtain the green body with high solid loading, which is in the solid state under static conditions but is in the fluid state if certain external force (such as vibration) are applied. Then, the suspension will solidify *in-situ* under static conditions after it being poured. The advantage is the use of the thixotropic property when the concentrated suspension has a high ionic strength. Moreover, the concentrated suspension don't need to prepare with high solid loading. However, the green bodies have poor strength and are prone to cracking and deformation.

Direct coagulation casting was invented by the research team of Gauckler at the Swiss Federal Institute of Technology in Zurich, Switzerland in 1994. In this method, first, a biological enzyme and a substrate are introduced into the ceramic concentrated suspensions at low temperature (~5℃). At this time the enzyme is in the inactivated state and does not almost react with the substrate. Then, the temperature of the suspension rises from 0 – 5℃ to 20 – 40℃, and the enzyme is activated and reacts with substrate. By adjusting the pH to the isoelectric point or by increasing ionic strength, the suspensions is coagulated *in- situ*. This method results in wet bodies with enough strength to be demoulded. In the process, it is necessary that the solid loading of suspension is more than 50 vol%, but the strength of the wet green body is low. However, the body is extremely uniform and does not contain any organic binder. Thus, it is suitable for preparing high reliability ceramics, for which the Weibull modulus can reach 40.

Thus, after the 1990s, the research on the *in-situ* forming technique had become a new hotspot in the field of high-performance ceramics. It is gradually realized that the forming technique take a very important position in the study of ceramic materials, and that the industrialization process of high-performance ceramics would be promoted greatly through the in-depth study on the forming technique and its basic theory. In the mid-1980s, the injection molding and pressure filtration had been studied at Tsinghua University in China. After the 1990s, the study of gel-casting and direct coagulation casting had been progressed significantly, especially with the invention of the novel injection molding for water-based nonplastic slurry of ceramic, also called as colloidal injection molding. A colloidal injection molding machine was also developed. The development of gel-tape-casting process and gel-casting with low toxicity system was studied in detail in about five years. Recently, freeze-gel-casting technique and colloidal forming of ultra-light and high strength porous ceramic have been invented, which further enrich the theory and technique of ceramic forming. The industrialization process of high-performance ceramics is also being promoted greatly. In China, other universities and institutes such as Shanghai Institute of Ceramics of Chinese Academy of Sciences and Tianjin University, etc., are doing the similar research. The key technology and development trends, regarding the colloidal forming of ceramics, include the following aspects:

(1) The preparation of a concentrated suspension with low viscosity and high solid loading is the precondition to guarantee the density, uniformity, and strength of the green bodies. With high density, the shrinkage of the green bodies can be

decreased in drying, and the deformation and cracking can be avoided in the sintering process. Therefore, the foundation of colloidal forming is to prepare the suspension with high solid loading.

(2) The *in-situ* consolidation technique. In a sense, a new consolidation technique of suspension means a new forming process. It is very important for colloidal forming to look for a new solidification method of suspension. For *in-situ* forming, the relative position of particles does not change during the solidification of suspension. It is a necessary condition to guarantee the body uniformity, which is also a key factor to improve the reliability of ceramic materials. Currently, the development of an *in-situ* consolidation technique requires the suspension to contain few or no organic substances. By varying the charged and dispersant characteristic between colloidal particles, the viscosity of suspension is increased and this leads to *in-situ* solidification of the green bodies. At the same time, with the enough strength to be demoulded, the ceramic green bodies can be easily transported during a large scale production.

(3) It is very important to avoid the residual stress in green body by using the colloidal forming. Normally, there is no shrinkage during forming suspension to green body, and no residual stress in green body. So we believe that the colloidal forming without shrinkage will become an important trend in the next 10 – 20 years.

(4) Near net size forming. Ceramics are the hard-machined material with high hardness and brittleness. The sintered bodies should be close to the actual size of the final parts in order to decrease machining of ceramics.

The preparation process of high-performance ceramic materials and parts is very important for development and application of ceramics. It involves the preparation of high-performance ceramic powders, and the forming and sintering processes. The quality of green and sintered body is directly impacted by the quality of the powders because the rheological behaviors and solid loading of concentrated suspension are directly determined by the properties of the powder during colloidal forming. Moreover, the quality of green bodies guarantees the quality of sintered body. The preparation of powders, and the forming and sintering of ceramics are mutual constraints and complementary. However, according to the current development of ceramic forming, the forming process, which plays an important role in the preparation process of ceramic materials and is related with the industrialization and scale of production, is the critical step to guarantee the performance reliability and repeatability as well as the yield of ceramics. Hence, it is of great significance to research on the forming technique of ceramics with low cost and high reliability, and it will promote the industrialization of high-performance ceramics in the world. This is not only the requirement of governments around the world and of industries, but also the urgent need of ceramic scientists.

To sum up, the green bodies prepared by an ideal forming process should have a good uniformity, a high green density, enough strength to be demoulded, and no residual internal stress. The ceramic materials with high performance can be obtained by using the forming process, the homogeneous of shrinkage would be guaranteed and deformation avoided during the sintering process. In order to meet the above requirements, the forming process from a technical perspective should have the following characteristics:

(1) Adding no or as few as possible organic polymers (0.1 – 4 wt%).

(2) The solid loading of suspension is more than 55 vol%.

(3) The solidification of suspension is a rapid and *in-situ* technology.

(4) The suspension has no shrinkage and internal stress during the transformation from liquid to solid state.

(5) A kind of near net size forming process.

The first point is to avoid the problem of debindering. The second point is to improve the density and the strength of the green bodies. The third point is to guarantee the uniformity and the dimensional accuracy of the green bodies. The fourth point is to avoid deformation and cracking of the green bodies during burn-out and sintering, in order to improve the reliability. The fifth point is to guarantee the dimensional accuracy of sintered products, so that there is no or less machining.

There are seven chapters in the book. The main contents involve our research achievements and developments in two decades. Recent international research results in the field of ceramic forming are also introduced. Finally, the basic problems and the future development trends of forming process are proposed and discussed in detail.

The book writing was completed by Yong Huang and Jinlong Yang.

It should be noted that professor Zhipeng Xie and his graduate students achieve the excellent results in the field of colloidal forming process, which enriched the content of this book. Meanwhile he was also personally involved in the compilation and modification of Chapter 3 in this book. Thus we are appreciated his positive contribution for the publication of this book.

Besides Prof. Xie, Xiaoqing Xi, Tian Ma and Yali Wang helped to compile and check for some chapters of the book. At the writing period of this book, Dr. Tao Zhang (Kingston University, United Kingdom), Dr. Hui-rong Le (the United Kingdom Dundee University), Chang-an Wang, Yang Sun, Liguo Ma et al. proposed some revision opinions to some chapters of the book. The special thanks are given to them for their support and help during finishing this book.

Yong Huang, Jinlong Yang

Contents

Chapter 1 Aqueous Colloidal Injection Molding of Ceramics Based on Gelation

Abstract In this chapter, an aqueous colloidal injection molding of ceramics (CIMC) which is based on gelation of monomer polymerization, like a reactive injection molding of double-slurry of polyester, is discussed systematically. The solidification mechanism is based on gelation of monomer polymerization. It was found that pressure can induce the polymerization of monomer. The gelation time can be effectively controlled by temperature and pressure. The testing equipment of gel point was developed and used to test gel reaction of monomer in suspension under different pressures. A machine for ceramic colloidal injection molding was developed. The copper was proved to accelerate the polymerization of monomer. Therefore, the copper parts are forbidden for use in a CIMC machine.

Keywords gel-casting, injection molding, colloidal injection molding, pressure induced forming, fast mixing with double suspensions, ceramic slurry

High-performance ceramics are becoming increasingly more important in several fields, such as spaceflight, energy, mechanics, and bio-techniques. But the high cost and low reliability have prevented high-performance ceramics to be utilized on a large scale until now. To solve this problem, more attention is being paid to the forming process of ceramics. In the industrialization of high-performance ceramics, forming has become a bottleneck and needs to solve. Colloidal forming is an important forming technique. It includes slip casting, tape casting, direct coagulation casting (DCC), injection molding, gel-casting and so on. Among these techniques, gel-casting and injection molding are considered the two possible solutions to address the industrialization of high-performance ceramics. Though they both have many advantages, there are still several problems to be solved in the industrialization process.

Gel-casting is a new ceramic forming technique that is generating worldwide attention. The process is based on the casting of slurry, containing powder, water, and water-soluble organic monomers. After casting, the mixture is then polymerized to form gelled parts. Drying, burning out and sintering complete the manufacturing process. The process is generic and can be used for a wide range of ceramic and metallic powders. It is a suitable technique for the fabrication of near net shape prototypes or small series using cheap moulds. In contrast with slip casting, gelled parts are more homogeneous and have a much higher green strength.

Y.Huang et al., *Novel Colloidal Forming of Ceramics*
© Tsinghua University Press, Beijing and Springer-Verlag Berlin Heidelberg 2010

Gelcast parts contain only a low percent of organic components, thereby making binder removal much less critical compared to injection molding. The advantages of gel-casting can be summarized as follows:

(1) Capability of producing complex parts like injection molding;

(2) Ease of implementation owing to its similarity with other well-established processes like slip casting;

(3) Low capital equipment cost;

(4) Possibility to use low-cost mould materials;

(5) Capability of implementation for mass production;

(6) High green strength;

(7) Excellent green machinability that allows machining much finer details that in cold static pressing (CIP) parts;

(8) Highly homogeneous material properties;

(9) Low organic content that translates into easy binder removal;

(10) Generic method applicable for both ceramic and metal powders.

Meanwhile, there are still some disadvantages that make it difficult to realize the industrialization by gel-casting. Most important of all, low automation prevents gel-casting from being used in the industrialization of high-performance ceramics (Omatete O. O., et al., 1991; Gilissen R., et al., 2000).

Injection molding, however, has been used in the ceramic industry for several decades for its high automation. Ceramic injection molding is a well-established shaping technique, which involves the mixing of ceramic powder with a large concentration of a melt polymer (up to 50 – 60 vol%). The carrier polymer provides very high viscosity, and so very high pressures ((150 ± 10) MPa) and temperatures ((200 ± 120)℃) are needed for injection. In addition to the high cost derived from the use of organics, the major problem arises from the debindering step that can easily lead to defects and failure of the sintered body. In the preparation of products with complex shape and big cross section, the problem is more obvious. These disadvantages prevent injection molding from being used in the industrialization of high-performance ceramics (Novak S., et al., 2000; Wei Wen-Cheng J., et al., 2000; Liu D. M., 1999; Krug S., et al., 2000).

The aim of this study is to develop a new forming technique to meet the needs of industrialization. The new technique is called colloidal injection molding of ceramics (CIMC).

1.1 Colloidal Injection Molding

1.1.1 The Concept of CIMC

Traditional injection molding can be used to form complex ceramic parts with high-dimension precision, so that it can be applied to realize the automation of large-scale production of ceramic products in industries. The problem with this technique is that, due to the high organic content, debindering becomes very difficult. The colloidal forming processes such as gel-casting and DCC can avoid this disadvantage, and improve the uniformity of green bodies and the reliability of the ceramics products. These approaches, however, have other shortcomings. For example, they need manual operation and thus are difficult to be used in the ceramic industries where automation is a necessity. An optimal approach would be a combination of both injection molding and colloidal forming processes, bringing their advantages together. The problem is that the injection molding is a plastic forming process, and the colloidal forming process starts from slurry. Since the slurry has no plasticity in colloidal forming, it is normally impossible to apply injection molding.

We have developed a new technique, namely CIMC (see Fig. 1.1; Yang Jinlong et al., 2002; 2005), combining the injection molding and the colloidal forming process together, to produce complex ceramic parts of high reliability, by using a colloidal injection molding machine invented by the authors. The new approach (with our machine) provides high automation of injection molding, high homogeneity of products, and low content of organics in the colloidal forming. It is possible to produce high-performance ceramic parts of complex shapes with high reliability by this newly developed technique, and thus it can be applied to ceramics industries.

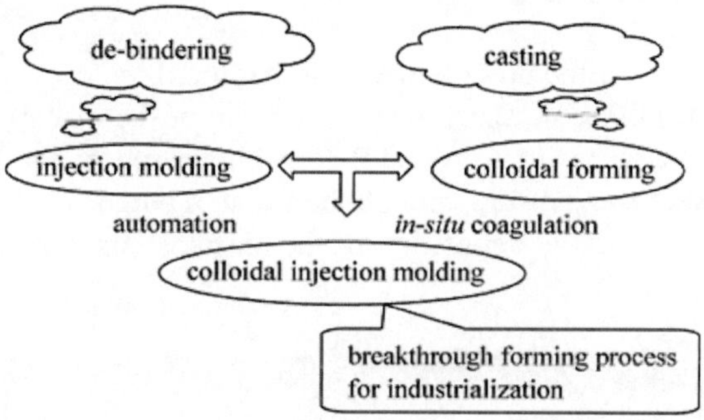

Figure 1.1 The idea for colloidal injection molding of ceramics

1.1.2 The Flowchart of CIMC

The fast uniform mixing and controllable colloidal forming process of ceramics could be used to divide ceramic suspension into two components A and B. The monomer was added into A, and the initiator was added into B. In this case, there was no reaction in either A or B because of segregation of the monomer and the initiator. The suspensions maintained good fluidity until they were mixed quickly and uniformly. Once A and B were mixed together, the encounter of monomer and initiator resulted in gel reaction in the suspension. The initiator quantity and the applied pressure were the factors dominating the solidification reaction. Figure 1.2 shows the schematic graph of this process. The catalyst in suspension A could not generate a reaction with the monomer. Its main function was to accelerate the reaction when the monomer and initiator mixed.

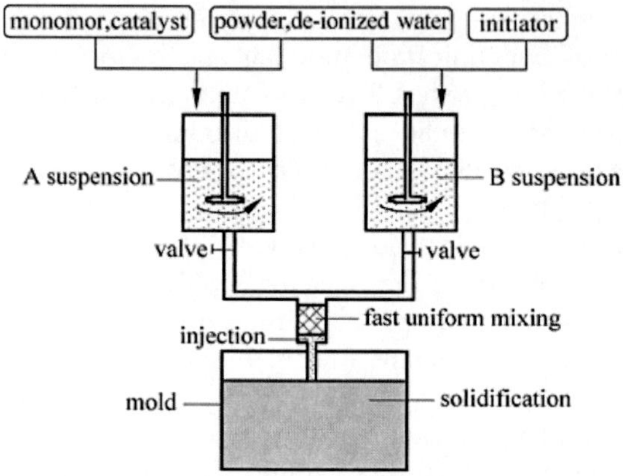

Figure 1.2 Schematic graph of colloidal injection molding of ceramics

1.1.3 The Machine of CIMC

A special equipment for the process was manufactured.

Figure 1.3 shows the flow chart for the working of the machine. The prepared suspensions were transferred into A and B container, respectively. In the containers, the suspensions were vacuumized through the vacuum pumps to completely remove the gas inside. To avoid the deposition of the suspensions, a stirring paddle was

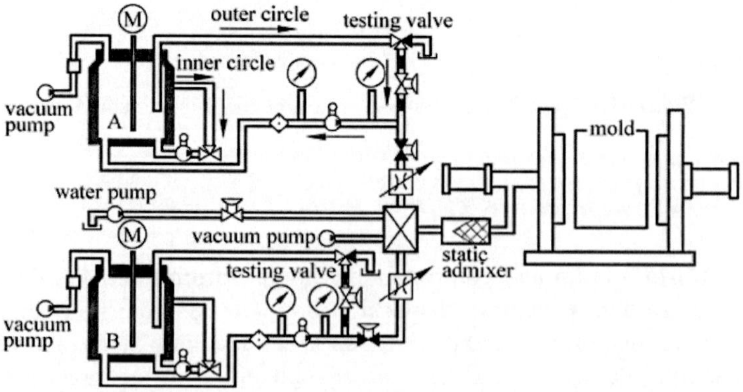

Figure 1.3 Flow chart for the working of the equipment for colloidal injection molding of ceramics

used in each of the containers to continuously intermix the slurries. Furthermore, with the use of bellow pumps, the slurries could circulate in two routes: inner circle and outer circle. This circular transportation in addition to the stirring action on the slurries could completely prevented the occurrence of deposition. The suspensions state was monitored through the testing valves, which were positioned on the outer circle line.

Before injection molding, the metering pumps quantified two slurries of the same volume. The two slurries were simultaneously and quickly injected into the static admixer with a high pressure of 5 – 8 MPa. The core status of the static admixer can be seen in Fig. 1.3. It is similar to a bridge connecting the machine and the mold. Its main function was to realize fast uniform mixing of A and B suspensions. The inner structure of the device can be seen in Fig. 1.4. From this chart, we can observe that the mixing process was realized by a series of mixing units with different shapes in the hollow tube. These units made the penetrating slurries rapidly levo-rotate and dextral-rotate alternately. These frequent changes of the slurries' flowing directions could help in obtaining a good mixing effect, which is called fast uniform mixing.

These mixed suspensions were injected into the mold under a pressure of 3 – 8 MPa. The demoulding agent is usually pre-coated onto the wall of the stainless steel mold. The initiator volume and injection pressure are the dominative factors to control the solidification process. The separation of monomer and initiator guaranteed no reaction in separated suspensions, which had gained enough operating time for the process. By adjusting the initiator volume and injection pressure in the process, the solidification speed of the suspension in the mold could be effectively controlled, which explains the meaning of controllable colloidal forming. In the process, the crucial impact of temperature on gel reaction was completely avoided.

It is well known that the pressure distribution is homogenous in flowing suspensions. And the fast uniform mixing results in the physical uniformity in the suspensions. The combination of these two points can ensure simultaneous solidification in different parts of the suspensions. The synchronous solidification considerably decreased the inner stress in the green bodies.

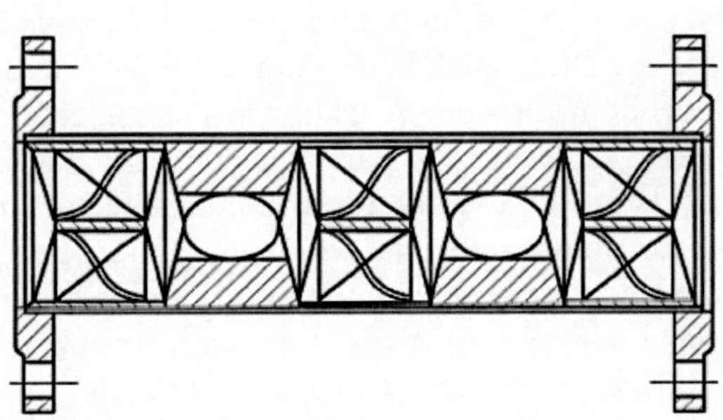

Figure 1.4 The inner structure of the static admixer

1.2 Pressure Induced Forming

1.2.1 Effect of Hydrostatic Pressure on Solidification

A pressure induction unit was then added to apply an external pressure after injection. The gelation curves of Al_2O_3 slurry under different pressures with a mould temperature of 25 and 36℃ are shown in Fig. 1.5. Under increased pressures, the onset time of gelation becomes shorter and the gelation speed increases significantly. At higher temperature, the influence of the pressure becomes more significant and the gelation finishes almost immediately. Pressure-induced solidification has distinct advantages over temperature-induced solidification.

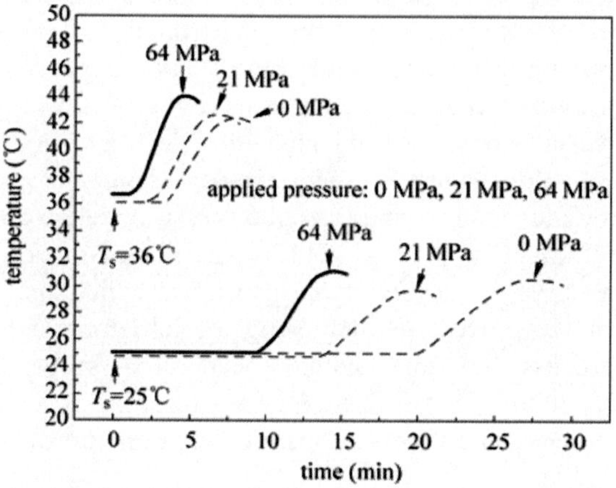

Figure 1.5 The effect of external pressure on the gelation of Al_2O_3 slurry with a solid loading of 50 vol% (T_s is the starting temperature)

1.2.2 Homogeneity of the Green Bodies

An alumina (Al_2O_3) green body of turbine shape with a diameter of 105 mm was prepared by this process. The spatial distribution of density is compared with that obtained by the conventional process by using the same injection molding system shown in Fig. 1.6. The value in each part represents the density in g/cm^3. The relative standard deviation of density for colloidal injection molding is 0.2% compared to 0.7% for conventional injection molding. The results show that colloidal injection molding process significantly improves the homogeneity in density of the green body. Nevertheless, the solidification on this process is associated with the surface temperature of the mould. The gradient of the temperature will inevitably cause non-uniform solidification and inner stress in the green body. A new strategy has been developed to minimize the inner-stress in green bodies prepared in this process.

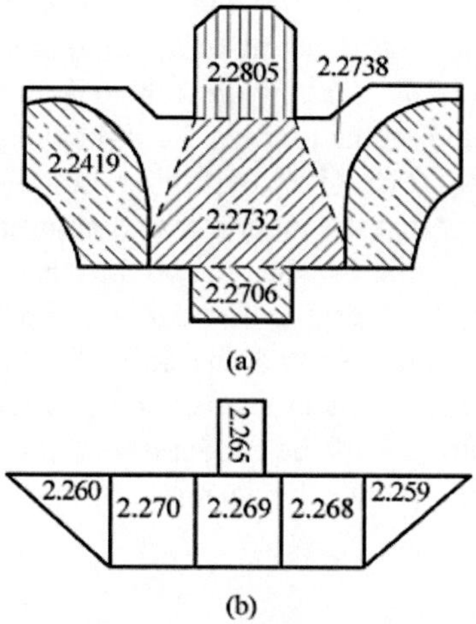

Figure 1.6 The comparison of density distribution in Al_2O_3 green bodies of turbine prepared by (a) conventional injection molding and (b) colloidal injection molding on the same system

1.2.3 Controlling the Inner Stress in the Green Body

The inner stress in green body is often responsible for the initiation of microcracks during the subsequent drying and debindering processes. During colloidal injection molding or other gel-casting processes, monomers in ceramic suspension polymerize and then form gel networks to hold the ceramic particles. The solidification speed increases with increasing temperature for a given composition. Non-uniform solidification occurs due to the temperature gradient in the ceramic suspension and results in the development of inner stress in the green body. Theoretical analysis has indicated that the magnitude of the inner stress increases with the stiffness of the green body. Based on this finding, a fraction of the monomers is replaced by a moderator, hydroxylethylacrylate in the suspension during gel-casting or colloidal injection molding to control the stiffness of the gel network. Test bars with dimensions of $6\times6\times42\ mm^3$ were prepared with 50 vol% solid loading and various amounts of the moderator from 0 to 100 wt% in the total monomers. The samples were dehydrated at about 70℃ in an electrical furnace. The flexural strength and elastic modulus of the green body are evaluated using three-point bending. The results are shown in Fig. 1.7 against the amount of the moderator. Both the flexural strength and elastic modulus of green body decrease simultaneously with the amount of the moderator. This reveals that the strength of polymer network is reduced when the harder polymer chain is relaxed by the incorporation of the shorter chain molecules of the moderator. Five alumina disks, a, b, c, d, and e, were produced under the same conditions. The diameter of the samples is about 50 mm and the thickness is about 4 mm. The identifiers of the disk samples are marked above the corresponding amount of the moderator in Fig. 1.7. Figure 1.8 shows the surface patterns of the alumina green bodies after drying and debindering to remove all the water and organic binders. Radial cracks were found on samples (c, d, and e), while samples (a and b) with higher amount of moderator were free of visible cracks. It is likely that the inner stress was initiated at the forming stage and magnified during drying process when the preform becomes harder. The intrinsic strength of the green body after drying and debindering is mainly determined by the nature of the ceramic powder and the solid loading. Observations of the surface patterns confirm that the inner stress and the elastic modulus of the green body decrease with the amount of the moderator. Choosing a proper amount of the moderator can effectively reduce the occurrence of cracks during the subsequent processes.

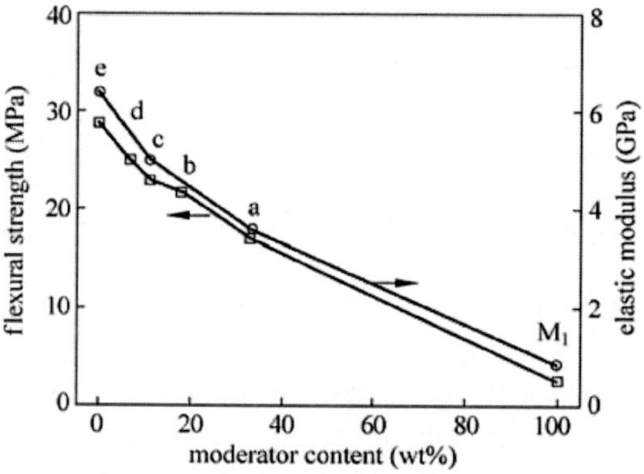

Figure 1.7 The strength and elastic modulus of alumina green bodies versus the amount of moderator

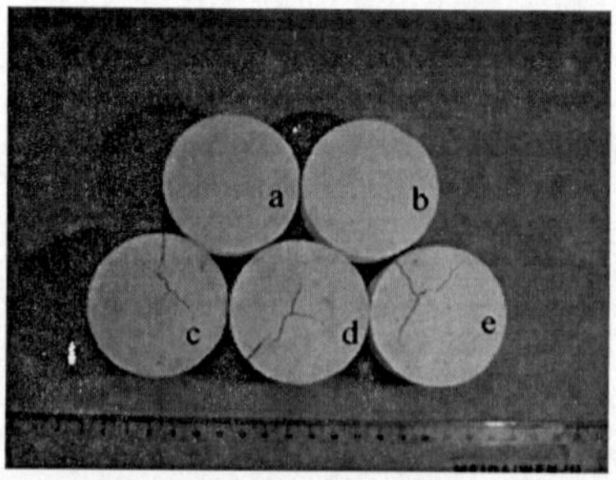

Figure 1.8 The surface patterns of alumina green bodies after drying and debindering

1.3 Storage Stability of Ceramic Slurries

1.3.1 The Importance of Storage Stability of Slurry

In this section, the induced time of polymerization was used to judge the store stability of slurry. Because acrylamide polymerization is an exothermic reaction, gelation process can be determined by the change of temperature.The metering system (Ma Liguo, et al., 2002) is shown in Fig. 1.9. At first, ceramic slurries with high solid loading were prepared by long-term ball milling. Then the slurry was poured into the mold, and the mold was placed in constant temperature water until its temperature became equal to scheduled temperature. The initiator and catalyst were added to slurry later. The interval from this addition to the beginning of polymerization was defined as induced time. When monomer was polymerized, the temperature of slurry rose because of exothermal polymerization. During this process, temperature of slurry was detected by a temperature detector and recorded by a computer. Based on this temperature, the polymerization process could be analyzed.

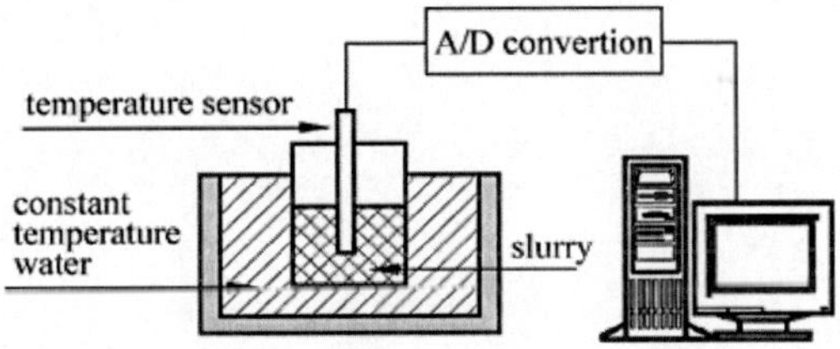

Figure 1.9 Metering system for polymerization process

1.3.2 Chemical Stability

During the research on gel-casting it was found that after long-term storage, even though the slurry had not been polymerized, the solidification properties of this ceramic slurry had been modified greatly. Sometimes the induced time was prolonged, and sometimes it was shortened. This affected the repeatability and stability of forming process. Hence, it was an obstacle for industrialization of gel-casting.

It is known that acrylamide in the slurry is prone to hydrolysis and thereby acrylic acid is generated. If the slurry was stored under strong light or there were some impurities in the slurry, the process of hydrolysis would be accelerated. In our research, it was found that acrylic acid has great effect on solidification of ceramic slurry. The influence of minute amounts of acrylic acid on the induced time and reaction time of acrylamide polymerization is shown in Fig. 1.10.

From the Fig. 1.10, it can be seen that the solidification process is accelerated with very low concentration of acrylic acid. However, the reaction is decelerated when acrylic acid concentration is more than 15 mmol/L. Both its induced time and reaction time were prolonged. The mechanisms may be explained as follows. Acrylic acid is weak acid. Even if there is only minute amounts of acrylic acid in the slurry, the acidity of slurry still increases. When acrylic acid concentration is 3 mmol/L, the pH of slurry changes to about 3.5. More acrylic acid leads its pH change to 3.1 slowly. It has been found that pH of slurry has great effect on polymerization initiated with ammonium persulphate (APS) and TEMED. On the one hand, hydrogen ion acted as catalyst for decomposition of APS; on the other hand, hydrogen ion could react with TEMED and generate stable ammonium, then decomposition speed rate of APS decreases. Thus, when the concentration of acrylic acid is greater than a certain value, 15 mmol/L, the rate of polymerization decreases and the induced time becomes longer. Therefore, the ceramic slurry for gel-casting should be protected from light because the light can lead to acrylamide decomposing into acrylic acid. Before the slurry is used after long-term storage, its solidification properties have to be tested.

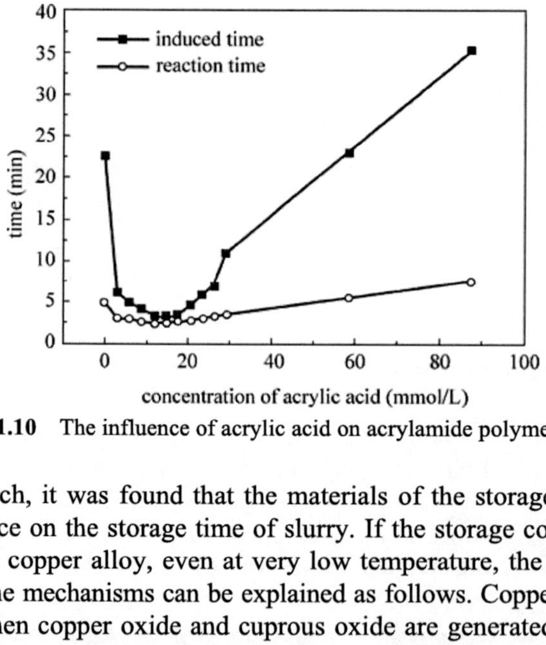

Figure 1.10 The influence of acrylic acid on acrylamide polymerization

In this research, it was found that the materials of the storage container also have an influence on the storage time of slurry. If the storage container is made up of copper or copper alloy, even at very low temperature, the slurry was also polymerized. The mechanisms can be explained as follows. Copper in air is prone to oxidation. Then copper oxide and cuprous oxide are generated on the surface of storage container. Copper oxide has no effect on stability of slurry. Nevertheless, the reducibility of cuprous oxide is strong, so it can react with the impurities present in the ceramic slurry thereby generating freeradicals. When all inhibiting agents are consumed, polymerization reaction occurs. Thus, the storage container or valve for gel-casting should not be made up of copper or copper alloy.

1.3.3 Inhibitor for Slurry Storage

The activity of acrylamide is so strong that it can self-polymerize under certain conditions. If the season was summer or large milling balls were used, the temperature of ceramic slurry could reach more than 50℃. The slurries are often polymerized during ball milling or storage. Even if the initiator and catalyst were not added into slurry, some free radicals would still be generated in the slurry. These free radicals reacted with oxygen in ceramic slurry. The peroxidized free radical is so stable that it cannot initiate polymerization. However, there is only about 20 wt% water in high solid loading ceramic slurry, so there is only little oxygen dissolved in slurry. The higher the environment temperature was, the faster the consumption rate of oxygen was. When all oxygen dissolved in ceramic slurry is exhausted, the slurry will be solidified unexpectedly.

In order to avoid this problem, inhibitors for gel-casting were studied. Many kinds of inhibitors, such as ferric chloride, phenol, diphenylamine etc., have been investigated. A compound inhibitor of phenothiazine and catechol has the best effect among these inhibitors at room temperature. The best mixture ratio for phenothiazine and catechol was about 1:5. Table 1.1 shows the inhibiting effect of this compound inhibitor on acrylamide polymerization.

Table 1.1 Inhibiting effect of phenothiazine-catechol on acrylamide polymerization

No.	temperature (℃)	phenothiazine (mmol/L)	catechol (mmol/L)	induced time τ (min)
1	30	0	0	2.6
2	30	0.025	0.125	5.8
3	30	0.050	0.250	7.2
4	30	0.075	0.375	12

([APS] =2.76 mmol/L, [TEMED] =40 mmol/L)

It has been observed that the induced time was extended greatly after the compound inhibitor was added into slurry. Now, the ceramic slurry can be stored for a long time even at higher temperatures. Because phenothiazine and catechol are organic substances, no impurity was left in ceramic components after sintering. The influence of phenothiazine and catechol on the rheological property of ceramic slurry was also discussed. The viscosity of slurry decreased with the addition of phenothiazine and catechol. Therefore, this compound inhibitor has not influence on the viscosity of ceramic slurry and is suitable for gel-casting.

1.4 To Prepare High Reliability Ceramic Parts with Complex Shapes: Aqueous Colloidal Injection Molding

The repeatability of the mechanical properties of final products has been greatly improved. A typical zirconia-toughened alumina (ZTA-50) and several grades of silicon nitride-based ceramics (Si_3N_4) were formed by colloidal injection molding. The ZTA was sintered at 1580℃ in air for 2 h. The Si_3N_4 green bodies were further cold isostatic pressed to improve the green density prior to a two-stage gas pressure sintering process. The first stage was performed at 1750℃ for 1.5 h under a nitrogen pressure of 0.3 MPa and the second stage at 1900℃ under a pressure of 6 MPa in N_2 for various holding time from 1 to 2.5 h. The flexural strength was measured by a standard three-point bending method for ceramics. The Weibull modulus was derived from 20 such tests on each material. As shown in Fig. 1.11, the Weibull modulus of ZTA-50 materials is as high as 24 and the optimum of Si_3N_4-based materials is 34 for the optimum holding time (2 h). As a reference, the typical value of these materials prepared by conventional forming processes is about 15. Therefore, the repeatability of the mechanical properties has been improved greatly. The microstructures of the sintered ZTA-50 and Si_3N_4 bodies were observed by using the scanning electron microscopy (SEM) as shown in Fig. 1.12. The micrograph of ZTA-50 confirms that the ZrO_2 particles are uniformly dispersed. Similarly, in the sintered Si_3N_4, the elongated β-Si_3N_4 grains are also well distributed. The improved uniformity of the microstructure is responsible for the increased repeatability of the fracture strength.

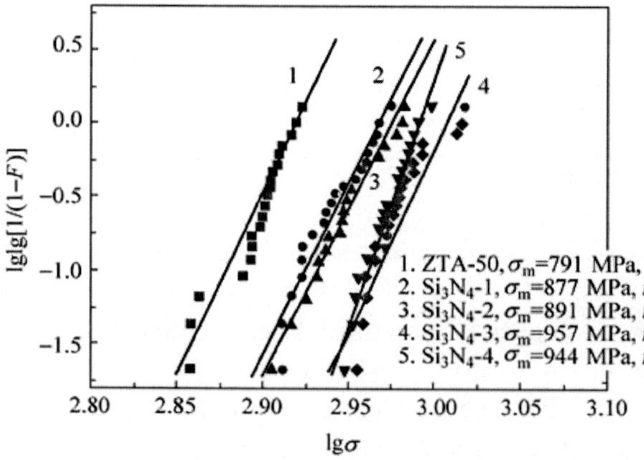

Figure 1.11 The strength and Weibull modulus of ceramics (σ is the flexural strength, F is the probability of rupture, σ_m is the mean flexural strength, m is the Weibull modulus)

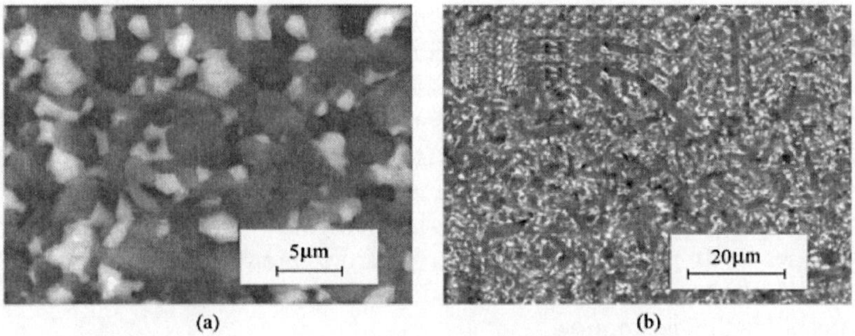

Figure 1.12 The microstructure of the sintered bodies by SEM
(a) ZTA-50; (b) Si_3N_4

As a new technique of ceramic forming, CIMC combines the advantages of gel-casting and injection molding, which helps to realize the injection molding of a ceramic aqueous suspension. It not only solves the problem of low automation in gel-casting, but also reduces the difficulty of debindering in injection molding. However, CIMC is not a simple combination of gel-casting and injection molding. Pressure induced forming is used in CIMC in order to eliminate the inhomogeneous polymerization due to the temperature gradient. The pressure-induced solidification occurs simultaneously within the slurry, by which the homogeneity of the green body can be improved and that the inner stress be reduced. The hydrostatic pressure helps to suppress the structural defects such as delamination or cracks. The solidification process can be achieved at room temperature so that the process can be simplified and easily automated for continuous production of a large batch. The pressure can be controlled precisely to optimize the speed of solidification and minimize the structural defects.

The inner stress in green bodies caused by nonuniform solidification in gel-casting processes can be significantly reduced by the addition of a moderator in the ceramic suspension. Combining this strategy with the colloidal injection molding process is a promising solution to form inner-stress free ceramic parts of complex shapes. These processing strategies have led to the improvement of the reliability of the products.

The process of fast, uniform mixing and controllable colloidal forming of ceramics was invented to completely avoid the crucial effect of temperature on colloidal forming of ceramics. The ceramic suspension for this process is divided into components A and B. The monomer is added into A, and the initiator is added into B. Under this condition, there is no reaction in either A or B because of segregation of the monomer and the initiator. The suspensions maintained good fluidity until they were mixed quickly and uniformly. The mixture solidified rapidly under the influence of the dominating factors of initiator volume and injection pressure. The new process has enough operating time and can be controlled expediently. The highly automatic equipment also had been developed for the process. The Al_2O_3 grinding balls with perfect physical properties were prepared by the corresponding equipment.

Storage of ceramic slurry has been one of the key problems in the industrialization of gel-casting. Except for the sedimentation of the ceramics particles, solidification properties of slurries were also affected after long-term storage. The induced time was prolonged or shortened because the acrylamide in the slurry hydrolyzed into acrylic acid. Even minute concentrations of acrylic acid has great influence on the polymerization of acrylamide. The material of the storage container also has influence on the storage time of slurry. The storage container should not be made up of copper or copper alloy. In order to store the ceramic slurry much safely, an inhibitor should be added into the slurry. A compound inhibitor of phenothiazine and catechol is the best option.

The advantages of CIMC can be summarized as follows:

(a) Easy de-bindering;

(b) Good homogeneity;

(c) Complex-shape parts with all kinds of cross section;

(d) Near net size forming;

(e) High automation.

With these advantages, CIMC can meet the needs of industrialization well. High performance ceramics parts of complex shapes with low costs can be prepared by CIMC.

References

Dai Chunlei, Yang Jinlong, Huang Yong, 2005. Storage stability of ceramic slurries for gel-casting. Key Engineering Materials, 280 – 283: 743 – 746.

Gilissen R., Erauw J. P., Smolders A., et al., 2000. Gel-casting—a near net shape technique. Materials and Design, 21: 251 – 257.

Huang Yong, Ma Liguo, Le Huirong, et al., 2004. Improving the homogenerity and realiablity of ceramic parts with complex shapes by pressure assisted gel-casting. Materials Letters, 58: 3893 – 3897.

Huang Yong, Ma Liguo, Yang Jinlong, et al., 2003. A new approach to prepare high performance ceramic parts with complex shapes: aqueous colloidal injection molding. Advanced Ceramics and Composites, 247: 39 – 44.

Huang Yong, Zhang Liming, Yang Jinlong, et al., 2007. Research progress of new colloidal forming processes for advanced ceramics. J. China Ceram. Soc., 35 (2): 1 – 10.

Krug S., Evans J. R. G., ter Maat J. H. H., 2000. Residual stresses and cracking in large ceramic injection moldings subjected to different solidification schedules. Journal of the European Ceramic Society, 20: 2535 – 2541.

Liu D. M., 1999. Control of yield stress in low-pressure ceramic injection moldings. Ceramics International, 25: 587 – 592.

Ma Liguo, Huang Yong, Yang Jinlong et al., 2002. Solidification course and its influence factors for gel-casting—gel point measurement of ceramics slurry and its iufluence factor. Jounal of Chengdu University (natural science). 21(2): 5 – 10.

Ma Liguo, Huang Yong, Yang Jinlong, et al., 2006. Control of the inner stresses ceramics green bodies formed by gel-casting. Ceramics International, 32 (2): 93 – 98.

Ma Tian, Huang Yong, Zhang Liming, et al., 2005. The 5th Pacific Rim International Conference on Advanced Materials and Processing, PTS1-5, 475 – 479: 1305 – 1308.

Novak S., Dakskobler A., Ribitsch V., 2000. The effect of water on the behavior of alumina-paraffin suspensions for low-pressure injection molding (LPIM). Journal of the European Ceramic Society, 20: 2175 – 2181.

Omatete O. O., Janney M. A., Strehlow P. A., 1991. Gel-casting—a new ceramic forming process. Am. Ceram. Soc. Bull., 70(10): 1641 – 1649.

Wei Wen-Cheng J., Wu Rong-Yuan, Ho Sah-Jai, 2000. Effects of pressure parameters on alumina made by powder injection molding. Journal of the European Ceramic Society, 20: 1301 – 1310.

Yang Jionlong, Dai Chunlei, Huang Yong, 2005. Controllable forming technology in gel-casting. Materials Science Form, 475 – 479: 1325 – 1328.

Yang Jinlong, Dai Chunlei, Su Liang, et al., 2005. Technology and equipment on colloidal injection molding of ceramics. Key Engineering Materials, 280 – 283: 739 – 742.

Yang Jinlong, Su Liang, Ma Liguo, et al., 2002. Colloided injection molding of ceramics. Key Engineering Materials, 224 – 227: 667 – 671.

Chapter 2 Gel-Tape-Casting of Ceramic Substrates

Abstract In this chapter, a new aqueous gel-tape-casting processing is introduced systematically. The solidification mechanism and influence factors of the gel-tape-casting are analyzed. Alumina green tapes have been prepared through gel-tape-casting successfully. Experimental results showed that the efficiency of styrene-acrylic binder is obvious for water-based tape casting. Sodium alginate has been successfully used to consolidate alumina tapes from aqueous suspensions. Based on gel-tape-casting, another novel process, spray inducing instant solidification of gel-tape-casting, was developed to overcome the limitation of the operating time. This process is fit for forming of ceramic sheets with thickness less than 0.3 mm. The researched works have demonstrated that the ceramic sheet products fabricated by gel-tape-casting process have good quality, low pollution, low costs, and significant social and economic benefits.

Keywords gel-tape-casting, slurry, ceramic sheet, substrate, styrene-acrylic latex, sodium alginate

Ceramics substrates (ceramic sheets) are important electrical component materials that possess many outstanding characteristics such as the high-temperature resistance, wear-resistance, corrosion-resistance, anti-oxidation, light weight and insulation, and so on. So they are widely applied in integrated circuits, ceramic capacitors, piezoelectric ceramic devices, layered composite materials and many other areas. Generally ceramic green sheet can be produced through various forming methods (Qin Zhengqi, et al., 1996) as follows: ① extrusion forming; ② dry-pressing; ③ rolling film forming; ④ screen printing forming; ⑤ tape casting, etc. However, all the methods used presently have their limitations for mass production of ceramic sheets even though some new forming methods and technologies will be further developed and perfected. The tape casting is considered as the main manufacturing method for large-scale ceramic sheet production.

Tape casting technology first appeared in the 1940s during the second world war when there was a serious shortages of the quartermaster materials to produce mica capacitors. In order to alleviate this problem, people began using tape casting to produce ceramic sheets for ceramic capacitors.

Tape casting (cast extension molding) is also known as doctor-blading (or knife-coating). During the period 1943 – 1945, Howatt from the Massachusetts

Y.Huang et al., *Novel Colloidal Forming of Ceramics*
© Tsinghua University Press, Beijing and Springer-Verlag Berlin Heidelberg 2010

Institute of Technology at Fort Monmouth Signal Laboratory had pioneered the tape casting and at that time they built the world's first-cast extension molding machine. This technology of ceramics forming was reported publicly in 1945 (Howatt G. N., et al., 1945) thereafter.

Tape casting process was formally applied to industrial production of ceramic capacitors in 1947, and G. N. Howatt received the franchise in 1952. This was the first ceramic patent (Howatt G. N., 1952) for tape casting for the production.

In 1967, an Al_2O_3 film was successfully prepared by using tape casting by H. N. Stetson and W. J. Gyurk (Stetson H. N., et al., 1967). In the same year, IBM corporation announced that layer packaging materials for use of computer had been made by tape casting technology (Schwartz B., et al., 1967). In the 1970s, ultra-fine powder tape casting began to appear. As the technology was advanced, many new products were successfully developed, and a number of tape casting applications were grown. In 1996, capacitors with 5 μm film were successfully fabricated by tape casting in Japan. By 1997, tape casting machinery, which could form film of 5 μm thickness, began to appear in Japan and the U.S.A. markets. In 1998 it had been announced by researchers that the film with a thickness of 3 μm had been obtained through tape casting technology (Rechard E., et al., 1998).

Tape casting, as a processing method, is a multidisciplinary technique (Rechard E., et al., 1998) which involves ceramic technology, powder surface physical chemistry, gel chemistry, organic and polymer chemistry and so on. As tape-forming process had been put forward for several decades, and extensive studies of the process were subjected to the attention of the scientific community, it has undergone continuous improvement and enhancement.

Although the traditional tape casting technology for ceramic sheets has many advantages including suitability for large scale continuous forming of ceramic sheets with simple and easy to operate equipment, several organic compounds are necessary, such as solvents, binder agent, plasticizers, and other additives. Most of these organic compounds are to some degrees of toxicity, especially organic solvents. They are volatilized during the forming process and cause environmental pollution. Since the 1970s, increasingly stringent environmental requirements have been established by countries across the world. Therefore, more and more researchers have engaged in investigation of water-based tape casting system.

Although water-based tape casting is a less cost, non-toxic and non-flammable technology, there are also a number of issues to be addressed. These issues include low water evaporation rate, less conductive to drying, high concentration binder in slurry, sensitive slurry to the process parameters and difficult to form smooth dense surface on the ceramic film, prone to dry crack. In addition, the ceramic powder in the water reunite flocculation easily and difficult to form a stable slurry of good dispersion.

A water-based system is not available for industrial scale forming today due to the existing many difficulties mentioned above to be overcome. The slurry curing principle should be studied thoroughly in order to make the water-based tape casting mass production a reality. For this purpose, gel-casting is introduced into tape casting process and then a fast gel curing water-based tape casting or aqueous gel-tape-casting process was proposed by the structural ceramic research team at the Tsinghua University.

In this chapter, the novel gel-tape-casting technology is introduced comprehensively covering the curing mechanism and kinetics, slurry preparation and characteristics, and the slurry rheological behavior and other research aspects.

2.1 Fundamental Principle and Processing of Aqueous Gel-Tape-Casting

2.1.1 Tape Casting Types and the Raw Materials

1) Tape Casting Types

(1) Traditional Tape Casting

Tape casting is a well-established technique used for large-scale fabrication of ceramic-based substrates and multilayered structures, such as multilayered ceramic capacitors and multichip ceramic modules (Hotza D., et al., 1995; Mister R. E., 1998). The applications of this technique also have been tested in many other fields, including solid fuel cells, photovoltaic cells and laminated object molding (Mister R. E., 1998).

Tape casting is the basic process for the fabrication of high-quality ceramic sheets such as Al_2O_3 or AlN substrates for circuitry, $BaTiO_3$ for multilayer capacitors, and so on (Mister R. E., 1998, 1990; Fiori C., et al., 1986; Hyatt E. P., 1986). The tape casting process can be divided into non-aqueous and aqueous systems according to the types of solvents used (Hotza D., et al., 1995a, b; Moreno R., 1990; Nahass P., et al., 1990). Non-aqueous solvents typically have lower boiling points and can avoid hydration of the ceramic powder, but require special precautions concerning toxicity and inflammability. Aqueous solvents, on the other hand, have the advantages of incombustibility, non-toxicity and low cost, and are therefore a good substitutes for non-aqueous solvents.

However, the aqueous system is subject to high sensitivity to minor process variables changes such as drying conditions, casting composition, and film thickness (Nahass P., et al., 1992; Mister R. E., et al., 1978). Therefore, cracks, bubbles, and other defects usually appear in products. Due to these issues, aqueous systems are seldom used for industrial applications up to today. To overcome the drawbacks of tape casting mentioned above, some researchers have studied alternative routes to improve the process (Carisey T., et al., 1995; Landham R. R., et al., 1987; Grader G., 1993; Yoshikawa S., et al., 1988; Lee H. D., et al., 1993; Smith D. J., et al., 1990; Waack R., et al., 1988). A possibility to form tapes by gel-casting is reported in this chapter.

(2) New Aqueous Gel-Tape-Casting

Based on the advantages and shortcomings of the traditional tape casting technology, a structural ceramic research team from Tsinghua University (Xiang Junhui, 2001; Huang Y., et al., 2000a, b) has combined the techniques of gel-casting and the tape casting, and proposed a new aqueous gel-tape-casting process. The gel-tape-casting with a water-based ceramic slurry was achieved through water-soluble organic monomer polymerization and curing under certain conditions. In the water-based gel-tape-casting process, ceramic powder and some additives are added into premixed solution that is prepared with organic monomer and cross-linker dissolved in deionized water. The ceramic slurry with a solid loading of greater than 55 vol% or 60 vol% containing initiator and catalyst shows good liquidity and moderate viscosity. After tape casting, the casted slurry is heated at appropriate temperature, the organic monomer is polymerized due to existing initiator and catalyst. Therefore, the slurry is cured and formed the green tape.

This technology can be applied for different ceramic powders. The green tapes have less organic content that can be removed through special degreasing processes and then directly sintered sequentially. The shaped green tapes have smooth surface, high strength, and good flexibility. However, oxygen impedes free radical polymerization of organic monomer. As a result, the gel-tape-casting process adopts either inert gas protection or other oxygen isolation devices.

2) Effects of Solvent Systems on the Tape Casting Process

Owing to its lower boiling point, the organic solvent system is more favorable to dry green tapes, during which the organic solvents will not subject to hydrolysis reaction with ceramic particles so as to ensure the stability of slurry. The disadvantages of this process include the followings: ① Most solvents used, such as toluene, xylene, and trichloroethylene, are toxic to a certain extent and thus lead to deterioration in production conditions, cause environmental pollution and increase production costs; ② It is more difficult to prepare the high solid loading of the slurry which requires high dose of the binder, plasticizers, and other organic compounds to ensure a high liquidity demand of the slurry; ③ Green tapes are prone to crack and deform due to more organic binders and lower density. It is obvious that the consequent poor quality of products causes low production efficiency and high energy consumption (Medowski G. O., et al., 1976; Schuetz J. E., et al., 1987; Kita K., et al., 1982; Burnfield K. E., et al., 1992; Nagata K., et al., 1993; Hotza D., et al., 1995a, b; Theodore P., 1995).

Water-based system uses water as solvent that takes on the lower dosage of organic matters, and it is noncombustible, non-toxic and low cost. However, because water is less volatile, a water-based system is not conducive to the drying of green tapes, and has low production efficiency. During the drying process the water evaporation is often accompanied by organic compounds and fine particle segregation, leading to nonuniform microstructure of the green tapes. Drying shrinkage with anisotropy in the clear will easily lead to cracking of products. In addition, hydration reaction will take place on some ceramic powders such as AlN, affecting the rheological behaviour of slurry and the final green body composition (Egashira M., et al., 1991; Greil P., et al., 1994; Hotza D., et al., 1995a, b). Because of these problems existing above, water-based system has been studied for many years, but it has not yet been widely applied.

The slurry formulas used commonly in organic solvent and water-based slurry system are listed in Table 2.1 and Table 2.2 (Qin Zhengqi, et al., 1996). It can be seen from Table 2.1 and Table 2.2 that there are organic compound content of 40.3 wt% and ceramic powder content of 59.7 wt% in organic solvents; while

Table 2.1 The slurry formula used commonly in organic solvent (Qin Zhengqi, et al., 1996)

raw materials	functions	dosage (g)	process
alumina	ceramic powder	194.40	first: mixing by ball milling for 24 h
magnesia	graingrowth inhibitor	0.49	
menhaden fish oil	suspending agent	3.56	
trichloroethylene	solvent agent	75.81	
ethanol (alcohol)	solvent agent	29.16	
polyethylene butyral	binder	7.78	second: blending for short time after joining additives
polyethyleneglycol(PEG)	plasticizer	8.24	
octyl phthalic acid	plasticizer	7.00	

Table 2.2 The slurry formula used commonly in water-based system (Qin Zhengqi, et al., 1996)

raw materials	functions	dosage (g)	process
distilled water	solvent agent	31.62	premixing
magnesia	graingrowth inhibitor	0.25	
polyethyleneglycol	plasticizer	7.78	
butyl benzyl phthalates	plasticizer	57.02	
non-ionic octyl phenoxy ethanol	wetting agent	0.32	
propylene-based sulfoacid	suspending agent	4.54	
alumina	ceramic powder	123.12	milling for 24 h after joining to the premixed solution
acrylic resin emulsion system	binder	12.96	milling for 0.5 h after joining to the above mentioned premixture
paraffin emulsion	defoamer	0.13	milling for 3 min after joining to the above mentioned premixture

34.8 wt% and 51.9 wt% as well as moisture 13.3 wt%, respectively, in water-based system. Although the amount of organic matter was reduced in water-based system, but at the same time, the content of ceramic powder was also decreased. In general, the strength of a water-soluble adhesive is significantly lower than that of binder used in the organic solvent system (Morris J. Jr., 1986). To ensure the green body strength and to avoid cracking, the high dosage of the binder in water-based system is necessary, so the content of ceramic powders will be reduced accordingly.

On the other hand, organic solvents have lower the surface tension and better wettability on the surface of the ceramic particles. Therefore, the ceramic slurry with stable, and good dispersion is easier to be formed. However, the water surface tension is much higher than that of common organic solvents, and poor wettability on the surface of ceramic particles leads to decrease solid loading in suspensions. By adding a wetting agent in slurry system, the wettability between water and ceramic particles can be improved and the solid loading should be also increased in slurry.

To explore the molding quality of green tapes to different solvent systems, the contrast research for tape casting of the Al_2O_3 slurries, which were prepared by water-based system and an organic solvents system, were carried out by Paul Nahass, et al. (Nahass P., et al., 1990). The results showed that the slurry with organic solvents had lower sensitivity on the change of the molding process parameters, and it was relatively easy to receive high intensity, uniform structure of the green tape. While the slurry with water-based system was sensitive to the dry conditions, the composition of slurry and the film thickness for minor changes of technical parameters. The green tape with a uniform, crack-free zone can be fabricated only when the conditions are accurate and the control is stable during the forming process (Nahass P., et al., 1990; Hotza D., et al., 1995a, b).

Drying of ceramic green tapes is the most difficult procedure in the aqueous tape casting process (George W. Scherer, 1990; Briscoe B. J., et al., 1997; Ford R. W., 1986). Water-based slurry in the drying process is easier to have cracking, curling and to tend segregation of organism and small particles leading to peeling and other defects. Especially it is more obvious that the green bodies dried faster. Therefore, process conditions such as temperature, relative humidity, and air fluid speed should be controlled precisely according to the thickness of green tape, and compositions of slurry, so that the water evaporation rate could be slowed down. The pores in the full tape must be eliminated as much as possible to avoid uneven regional contraction caused by deformation or cracking and green tape curly (Briscoe B. J., et al., 1997; Ford R. W., 1986; Michel Descamps, et al., 1995). The drying rate of the thick tape should be controlled especially in the later stage of drying.

For a water-based slurry system, the green body will become brittle with water evaporation. The moisture content of the green body after drying should generally be remained at 2 – 5 wt% around (Michel Descamps, et al., 1995).

In recent years, tape casting has received special attention due to those reasons for the environmental and health aspects. Organic solvents have lower boiling points, but need special precautions concerning flammability and toxicity. Therefore, slurries using water as the solvent have appeared (Mister R. E., et al., 1990; Chartier T., et al., 1993; Kristoffersson A., et al., 1997). Aqueous-based systems have advantages of nonflammability, non-toxicity and lower cost (Hotza D., et al., 1995a, b).

However, some problems occur in water-based systems. For example, it is difficult to obtain stable slurry with appropriate rheological properties and high solid loading, which would be suitable to give uniform tapes with high green density. Furthermore, the drying procedures of the slurry are not easily controllable to obtain a favorable range of thickness (Chartier T., et al., 1993) and to decrease green tape defects that result from the drying process (Hotza D., et al., 1995a, b).

Therefore, a novel process, gel-tape-casting, has been developed to overcome those problems described. This process is a combination of tape casting and gel-casting, based on the mechanism of polymerization of soluble monomers. The slurry prepared using this process contains a solid loading of >50 vol%, and it can solidify in a few tens of minutes under a nitrogen atmosphere, at a specific temperature.

In general, an aqueous slurry has less tolerance to minor changes in type and content of additives, drying conditions, casting composition, and film thickness (Hotza D., et al., 1995a, b). Therefore, more attention must be paid to how additives and other conditions influence the properties of the slurry and green tape.

3) Starting Materials Used

The dispersants used usually are tetramethylammonium hydroxide and ammonium citrate. The plasticizer used usually is glycerol and polyethylene glycol (PEG). Surfactants used usually are dodecylbenzene sulfonic acid, sodium salt, tween 80, and tributyl phosphate.

4) Organic Compounds Used Commonly in Slurry for Tape Casting Process

The organisms commonly used in slurry are shown in Table 2.3.

2.1.2 Polymerization of the Monomer

1) Gel-Tape-Casting Mechanism

For organic tape casting, a green tape is prepared by the evaporation of an organic solution. However, the preparation of a green tape by the gel-tape-casting is based on the mechanism of polymerization of aqueous monomers. During the casting, the polymerization of monomers is initiated at elevated temperature, and a polymer network is formed (Wayne N. J., 1969).

Table 2.3 Organic compounds commonly used in slurry for the tape casting (Qin Zhengqi, et al., 1996)

	solvents	binders	plasticiser	suspending agent	wetting agent
organic solvent systems	acetone; butanol; benzene; bromochloromethane; di-acetone; ethanol; propanol; acetone ethyl toluene; trichloroethylene; xylene; dimethybenzene	cellulose acetate butene; cellulose ether; petroleum resin; polyethylene; polyacrylic ester; poly-propylene; polyvinyl alcohol (PVA); polyethylene butyral; ethylene chloride; polymethacrylate; ethyl cellulose; rosin resin acid	butylbenzyl phthalic acid; dibutyl phthalate acid; dimethyl phthalate acid; phthalates mixture; polyethylene glycol (PEG)-dielectric; tritolyl phosphate	fatty acid; natural fish oil; synthetic surfactants; benzene sulfonic acid; fish oil; oleic acid; methanol; octane	poly-propylene-based ethanol; polyethylene glycol ethyl ether; ethyl benzene glycol; polyoxyethylene(POE); single-oleic acid glycerol; three oleic acid glycerol; ethanol category
water-based systems	water	acrylic polymer systems; acrylic polymer emulsion systems; ethylene oxide polymer; hydroxyethyl cellulose; methyl cellulose; polyvinyl alcohol (PVA); isocyanate resin; paraffin wax lubricant; urethane (water-soluble); methyl acrylate copolymer salt; paraffin wax emulsion; ethylene vinyl acetate copolymer emulsion	butyl benzyl titanate; 2-butyl titanate; ethyl toluene sulfonamide; glycerol; poly-alkyl-glycerol; triethylene glycol; three-*N*-butyl phosphate; hydrochloric acid; gasoline; polyols	phosphate; phosphate complex salt; allyl sulfoacid; natural sodium salt; acrylic copolymer	non-ionic octyl phenoxy ethanol; ethanol-type nonionic surfactants

When the ambient temperature increases, the initiator in the slurry is decompounded into initiating radicals:

$$(NH_4)_2S_2O_8 \longrightarrow 2NH_4^+ + 2SO_4^{2-} \tag{2.1}$$

The initiating radical reacts with the monomer, and a monomer-free radical is produced:

$$SO_4^{2-} + CH_2{=}\underset{\displaystyle \mathrm{CONH_2}}{\underset{|}{CH}} \longrightarrow SO_4CH_2\underset{\displaystyle \mathrm{CONH_2}}{\underset{|}{CH}}{-} \tag{2.2}$$

That radical can react with other monomers and result in a propagating radical:

$$SO_4CH_2\underset{\displaystyle \mathrm{CONH_2}}{\underset{|}{CH}}{-} + nCH_4{=}\underset{\displaystyle \mathrm{CONH_2}}{\underset{|}{CH}} \longrightarrow SO_4CH_2\underset{\displaystyle \mathrm{CONH_2}}{\underset{|}{CH}} + CH_2\underset{\displaystyle \mathrm{CONH_2}}{\underset{|}{CH}} + nCH_2\underset{\displaystyle \mathrm{CONH_2}}{\underset{|}{CH}}{-} \tag{2.3}$$

As the reaction proceeds, monomers polymerize into polymer chains. In most cases, the C=C double bond of the cross-linker, MBAM, breaks down and adheres to the heads of two polymer chains, so that these polymer chains join. Finally, cross-linked polyacrylamide (PAM), which fixes ceramic powders in its framework structure, is formed.

The slurry cannot be solidified to green tape when exposed to air because of the antipolymerizing effect of oxygen. Oxygen has a greater tendency to react with a radical formed in the process than AM and hence it produces an inert radical. Thus, the polymerization stops. Therefore, a cover filled with nitrogen is required to keep out the oxygen.

However, the green tape formed from monomer polymerization is too brittle to be rolled. Plasticizer glycerol (G) containing hydroxyl is introduced into this system. Hydroxyl in plasticizer G tends to form a hydrogen bond with oxygen/nitrogen atoms in polymer chains, thus substituting for a part of the cross-linker molecules to build a soft organic bridge between the polymer chains. Thus, the flexibility of the green tape is improved (Moreno R., 1992a, b).

In the gel-tape-casting process, an AM monomer solution is used as the solvent, and the slurry is prepared by adding ceramic powders and additives to the solvent. During casting, initiator and catalyst are added to the slurry and polymerization proceeds. The slurry solidifies and a green tape is obtained. The structural formula of AM is

$$CH_2{=}CH{-}\underset{\displaystyle \mathrm{NH_2}}{\underset{|}{C}}{=}O \tag{2.4}$$

Polymerization of AM may occur as follows. Firstly, the initiator is decomposed into primary free radicals, as shown in reaction (2.1).

Primary free radicals can be combined with the monomer molecule, which produces monomer free radicals, as shown in reaction (2.2).

These free radicals can combine with monomers and result into chain free radicals,

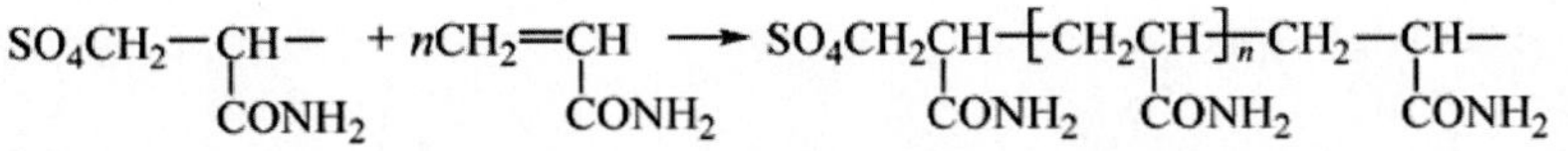

(2.5)

This reaction proceeds continuously and AM molecules polymerize into polymer chains. A network is formed by polymer chains by two ways, crosslinking (2.6) of polymer chains and bridging (2.8) between polymer chains by crosslinker molecules.

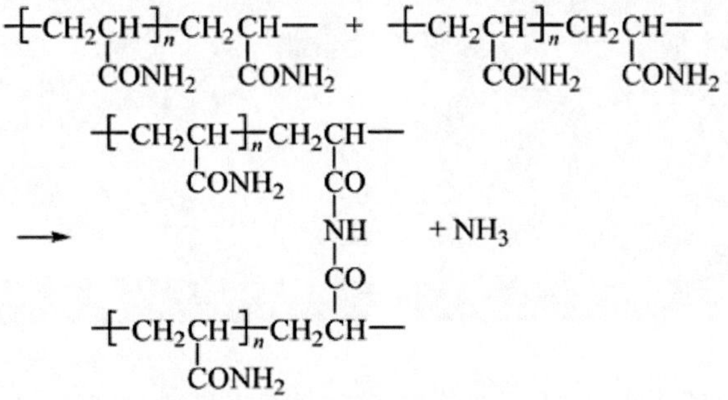

(2.6)

In order to make monomers polymerize completely, MBAM (molecular formula is shown in (2.7)) is added as crosslinker into the monomer solution.

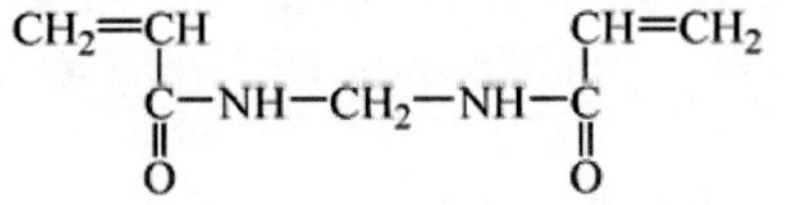

(2.7)

MBAM has two C=C double bond and can be connected between two polymer chains by a bridge effect, which is beneficial for network formation (2.8).

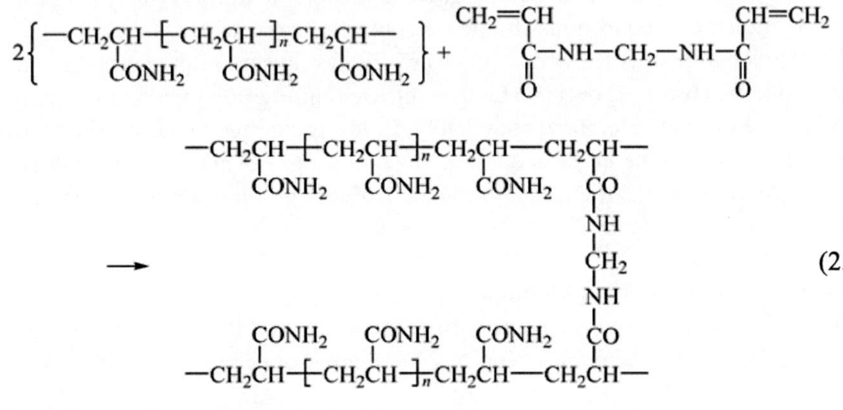

(2.8)

The ceramic slurry solidifies by polymerization of AM, and thus a ceramic green tape is obtained. The network of PAM acts as a frame that fixes ceramic powders by adsorption. The details of the combination between the ceramic powder and PAM are shown in Fig. 2.1.

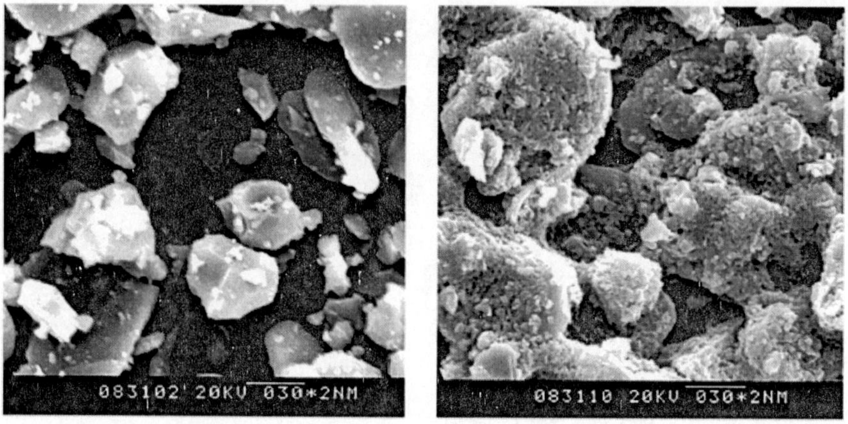

(a) (b)

Figure 2.1 The SEM image of ceramic particles before and after polymerization (a) before polymerization; (b) after polymerization

It should be mentioned that the slurry exposed to air could not be solidified to a green tape because of the anti-polymerizing effect of oxygen, therefore, during the casting process the slurry should be protected by an inert atmosphere of nitrogen or argon.

2) Study of Polymerization Kinetics

The polymerization rate of ceramic slurry depends on the properties of slurry, such as the concentration and pH of the monomer solution, and the reaction conditions, such as polymerization temperature and amount of initiator and catalyst. To obtain a green tape with high strength and suitable flexibility, slurry preparation and polymerization conditions should be controlled precisely.

Experimental results show that pores are always present in the green tape if the slurry solidifies too fast, because the gas released during polymerization cannot be eliminated from the green tape completely. Pores in the green tape would decrease the strength of the green tape and of the sintered sample. Therefore, it is necessary to keep the polymerization rate within a suitable range. In other words, the slurry would be better polymerized to a greater extent within a few minutes than a few seconds.

(1) Concentration of the Monomer Solution

The relationship between the polymerization rate and the concentration of the monomer solutions is shown in Fig. 2.2. The higher the concentration, the shorter the polymerization time, i.e., the higher the polymerization rate.

The polymerization time decreases sharply with the increase of the monomer solution concentration up to 40 wt% after which it remains practically constant. Thus, the polymerization rate can be adjusted more easily by controlling the monomer solution concentration more easily if the concentration is less than 40 wt%. For the Al_2O_3 slurry, in order to obtain a suitable polymerization rate in the gel-tape-casting, the concentration of the monomer solution in the premixed solution should be kept in the range of 30 – 40 wt%.

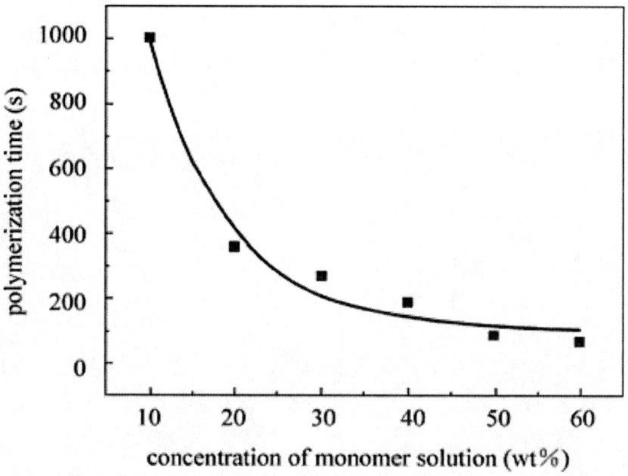

Figure 2.2 The relationship between concentration of monomer solution and polymerization time

(2) pH

The pH of the slurry affects the polymerization rate significantly because the initiator $(NH_4)_2S_2O_8$ decomposes more quickly in an acidic condition than in an alkaline condition. As a result, the polymerization rate decreases with the increase of the pH and the polymerization time increases almost linearly with pH increase (see Fig. 2.3). Besides influencing the polymerization rate, the pH also influences the microstructure of the green tape. At low pH, polymerization of AM tends to release NH_3 which forms pores in the green tape and leads to an inhomogeneous microstructure (see Fig. 2.4(a)). Experimental results show that a suitable pH of Al_2O_3 slurry lies in the range 7 – 9, a condition beneficial for both inhibiting pore formation in the green tape and improving the dispersibility of the ceramic powder in the slurry. As a result a green tape with a homogeneous microstructure can be obtained (see Fig. 2.4(b)).

(3) Polymerization Temperature

Temperature is one of the most important factors influencing polymerization rate. Although a high temperature is beneficial for improving polymerization rate, at an excessively high temperature, the water in the slurry will volatize so quickly that pores will be formed in the green tape.

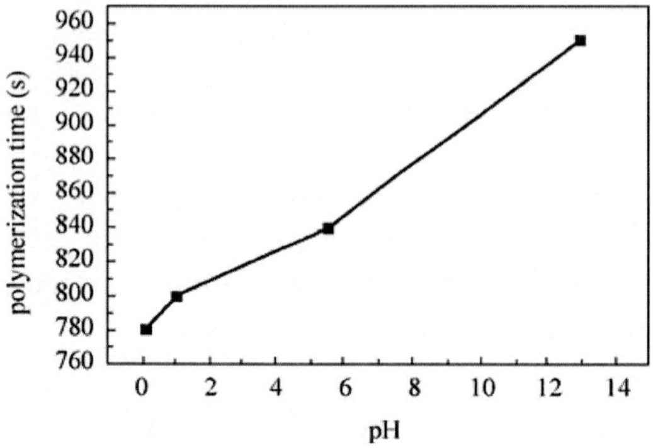

Figure 2.3 The relationship between the pH of the ceramic slurry and polymerization time

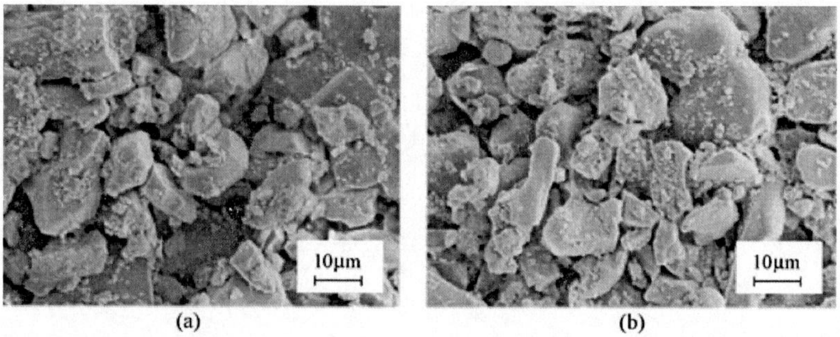

Figure 2.4 The SEM image of the broken section of green tape made from slurries with different pH

(a) pH = 3.5; (b) pH = 8.5

Figure 2.5 shows the relationship between temperature and polymerization time. With the increase of temperature, the polymerization time is reduced dramatically. A suitable temperature for polymerization of the system lies in the 40 – 80℃ range.

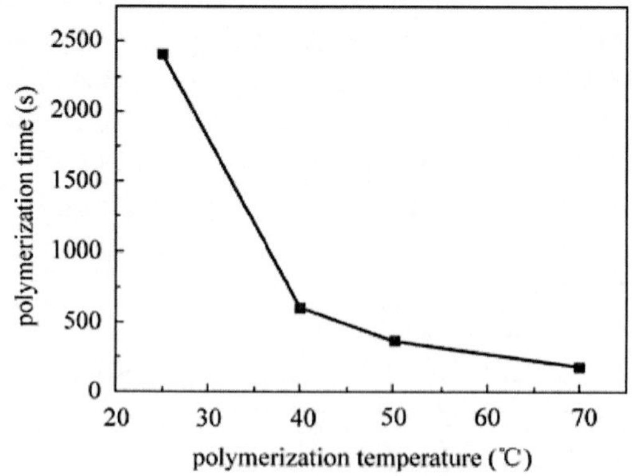

Figure 2.5 The relationship between temperature and polymerization time

During the forming process, it is expected that the slurry can be kept stable for a long time before casting, and it would polymerize very quickly after casting. According to the sensitivity of the polymerization rate related with temperature, the starting time of polymerization can be controlled accurately by keeping the ceramic slurry at low temperature (such as 25℃) before casting, and heating it to 40 – 80℃ to complete polymerization after casting.

(4) Initiator and Catalyst

It has been proved that the higher the polymerization rate, the more difficult it is to control the casting process. If the polymerization proceeds too quickly, the slurry may polymerize in the reservoir before passing of the blade. The role of the initiator is to create primary free radicals to promote the polymerization of the monomers. A large amount of initiator will increase the initiating rate and lead to a very fast polymerization. Figure 2.6 shows the relationship between the amount of initiator and the polymerization time of the Al_2O_3 slurry. A suitable amount of initiator lies in the range 0.5 – 1 vol% of slurry.

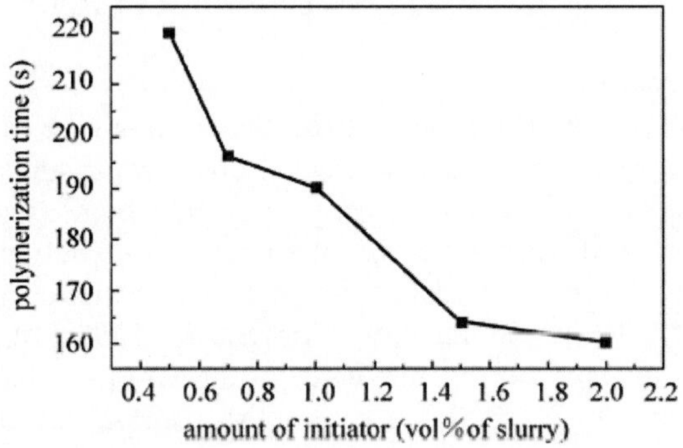

Figure 2.6 The relationship between amount of initiator and polymerization time

The catalyst can significantly lower the activation energy of the reaction and improve the polymerization rate. Experimental results show that the slurry cannot polymerize without a catalyst even if a certain amount of initiator is added. Nevertheless, a very high amount of catalyst makes the polymerization rate uncomfortable. For the studied Al_2O_3 slurry, an appropriate amount of catalyst is about 0.5 vol% of the slurry. The relationship between the polymerization time and the amount of catalyst is shown in Fig. 2.7.

It can be seen from above that the polymerization rate of AM depends on the properties of the slurry and the reaction conditions. The concentration of the monomer solution should be kept within 30 – 40 wt% to obtain a suitable polymerization rate. A suitable range of pH of the Al_2O_3 slurry and temperature are 7 – 9 and 40 – 80℃, respectively. The AM sensitivity to temperature can be used to control the starting time of polymerization accurately. For the studied Al_2O_3 slurry, the

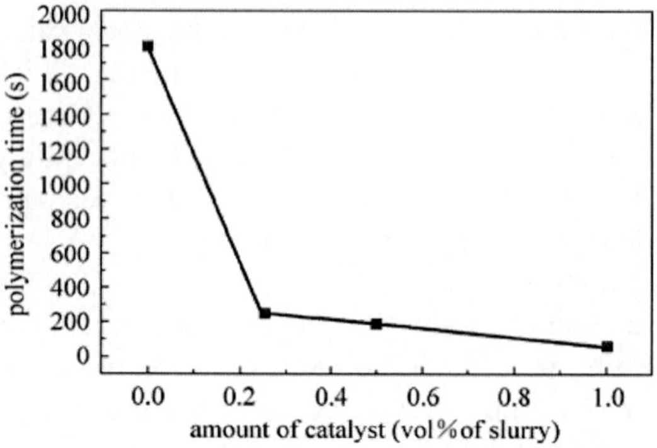

Figure 2.7 The relationship between amount of catalyst and polymerization time

suitable amounts of initiator and of catalyst are 0.5 – 1 vol% and about 0.5 vol% of the Al_2O_3 slurry, respectively.

2.1.3 Influence Factors on Polymerization of the Monomer

1) Effects of Dispersant

The ammonium citrate is used as a dispersant in this system to improve the dispersion of ceramic powders and to achieve a solid loading >50 vol%. Molecules of the dispersant are absorbed onto the surface of the ceramic powder when they are dissolved into a slurry; this changes the charge distribution on the particle surface. Generally, ceramic powders with high zeta-potential (absolute value) usually exhibit high stability in the slurry (Moreno R., 1992a, b). The addition of the dispersant can increase the absolute value of the zeta-potential of the slurry at pH = 8 – 10 (as shown in Fig. 2.8). Therefore, the dispersion of ceramic powders in the slurry is improved.

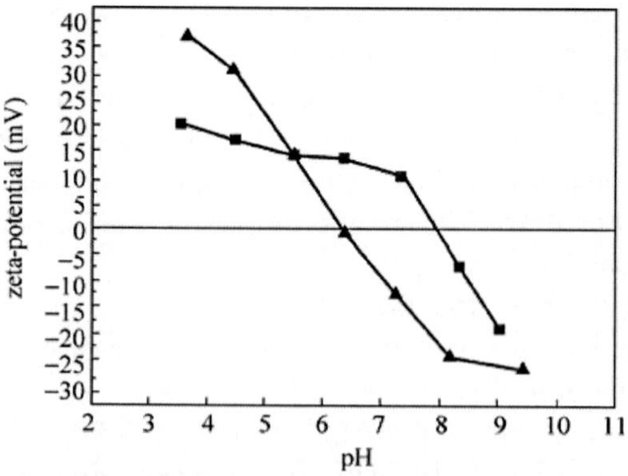

Figure 2.8 Absolute value of the zeta-potential of a slurry with ammonium citrate dispersant (-▲-) is higher than that of a slurry without dispersant (-■-) at pH = 8 – 10

An optimal amount of the dispersant is used in the slurry to achieve good fluidity. The apparent viscosity of slurries (solid loading 55 vol%) related with various contents of the dispersant has been tested. The viscosity of the slurry decreases with the increase of dispersant at relatively lower dispersant concentration; however, it increases when the dispersant content exceeds a fixed value (1 wt% in this system). Graule and Grauckler (Graule T., et al., 1993) reported the same phenomenon in an aqueous slurry by using Tiron as dispersant. For amounts of dispersant higher than a fixed value (as shown in Fig. 2.9), the increase in viscosity can be attributed to an increase in ionic strength. The diffuse electrical double layer around each ceramic particle is compressed by the ions, and the repulsive potential is decreased.

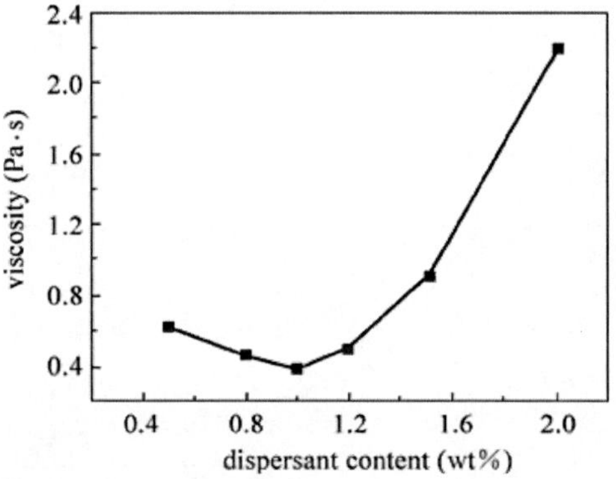

Figure 2.9 Viscosity at shear rate 50 s^{-1} of a slurry with a solid loading of 55 vol% vs. the dispersant content (Viscosity of slurry reaches the minimum value when the content of dispersant is 1 wt%)

2) Effect of Plasticizer

The solidified green tape made from the slurry without plasticizer is too brittle to be rolled. A plasticizer is introduced into the slurry to make the green tape soft. The addition of the plasticizer affects the rheological behaviors of the slurry. Three curves of viscosity have been compared that represent the slurries (55 vol%) with 10 wt% plasticizer G, 10 wt% PEG and without plasticizer (as shown in Fig. 2.10). The addition of the plasticizer somewhat increases the slurry viscosity to a certain extent.

The relationship between the content of plasticizer G and the slurry viscosity (solid loading 50 vol%) has also been studied. These viscosity curves indicate that the viscosity of the slurry increases with an increase of plasticizer G (as shown in Fig. 2.11).

The flexibility of the green tape is judged by the number of times that the tape can sustain bending without breaking. The experiment shows that the flexibility is improved (50 or more bends) gradually when the amount of plasticizer G increases.

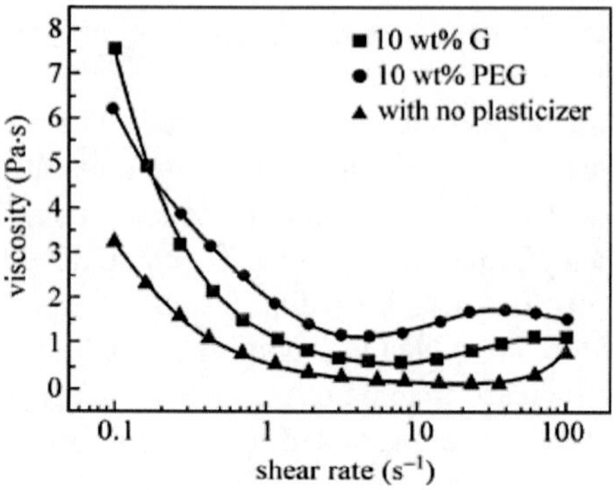

Figure 2.10 The effect of various plasticizers on viscosity of the slurry (solid loading 55 vol%). Viscosity of slurry increases after additions of plasticizer

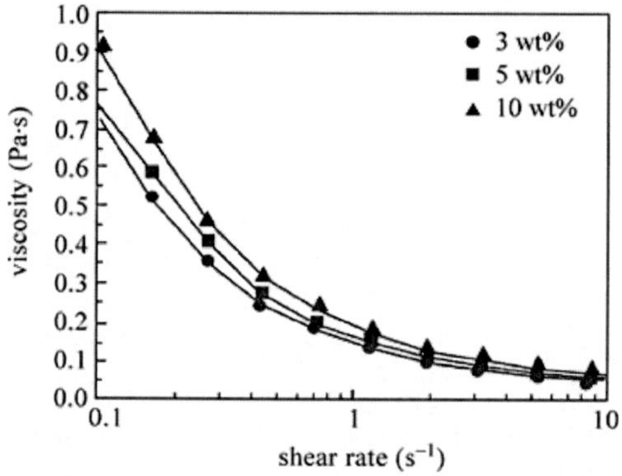

Figure 2.11 The relationship between the amounts of plasticizer G and viscosity of the slurry. Viscosity increases with increasing of amounts of G at relatively low shear rate. However, the difference disappears at shear rates >10 s^{-1}

However, after the plasticizer G amount reaches about 5 wt%, the flexibility does not seem to change. Meanwhile, the fracture strength of the green tape decreases at high amounts of plasticizer G addition. Therefore, 5 wt% of plasticizer is considered as the optimal amount in this system to balance the needs of flexibility and fracture strength.

3) Effect of Surfactant

During gel-tape-casting, it is difficult to wet the carrier film with the aqueous slurry, and, thus, a flat and homogeneous green tape cannot be formed. Therefore, a surfactant is needed in this system. It can be concluded from the wetting angles of slurries with surfactant that a small amount of surfactants in the slurry decreases surface tension significantly, which results in the decrease of the wetting angle of the slurry. For example, the wetting angle shifts from 109.8° to 72.2° after using surfactant D (as shown in Table 2.4).

Table 2.4 Wetting angles and surface tensions of slurries with various surfactants

solution	wetting angle (°)	surface tension (mN/m)
deionized water	109.8	70.86
slurry without surfactant	103.3	46.76
slurry with surfactant A	69.5	36.05
slurry with surfactant C	75.0	—
slurry with surfactant B	73.5	24.24
slurry with surfactant D	72.2	26.96

Note: The amount of surfactant in slurry is 1%. Addition of surfactant can decrease the wetting angle of the slurry and, thus, improve its wetting property. A–anionic surfactant; B–cationic surfactant; C–non-ionic surfactant; D–special type surfactant.

The releasing property of the green tape made from organic slurry is perfect because of its strong adhesion. On the contrary, the green tape from aqueous gel-tape-casting is difficult or impossible to release from the carrier film. However, the green tape with surfactant D added easily releases from the carrier film and has smooth surfaces. The good releasing property is attributed to the structure of surfactant D.

In general, surfactant molecules gather on the surface of the solution during processing when hydrophilic groups come into solution and hydrophobic groups come out. The common surfactant has an organic hydrophobic group, which has a strong attractive force to the plastic surface of the carried film. However, the hydrophobic group of surfactant D shows great electronegativity, and its radius is large enough to cover the entire carbon chain of the surfactant D molecule (Tadros T. F., 1984). Thus, the active force between surfactant D and the carrier film is so small that the green tape can be released easily.

4) Defoaming and Deairing

Foams are introduced when a surfactant is added to a slurry. The amount of foam increases with increasing the content of the surfactant. If foams are not removed from slurry before casting, pinholes and other defects will appear in the green tape. Some defoamers are used to eliminate the foam from the slurry. Small amounts (<1 wt%) of defoamer decreases the amount of foam.

There are three main mechanisms of foam removal: decreasing the solubility of surfactant; spreading over the interface and displacing the stabilizing surfactant; and causing molecules in the liquid membrane to drain quickly (Myers D., 1992). Defoamers used in this experiment all are oleaginous and follow the second mechanism.

Some foams remain in the slurry even after a defoamer is used. To eliminate the residual foam thoroughly, the slurry is deaired for about 30 min in a rotating flask under high vacuum. To avoid the polymerization of monomers, the deairing process is conducted in an ice-water bath. The green-tape surface contains no evident foam after using the defoamer and deairing under vacuum.

5) Debindering and Sintering

The green tape is dried at 120℃ for 24 h after it is released. The TGA/DTA of the green tape has been performed at 1 atm, as shown in Fig. 2.12. These analysis have been conducted at a heating rate of 5℃/min to a maximum temperature of 800℃. A large weight loss occurs between 285 and 579℃, which is associated with the burnout of most of the organic components. The decomposition of some reagents and organic components with large molecular weight results in another weight loss between 660 and 750℃.

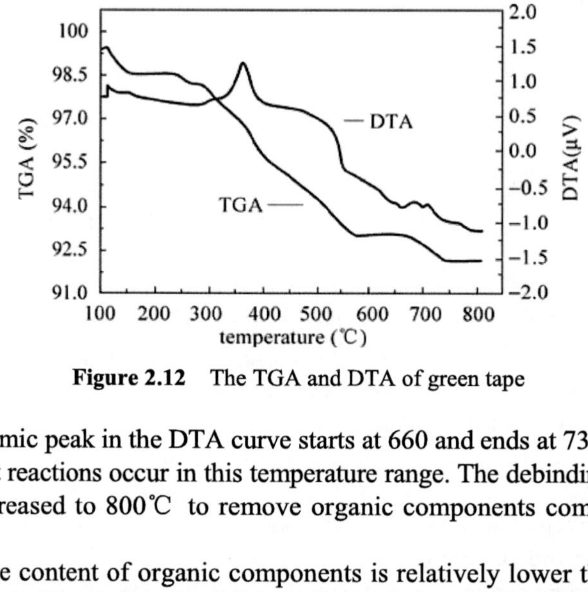

Figure 2.12 The TGA and DTA of green tape

An exothermic peak in the DTA curve starts at 660 and ends at 730℃. This peak illustrates that reactions occur in this temperature range. The debinding temperature has to be increased to 800℃ to remove organic components completely before sintering.

Because the content of organic components is relatively lower than in organic tape casting, it is not necessary to have an individual debindering process. Sintering systems with various debindering temperature and holding time have been conducted, and the properties of the sintered samples by these sintering systems are listed in Table 2.5.

The comparison of sintering results shows that a slow heating rate and a high sintering temperature are favorable to acquire sintered samples with high density and flat surfaces.

Debinding and sintering are performed and alumina substrates with high density (3.84 g/cm^3), high fracture strength (315 MPa) and flat surfaces have been obtained.

Most organic components are burned out between 280 and 580℃.

Weight loss that occurs between 660 and 750℃ is associated with an exothermic peak in DTA curve. Therefore, it is necessary to increase the debindering temperature to 800℃.

Table 2.5 The properties of samples sintered under various systems

sintering systems	line shrinkage (%)	fracture strength (MPa)	density (g/cm^3)	appearance evaluation
1	10.24	266	3.63	fluctuant
2	10.87	257	3.64	fluctuant
3	11.56	281	3.72	
4	12.31	284	3.75	
5	14.68	292	3.80	flat
6	15.13	296	3.83	flat
7	15.95	315	3.84	flat

Note: sintering systems

1. Room temp. $\xrightarrow{180'}$ 600℃ $\xrightarrow{60'}$ 600℃ $\xrightarrow{720'}$ 1560℃ $\xrightarrow{120'}$ 1560℃;
2. Room temp. $\xrightarrow{180'}$ 600℃ $\xrightarrow{60'}$ 600℃ $\xrightarrow{600'}$ 1250℃ $\xrightarrow{300'}$ 1560℃ $\xrightarrow{120'}$ 1560℃;
3. Room temp. $\xrightarrow{300'}$ 600℃ $\xrightarrow{60'}$ 600℃ $\xrightarrow{650'}$ 1250℃ $\xrightarrow{300'}$ 1560℃ $\xrightarrow{150'}$ 1560℃;
4. Room temp. $\xrightarrow{60'}$ 360℃ $\xrightarrow{30'}$ 360℃ $\xrightarrow{75'}$ 600℃ $\xrightarrow{60'}$ 600℃ $\xrightarrow{140'}$ 1150℃ $\xrightarrow{840'}$ 1580℃ $\xrightarrow{120'}$ 1580℃;
5. Room temp. $\xrightarrow{20'}$ 120℃ $\xrightarrow{60'}$ 120℃ $\xrightarrow{180'}$ 360℃ $\xrightarrow{60'}$ 360℃ $\xrightarrow{500'}$ 800℃ $\xrightarrow{240'}$ 800℃ $\xrightarrow{720'}$ 1580℃ $\xrightarrow{120'}$ 1580℃;
6. Room temp. $\xrightarrow{20'}$ 120℃ $\xrightarrow{100'}$ 120℃ $\xrightarrow{360'}$ 800℃ $\xrightarrow{600'}$ 800℃ $\xrightarrow{720'}$ 1590℃ $\xrightarrow{120'}$ 1590℃;
7. Room temp. $\xrightarrow{20'}$ 120℃ $\xrightarrow{100'}$ 120℃ $\xrightarrow{600'}$ 800℃ $\xrightarrow{720'}$ 800℃ $\xrightarrow{720'}$ 1590℃ $\xrightarrow{120'}$ 1590℃.

The SEM photographs of the green tapes on the surface and underside are shown in Figs. 2.13 and 2.14.

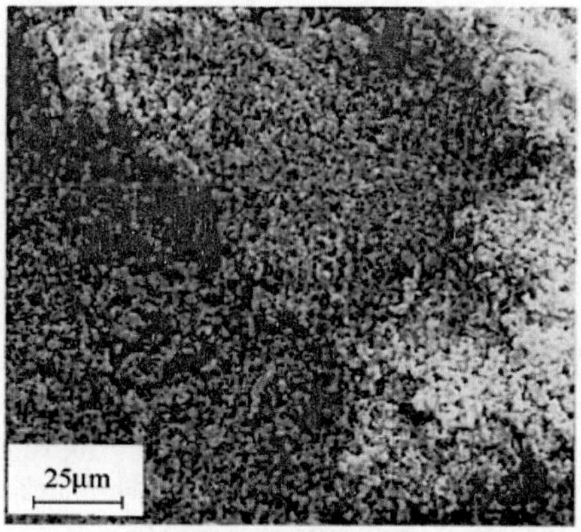

Figure 2.13 The SEM photograph showing defects on underside of the green tape due to ceramic particles adhering to carrier film

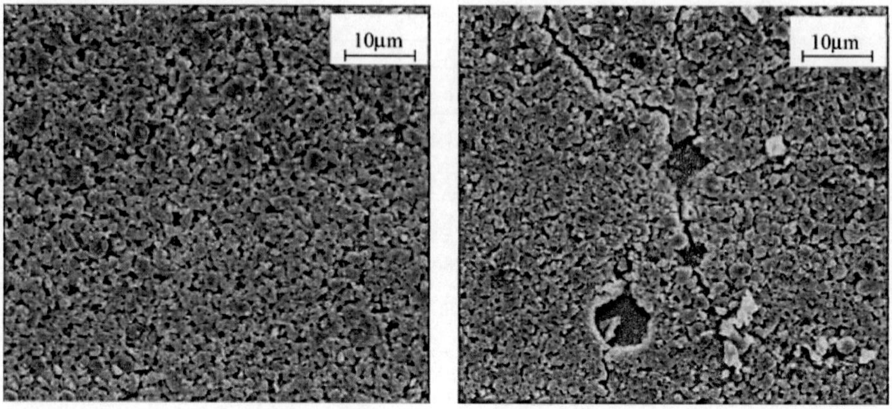

(a) (b)

Figure 2.14 The SEM photograph of the green tapes on the surface
(a) Green-tape surface without bubbles after defoaming and deairing;
(b) Surface with pinholes caused by foam

2.1.4 Processing of the Gel-Tape-Casting

1) Technological Characteristics and Device for Gel-Tape-Casting

The ammonium citrate was selected as a dispersant. The glycerol was used as a plasticizer (labeled as G). Several types of surfactants, including anionic A, cationic B, non-ionic C, and special type D, were chosen. Defoamers were olefin, tributyl phosphate, dipentylamine and 2-ethylhexanol. The content of all reagents were based on the premixed solution.

The ceramic powder, organic binder, plasticizer, decentralized agents and other additives added to solvents are mixed through the ball milling to form the slurry with uniformity, stability and certain liquidity. According to the different solvents used, cast extension molding system is divided into organic compounds and water-based system. The organic compounds and their amounts in different slurry systems are been varied also. The organic compounds used commonly in the two kinds of the systems are listed in Table 2.3 (Qin Zhengqi, et al., 1996).

Some additives were mixed with ceramic powders in the premixed solution to obtain a slurry with a solid loading of >50 vol%. The slurry was then ball-milled for 10 – 24 h to reach homogeneity. After the ball-milled process, initiator and catalyst were mixed into the slurry, and the slurry was deaired in an ice-water bath for 30 min. The slurry was then ready for casting.

The gel-tape-casting device was designed especially for this study as shown in Fig. 2.15. It included two heaters beneath the support and a protecting cover above the support. During casting, the slurry was poured into a reservoir and then drawn, passing the blade on the carrier film, which moved into the cover filled with nitrogen atmosphere. The baseband speed was constant at 1.2 cm/s, and the gap between the blade and the support was adjusted at several hundred microns. The polymerization of the monomers was then initiated, and the slurry was solidified to a green tape in <20 min under a nitrogen atmosphere and a temperature of 40 – 70℃. A green tape with adequate strength and flexibility could be prepared, with a thickness in the range 0.1 – 1 mm.

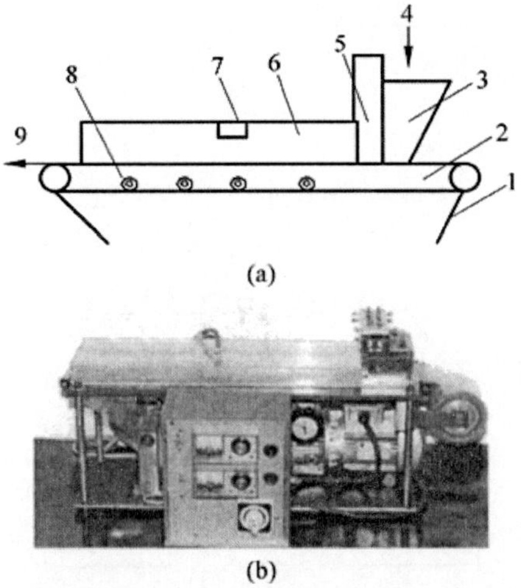

Figure 2.15 (a) Schematic of gel-tape-casting device; (b) gel-tape-casting machine designed and fabricated by authors
1–baseband; 2–baseplate; 3–slurry reservoir; 4–slurry; 5–doctor brade; 6–protective shield of atmosphere; 7–venthole; 8–heating wire; 9–green tape

The processes for preparing ceramic sheets are discussed below.

After milling and gas removal, the slurry can be used to cast extension molding. The slurry flow from the slurry reservoir 3 to the lower base-band 1 which can move to forward, tape film (tape band) thickness was controlled by the doctor brade 5. Blank film together with the base-band access to go around hot air drying room, the tape film after drying retain a certain solvent, together with the base-band together reel standby application. The solvent in the process of storage distributed evenly and eliminated moisture gradient. Finally, the dried green tape film was made into various products by cutting, drilling or stamping according to the required shape.

2) Flowchart for the Gel-Tape-Casting Process

The flowchart for the gel-tape-casting is presented in Fig. 2.16. A solution of monomer and cross-linker is prepared. Then alumina powder, dispersant, plasticizer and sintering assistants (Kaolin, $CaCO_3$ and SiO_2) are added into the solution to prepare slurry with solid loading more than 50 vol%. After 10 – 24 h of ball milling, the slurry is deaired under vacuum (0.02 atm). Then the initiator and catalyst are added in the slurry and the slurry is ready for tape casting.

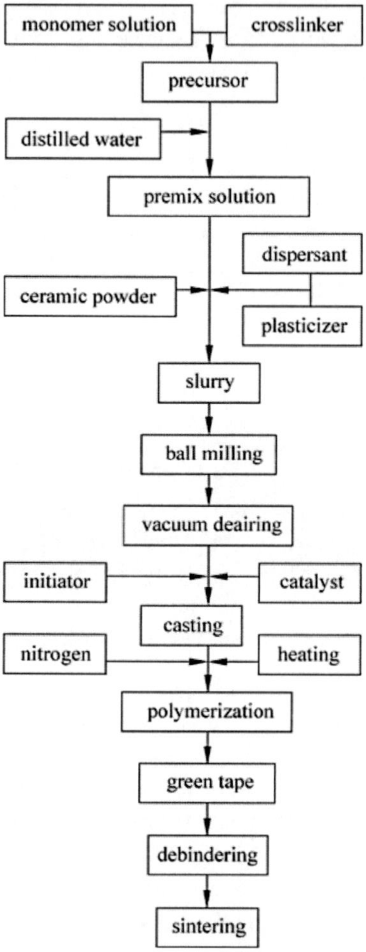

Figure 2.16 The flow chart of the gel-tape-casting process

The prepared slurry is poured into the reservoir and drawn past the blade by the moving of carrier film (baseband), and brought into the cover. The slurry polymerizes at about 50℃ under inert atmosphere condition. The cast tape has enough strength and flexibility to sustain shaping processes such as incision and punching. The cast tape is sintered by following stages: it is heated up to 1150℃ at a rate of 2℃/min, then for 0.5℃/min up to 1550℃ with a plateau for 2 h.

2.2 The Characteristics of Slurries Used for Aqueous Gel-Tape-Casting

2.2.1 The Properties of the Aqueous Ceramic Slurries with Binder

1) The Importance of Additives in Tape Casting

Tape casting is one of the main fabrication methods of thin ceramic substrates (Mister R. E., 1998; Hotza D., 1995a, b; Mister R. E., 1990). Traditional tape casting requires large amounts of organic solvent in the slurry, which may cause defects such as: ① cracks and skins resulting from anisotropic drying shrinkage; ② holes inside the green tape during binders burn-out; and ③ inhomogeneous microstructures along the thickness of the tape. Moreover, the organic components used in tape casting are highly volatile, toxic and expensive (Chartier T., et al., 1993; Doreau F., et al., 1998; Briscoe B.J., et al., 1997).

More and more emphases are focused on developing aqueous tape casting and new moulding mechanisms are being used to update traditional tape casting (Smith D. J., et al., 1990; Chartier T., et al., 1997). A new process, aqueous gel-tape-casting, is developed in this study. This process has the advantages of high solid loading (up to 60 vol%) and low-toxicity. The aqueous slurry can be rapidly solidified by the polymerization of aqueous organic monomers.

However, it was observed that the aqueous slurry hardly wetted the surface of the organic carrier film, and that it was also hard for the green tape to be released from the carrier film. Therefore, some additives, such as surfactants, are used to overcome those problems. As a result of adding surfactant into slurry, foams are also introduced, which may destroy the surface and form pores inside the green tape. To eliminate foams, defoamers are employed. The influence of above additives on the gel-tape-casting process was investigated.

2) The Features of Aqueous Ceramic Slurry with Binder

Tape casting is a low cost and useful technique for preparing thin ceramic sheets, and it has been used mainly to produce substrates, solid electrolytes, packages and multi-layers capacitors for the electronics industry, etc. (Hotza D., et al., 1995a, b; Zeng Y. P., et al., 2000; Cui X. M., et al., 2003; Kristoffersson A., et al., 1997). In general, tape casting slurry is a typical organic system. Because the volatility and toxicity of organic solvents are harmful to the environment and health, the water-based tape casting is already given much attention to the ceramics industry (Pagnoux C., et al., 1998; Doreau F., et al., 1998, 1999; Gutierrez C. A., et al., 2001; Briscoe B. J., et al., 1998).

An aqueous slurry must be adjusted in order to yield tapes that satisfy some quality criteria (Hotza D., et al., 1995a, b), such as: ① high solid loading in water; ② no defects during drying; ③ high strength to allow the manipulation of dried sheets; ④ microstructural homogeneity; ⑤ good thermo-compression ability (the object to laminate multilayer materials); ⑥ easy pyrolysis (burnout); and ⑦ high mechanical strength of green tapes. These require careful selection of the slurry additives together with accurate control of many processing parameters.

The water-based binder is very important because it provides strength to green tapes after evaporating the solvent water through organic bridges between the ceramic particles. Some water-soluble binders can be used in aqueous tape casting, e.g., cellulose ethers (such as hydroxyethyl cellulose), polyvinyl alcohols and latexes. All these three types of binders have been successfully used in aqueous tape casting (Kristoffersson A., et al., 1997; Pagnoux C., et al., 1998; Doreau F., et al., 1998, 1999; Gutierrez C. A., et al., 2001).

3) The Properties of Alumina Powder and Binder

The alumina powder used in this experiment had a median size of 3.84 μm as shown in Fig. 2.17(a), which was produced by Henan Xinyuan Corp. The binder used in the aqueous tape casting process was a commercial styrene-acrylic latex that had low foam and viscosity to produce flexible green tapes with good mechanical properties and lamination behaviors. As shown in Fig. 2.17(b), the styrene-acrylic latex was stable nonionic latex at pH = 7, which the IEP must be low at pH = 3.0 when the zeta potential arrived at zero. The latex was a dispersion of fine latex particles of $d_{50} \approx 300$ nm and a solid loading was about 50 wt%. The glass-transition temperature (T_g) of the latex was about 273 K, which allowed it to form a film at room temperature.

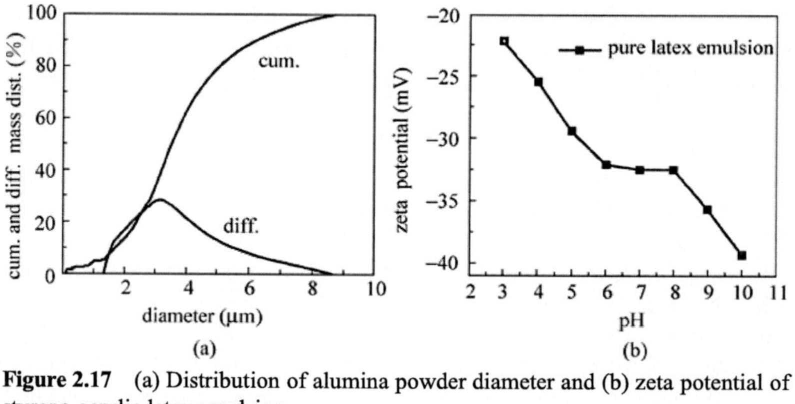

Figure 2.17 (a) Distribution of alumina powder diameter and (b) zeta potential of styrene-acrylic latex emulsion

The fabricated slurry was degassed in vacuum to remove gas bubbles and then it was casted on the polymer film substrate and dried naturally.

4) The Effect of pH and Dispersant on Characteristics of Slurries with Binder

The high-performance aqueous ceramic slurry can be acquired by adjusting the pH and selecting the appropriate dispersant (Hotza D., et al., 1995a, b; Zeng Y. P., et al., 2000). The zeta potential is a very important physical parameter which can decide the stability of aqueous ceramic slurry. The higher the zeta potential absolute value, the higher the repulsive energy and the more stable the slurries.

Figure 2.18 shows the relationship between the pH and the zeta potential of alumina in water. The curve is the zeta potential of Al_2O_3 without any additive at different pH; the isoelectric point at which the zeta potential was zero was at pH = 6.8. The IEP of emulsion was low than 3.0; therefore the pH of alumina slurry must be far from the range of pH = 3 – 6.8. Some research results showed the well-dispersed ceramic slurries at pH = 10 for aqueous slurry of alumina. In this experiment, the pH of latex and alumina solution were adjusted to 10, and then mixed together. It was reported that polyelectrolyte (PAA-NH_4, that is polyacrylic acid-NH_4) could not dissociate at pH<3.5, but could completely dissociate around pH = 10. The result shows that the pH would greatly affect the zeta potential and dispersive effects of dispersant. If the pH of alumina slurry was 10, the PAA-NH_4 polyelectrolytes could get completely ionization and acquire stable slurry. Therefore, the PAA-NH_4 was an appropriate dispersant for latex slurry.

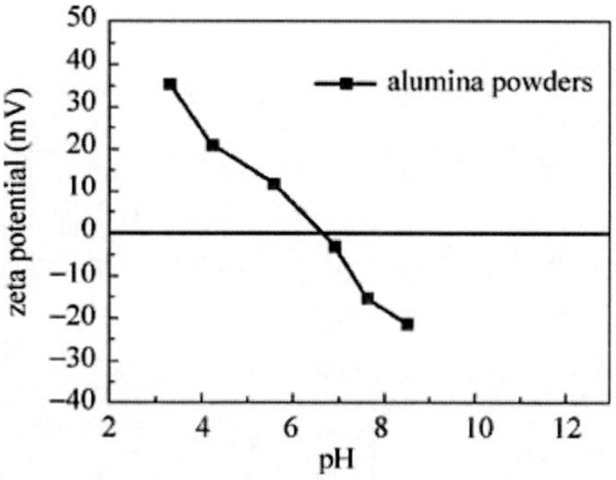

Figure 2.18 Zeta potential of alumina in water

Figure 2.19 shows the relationship between the concentration of dispersant and the shear viscosity of slurries at shear rate of 20 s^{-1}, and the function curve appears to be U-shaped. The reason is that PAA-NH4 is a long link molecule, which can adsorb on the surface of particles to prevent particles from contacting each other and thus stabilizing the slurries. At some point, the long chain molecule was bestrewed on the surface of alumina particles, which dispersant had the best effects in slurry. In this experiment, when the concentration of dispersant was around 1.0 wt% of alumina powders, the viscosity of slurry was the smallest at the same shear rate. When the concentration of dispersant was smaller than 1.0 wt%, the distribution of dispersant on the surface of alumina particles would be insufficient, and the dispersive effect would be not good. If the concentration of dispersant were more than 1.0 wt%, the long chain molecules maybe entwist or flocculate each other, on the contrary, it would bring rising on the viscosity of slurry. In other word, the main experimental factors which influenced the stability of the slurry, were the properties of dispersant and pH.

5) The Rheological Characteristic of Slurries with Binder

The rheological behaviors of the slurries prepared with latex binders always showed a typical shear-thinning behavior with different solid loading at first, and then showed shear-thickening from a critical point, which had hold the lowest viscosity (as shown in Fig. 2.20(a)). This behavior agreed well with the requirements for tape casting process when the shear rate was relatively low. At a shear rate of 21.7 s^{-1}, the viscosity of suspension remained approximately 0.2 – 1.0 Pa·s with fixed binder volume fraction of 30%. Hence, all the four slurries were suitable for tape casting (Pagnoux C., et al., 1998; Doreau F., et al., 1998, 1999; Gutierrez C. A., et al., 2001).

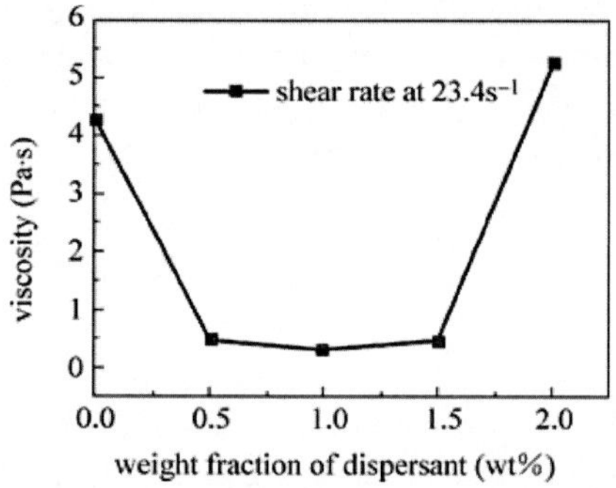

Figure 2.19 The relationship between the concentration of dispersant and the shear viscosity

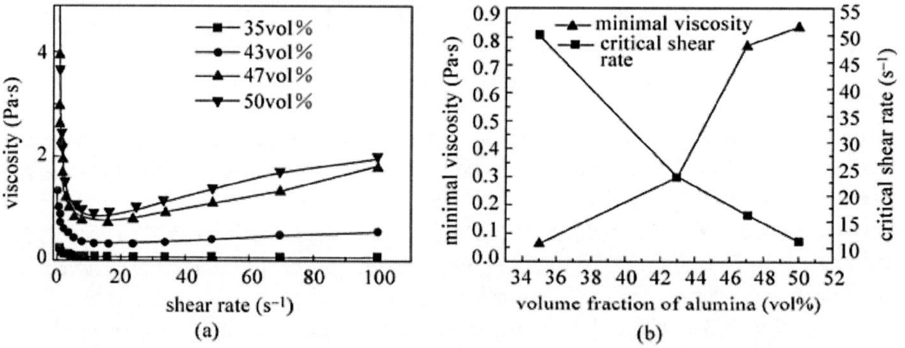

Figure 2.20 (a) The rheological behavior of the slurries with different solid loading and (b) The relationship of the lowest viscosity and corresponding shear rate

Figure 2.20 (b) showed that the lowest viscosity of suspensions and corresponding shear rate were increased with alumina solid loading by holding constant binder concentration (fixed solid loading 30 vol%). The variation curves showed the shear rate value at critical point, which decreased with the increase of solid loading. The result might be from the increasing of flowing resistance with increment of solid loading in suspensions. Therefore, the optimal alumina solid loading of slurry is less than 50 vol%.

6) The Drying Curves of Alumina Slurry with Binder

Figure 2.21 was drying curves of the suspension, which included different drying temperatures on the same slurry (solid loading was 40 vol%, latex binder content was 30 vol%). As shown in Fig. 2.21, the slope of drying curve was increased with the increase of drying temperature before transition point, which the water loss was hold invariability after this critical point. The slope was equivalent to drying rate which was solvent water losing at one time unit; therefore, the larger the slope of drying curve, the faster the drying rate. But the transition point was ahead of time that the water loss began to hold invariability with the increment of temperature from 20 to 30℃. In general, the drying stage before transition point was defined the first drying process. All weight-loss curves in Fig. 2.21 followed a linear relationship with time during the first initial drying process. Therefore, it might be concluded that the evaporation process from the slurry interface should control the first drying process of the suspensions before around 60 min in this slurry. However, the drying rate of surface of slurries was very high after the suspensions were cast on the support film. After some time, the drying rate became very low. The drying process might be controlled by the transportation of the water from inner to the surface of the suspensions after the surface of suspensions dried. Therefore, the resistance of diffusion of the solvent water was greatly decided by the slurry coating thickness, which was larger than evaporation resistance of the initial stage. The drying rate was slower than the initial drying stage, and Fig. 2.21 showed the results.

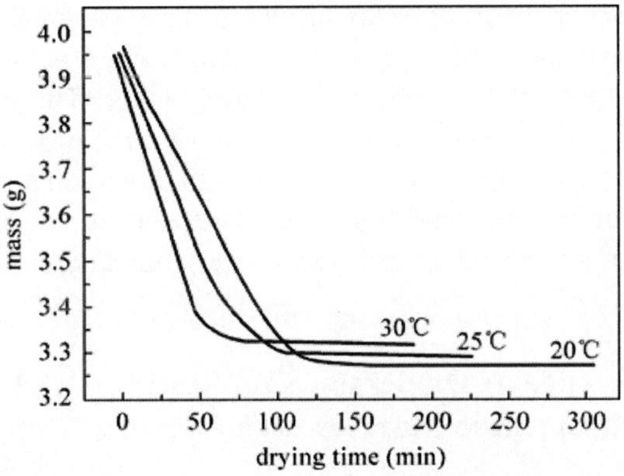

Figure 2.21 Drying curves with different temperature

7) The Microstructure of Green Tape Prepared by Slurry with Binder

During the drying process, the compressive force induced by evaporating would result in the latex particles distortion, and generate the cross-linking structures in contact latex particles of the solidified tapes. Figure 2.22 showed the cross-linking structures in SEM photograph of green tapes. Therefore, the drying process was one of the most important steps, which enabled to obtain a solid tape from slurry.

The film forming mechanisms usually include two necessary conditions: ① the compressive force for latex particles results from solvent evaporation, and ② the glass transition temperature of latex is lower than work temperature (Hong Xiaoyin, et al., 2001; Tan Mengkwang, 1996).

Figure 2.22 The cross-linking structure in SEM photograph of green tapes fracture

This study showed the efficiency of styrene-acrylic binder for water-based tape casting. The rheological behaviors of the slurries prepared with latex binders always showed a typical shear-thinning behavior with different solid loading at first, and then showed shear-thickening from a critical point. The PAA-NH_4 was found to be an effective dispersant for the slurry of latex binder. The main experimental factors that influenced the stability of the slurry were the properties of dispersant and pH. The research results showed that the glass transition temperature of latex (T_g) was also important to form film. The drying speed of slurries was mostly decided by the drying temperature and slurry coating thickness.

2.2.2 The Influence of Dispersants on Stability and Rheology of Aqueous Ceramic Slurries with Organic Monomer

1) The Effect of Dispersant on Powder Dispersion and Slurry Viscosity

In order to get a slurry with a higher solid loading, dispersants are used. A dispersant can change the charge distribution of the powder surface. Figure 2.23 shows the zeta potential of the slurries with and without ammonium citrate. It is clear that the dispersant (ammonium citrate) can raise the absolute value of zeta potential of the slurry when the pH is kept between 8 and 10, and with a shift in the IEP from 8 to 6.5.

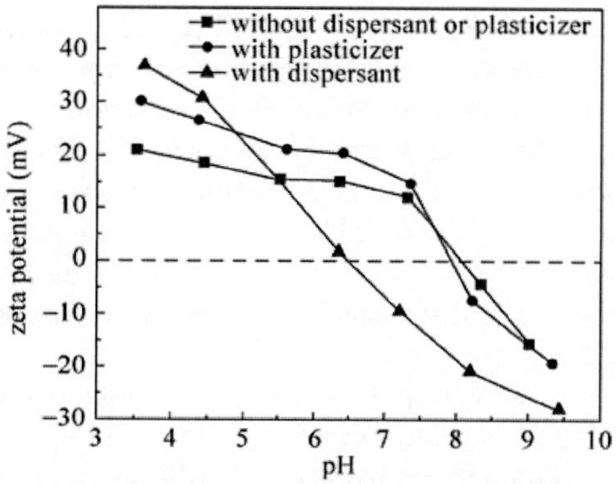

Figure 2.23 The zeta potential of slurry

The effects of ammonium citrate on zeta potential can be considered as follows. At pH = 5.3, alumina reacts with OH^-, leaving AlO_2^- with the negative zeta potential:

$$Al_2O_3 + 2OH^- \longrightarrow 2AlO_2^- + H_2O \tag{2.9}$$

When ammonium citrate dissolves in slurry, the concentration of protolyzable anions (citrate) is increased. Citrate anions tend to chemically bond to the surface of alumina particles. Because the absorption of citrate anions thickens the double layer around the alumina particle, the zeta potential increases. On the other hand, to acquire an equal number of positively charged sites and negatively charged sites, a higher concentration of cations, mainly H^+ in aqueous solution, are needed. So the pH curves shift down (Moreno R., 1992a, b).

The rheological behavior of slurries with different amounts of dispersant are tested at a shear rate of 50 s^{-1}, as shown in Fig. 2.9. The viscosity of the slurries decreases with increasing dispersant content at a relatively lower dispersant concentration. In this case, increasing dispersant content gradually augments the concentration of citrate anions around the alumina particle, thus it can enhance the zeta potential and reduce the viscosity of slurry. However, when the content of dispersant is greater than a certain value of approximately 1 wt% based on the alumina powder, the concentration of citrate anions around the alumina particle has reached its maximum value. The redundant citrate anions just reduce the thickness of the electriostatic double layer and the zeta potential. As the dispersion of alumina particle deteriorates, the viscosity increases rapidly. A slurry with larger than 55 vol% solid loading can maintain good fluidity with the help of 1 wt% dispersant.

2) The Effects of Dispersants on Rheological Property of Slurry with Monomer

The slurry preparation is onc of thc most important stages in the gel-tape-casting process. The expected slurry has high solid loading with good fluidity and stability. Al_2O_3 powder used in this study has a wide distribution of particle size, which is helpful to get high solid loading. The solid loading of the slurry prepared by this

Al_2O_3 powder is more than 55 vol%. The fluidity and stability of the slurry are influenced by those factors such as additives and pH in the slurry. The effects of these factors on rheological property of slurry are discussed.

The dispersants are added into slurries to improve the dispersibility of ceramic powders. For ceramic slurries, there is a relationship between stability and zeta potential of ceramic powders. The powders with high zeta potential (absolute value) usually exhibit higher stability in slurries than powders with low zeta potential do. Accordingly, the stability of slurries can be determined by testing the zeta potential of powders in slurries.

The testing results of the zeta potential of Al_2O_3 powders according to pH in slurries are shown in Fig. 2.24. Slurries (solid loading is 55 vol%) added ammonium citrate (AC), tetramethylammonium hydroxide (TMAH), respectively, as dispersant are compared with the slurry (solid loading is 45 vol%) without dispersant. The absolute value of the zeta potential of powders in slurries added dispersant is higher than that of powders in the slurry without dispersant. The curve of viscosity variation of these three slurries is shown in Fig. 2.25. It can be seen that the viscosity decreases dramatically after the dispersants have been added. This result

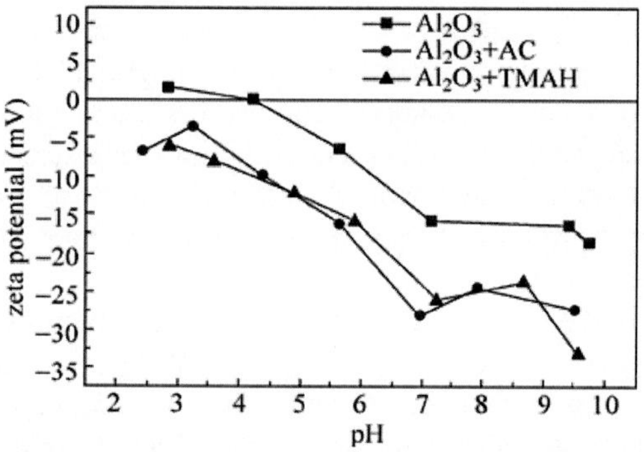

Figure 2.24 The zeta potential of Al_2O_3 with different dispersant

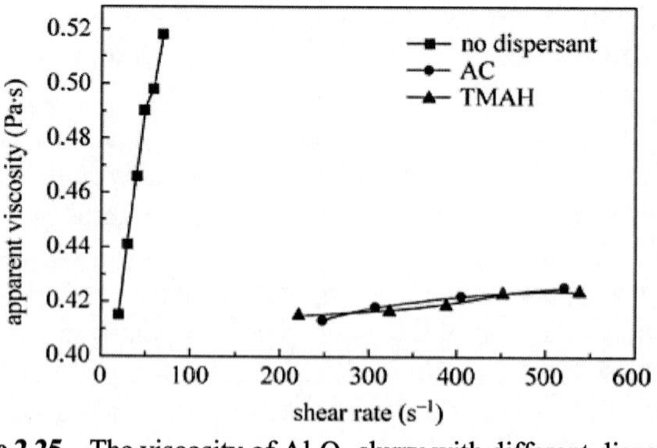

Figure 2.25 The viscosity of Al_2O_3 slurry with different dispersants

implies that AC and TMAH can effectively improve the dispersibility and stability of Al_2O_3 powders in slurries by enhancing the zeta potential, and will lead to a low viscosity and increased stability of slurries.

2.2.3 The Influence of Plasticizer on Properties of Aqueous Ceramic Slurry with Organic Monomer

In general, the polymerized green body is too brittle to be rolled and shaped. The plasticizer is used to make the green body soft enough to sustain machining and shaping processes. Glycerol (G) is selected as the plasticizer in this study. The relationship between the amount of G added and the viscosity of slurries is shown in Table 2.6. The testing results indicate that with increase of the amount of plasticizer, the viscosity increases, and the green tape is likely to be broken while being peeled from the carrier.

Table 2.6 The viscosities of slurries containing organic monomer with different content of G plasticizer (solid loading 30 vol%)

shear rate (s^{-1})	viscosity of slurries with different content of plasticizer (Pa·s)		
	6 wt%	10 wt%	20 wt%
0.0999			
0.215			
0.464	0.00201		0.0032
1.00	0.0016	0.000209	0.0027
2.15	0.00116	0.00286	0.00176
4.64	0.00149	0.00157	0.00231
10	0.00147	0.00184	0.00228
21.5	0.00157	0.00199	0.00236
46.4	0.00162	0.00195	0.0024
100.0	0.00174	0.00202	0.00254

A study is carried out to determine the optimal concentration of plasticizers that can yield a cast tape that is both soft and strong. The flexibility is determined by the number of times that the tape can sustain being bent without breaking. With the amount of plasticizer increased from 0 to 7 wt%, the bending time is increased from 1 to more than 50 (see Fig. 2.26).

The flexibility of the cast tape is improved gradually with the amount of plasticizer increasing and attains the optimal value when the amount is about 5 wt%. When the amount is more than 5 wt%, the flexibility is kept at normal state, but the viscosity of slurry increases obviously which results in a low fluidity of slurry.

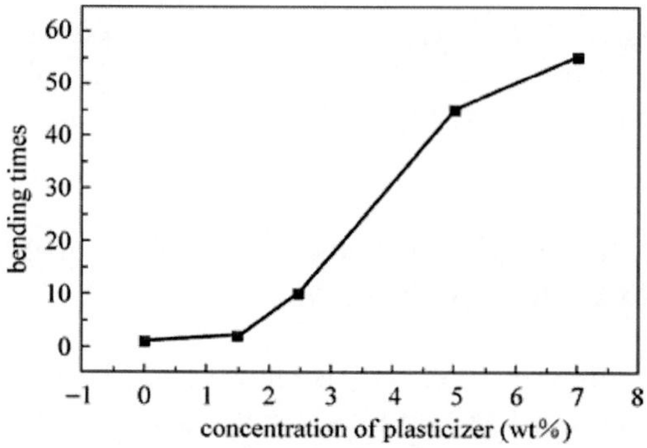

Figure 2.26 The relationship between bending time and concentration of G plasticizer

For this reason, it is optimal to keep the amount of plasticizer less than 5 wt% in slurry to achieve a balance between the flexibility and the strength of the cast tape.

Plasticizer has no evident influence on the zeta potential of the alumina particle, as shown in Fig. 2.23. However, it influences the viscosity of the slurry. The apparent viscosities of slurries with 6 wt%, 10 wt%, and 20 wt% plasticizer (based on premix solution) have been compared, and the results are shown in Fig. 2.27. Figure 2.27 illustrates that the viscosity of slurry increases with plasticizer content, although it does not obviously affect the solid loading and manoeuvrability of the slurry. The increasing viscosity may be attributed to aggregation of alumina particles at higher concentrations of plasticizer.

Addition of a plasticizer is accompanied by a decrease in tape strength. To balance the requirement of strength and flexibility, 2 wt% plasticizer is chosen for this study. In this case, the green tape can be bent to 90° for more than 50 times without cracking.

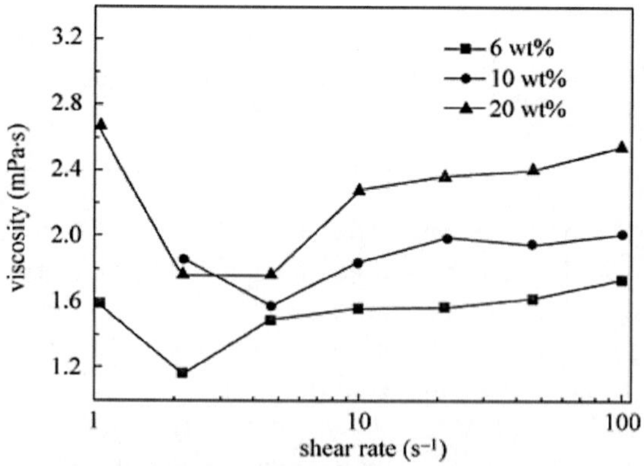

Figure 2.27 The viscosity of the slurries with 6 wt%, 10 wt%, and 20 wt% of plasticizer (G) based on the premixed solution (G–plasticizer containing hydroxyl)

2.2.4 The Influence of pH on the Properties of Slurries with Organic Monomer

The influence of pH on rheological properties of slurries is very important and a suitable pH is helpful for enhancing the solid loading of slurry. The pH is adjusted to review the variation of viscosity of the slurry. Table 2.7 shows the testing results of viscosity of Al_2O_3 slurries at different pH. It can be seen that the viscosity drops dramatically with the pH increasing from 2.14 to 10.11. With the increase of pH, the absolute value of zeta potential of Al_2O_3 powders increases significantly (see Fig. 2.24), resulting in an obvious improvement in dispersibility of the powders. So the slurries exhibit lower viscosity at higher pH.

It is obvious that the low viscosity is beneficial for making the forming process easily operable. It is also helpful for enhancing the solid loading of slurry. According to the experimental results, the suitable range of the pH for the Al_2O_3 slurry is about 7 – 9, and the solid loading of the slurry can reach more than 55 vol% with good fluidity and stability.

Table 2.7 Testing result of viscosity of slurries with monomer at different pH (solid loading 55 vol%)

shear rate (s^{-1})	viscosity (Pa·s)				
	pH = 2.14	pH = 3.9	pH = 6.05	pH = 7.8	pH = 10.11
0.1	29	13	6.61	3.26	1.71
0.164	17	8.26	4.07	1.95	0.916
0.268	12	5.69	2.39	1.22	0.554
0.439	9.02	4.01	1.83	0.783	0.354
0.72	6.13	2.86	1.14	0.506	0.236
1.18	4.31	1.96	0.813	0.342	0.168
1.93	2.98	1.33	0.541	0.231	0.125
3.16	2.14	0.904	0.409	0.165	0.089
5.18	1.47	0.629	0.278	0.118	0.065
8.48	0.965	0.443	0.196	0.0864	0.0482
13.9	0.652	0.319	0.133	0.0668	0.0377
22.8	0.483	0.235	0.0921	0.0553	0.0317
37.3	0.391	0.174	0.0792	0.0488	0.0289
61.1	0.312	0.131	0.0710	0.0462	0.0291
100	0.254	0.101	0.0658	0.0449	0.0314

2.2.5 The Effect of Surfactant on Wetting and Green Tape Releasing (Separating)

The aqueous slurry is hard to overspread on the organic carrier film. So a surfactant is required to improve the wetting property of the aqueous slurry. Introduction of even small amounts of surfactant into slurry can obviously reduce surface tension, which leads to the reduction of the contact angle at the interface of solution and carrier film. Table 2.4 shows the contact angles of slurries on carrier film when various surfactants are used. It can be concluded from these contact angles that the slurry without surfactant is a hard to wet carrier film because its contact angle is larger than 90°. All of these surfactants have an obvious effect on decreasing contact angle to below 90°.

It is also found that the bottom of green tape often sticks to the carrier film due to the high interface tension, as shown in Fig. 2.28. Various surfactants show different effects on this releasing property. Surfactants A, B, and C do not help to solve this problem, but make the interface tension stronger. While using surfactant D, the green tapes easily released from carrier film and held very smooth surfaces. The difference in releasing property results from the surfactant structures. In general, surfactant molecules are gathered on the surface of solution during the process—with the hydrophilic group in aqueous solution and the hydrophobic group on the surface—whilst most common surfactants have an organic hydrophobic group, which has a strong attractive force to the organic carrier film. But the hydrophobic group of surfactant D demonstrates great electronegativity, and the chemical bonds in surfactant D are like an electrovalent bond to some extent. Therefore, the attractive force between surfactant D and the plastic film is so weak that the green tape can be stripped easily from the carrier film.

Figure 2.28 Fracture surface of green tape that sticks to carrier film

2.2.6 Foam and Pore Elimination

It was observed that adding a surfactant into slurry may introduce foams, especially for surfactant D, there were many pinholes and cracked bubbles on the surface of green tape when the foams were not removed from the slurry before casting. Figure 2.29(a) shows a tape surface with pinholes related to surfactant D.

Regarding the foams, when surfactant is added into aqueous solution, the molecules of surfactant tend to aggregate at the liquid/air interface. As shown in Fig. 2.30(a), a layer of liquid film encloses a pocket of air at the surface of the slurry during mixing. The surfactant anions along both sides of the liquid film repel each other to maintain a constant thickness of the liquid film. Thus, a stable foam is formed.

Some methods are adopted to eliminate the foams in slurry. Various defoamers are mixed in slurry during ball-milling. When the defoamer molecules dissolve in slurry, they replace part of the surfactant molecules at the surface of the liquid film as shown in Fig. 2.30(b). Because the electrostatic repulsion in this local area decreases greatly, slurry in the liquid film descends by gravity until the film drains off. Then, the foam cracks and a small amount (<1 wt%) of defoamer shows a strong ability to reduce the number of foams.

Before tape casting, a vacuum de-airing process is required. Vacuum de-airing under 0.02 MPa for 25 min can remove almost all the residuary foams in slurry. The green tape after foam elimination has a highly homogeneous microstructure and no pinholes can be seen from the SEM photo, as shown in Fig. 2.29(b).

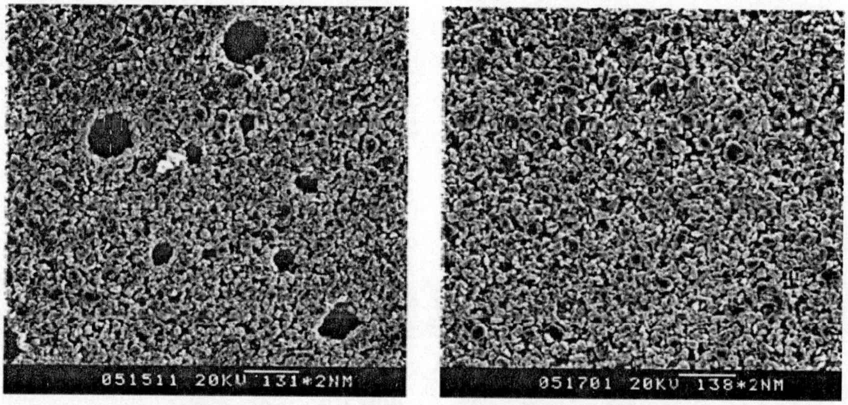

(a) (b)

Figure 2.29 The alumina green tape surface with pinholes (a) and without any pores (b)

Alumina green tapes have been prepared through gel-tape-casting, which combines the merits of gel-casting and tape casting. A slurry with adequate fluidity can sustain a solid loading of more than 55 vol% when 1 wt% dispersant is used. A plasticizer is used to improve the flexibility of green tapes. As the content of plasticizer in slurry increases, the flexibility of green tape is enhanced, but its strength decreases to a certain extent. A surfactant can decrease the surface tension of aqueous slurry,

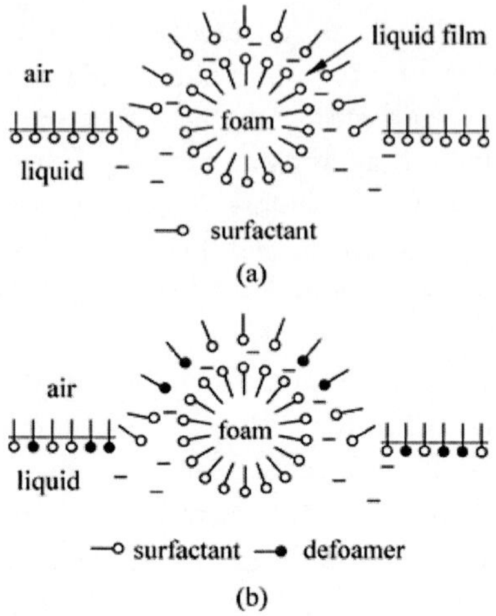

Figure 2.30 The sketch illustration of foam forming (a) and foam eliminating (b)

obviously due to its molecular structure, so that the slurry can easily wet the carrier film. Surfactant D can reduce the interface tension between the green tape and organic carrier film dramatically, thus making it possible for the green tape to be released from the carrier film. Foams are produced during mixing, which disrupt the microstructure of the green tape. The foam removal processes, including addition of defoamers in slurry and vacuum deairing, are adopted to eliminate the foams. A green tape with a high solid loading, high strength, flexibility, and good releasing property is thereby successfully fabricated.

2.2.7 Sintering of Green Tape Prepared by Slurry

To examine the performance of gel-tape-casting process, the cast tape is sintered at 1550℃ for 2 h. The little amount of organic compounds in cast tape makes it possible for the sample to be sintered directly without a special burnout process. No crack is found in the sintered sample.

The relative density of the sintered sample is 97.7%. The variation of solid loading of slurry (53 – 55 vol%) shows no influence on the density of the sintered sample. The SEM photograph of the section of the sintered sample is shown in Fig. 2.31. A homogeneous microstructure without exaggerated grain growth is observed.

Ceramic thin sheets of 0.1 – 1 mm have been successfully formed by gel-tape-casting process. In this process, polymerization of organic monomers is used to form the network in which a ceramic green body is solidified. Special drying and burnout procedures, which are critical stages in conventional tape casting, are omitted, and the production efficiency is highly improved.

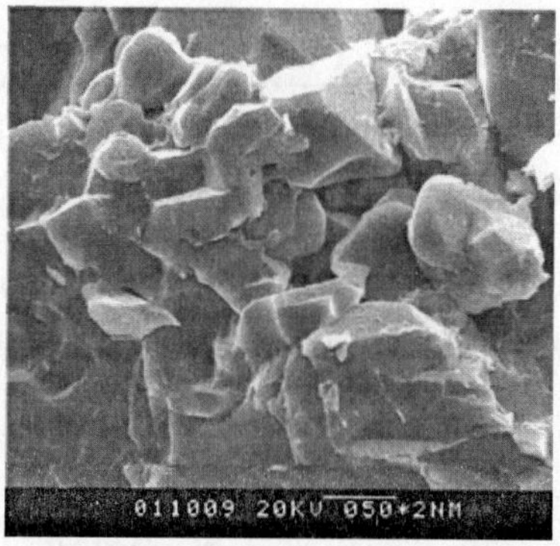

Figure 2.31 SEM image of section of sintered sample

Gel-tape-casting was performed with the protection of an inert atmosphere in order to overcome the effect of oxygen anti-polymerization. Al_2O_3 slurry is polymerized successfully in a few minutes, the temperature being around 50℃ with the protection of nitrogen gas. The relative density of the Al_2O_3 green tape is about 61%.

The Al_2O_3 sample is sintered at 1550℃ for 2 h and relative density is 97.7%. A homogeneous microstructure without exaggerated grain growth is observed. The variation of solid loading of slurry (53 – 55 vol%) shows no influence on the density of sintered sample.

2.3 Aqueous Gel-Tape-Casting with Styrene-Acrylic Latex Binder

2.3.1 The Importance of Binders in Gel-Tape-Casting Process

(1) Raw Materials

The alumina powder was produced by Xinyuan Company in Henan Province (China). The mean particle size was D_{50} = 3.84 μm. The binder was styrene-acrylic latex (Shandong Daming Water-Proof Materials Company) that had low foam, low viscosity and good forming-film properties to produce green tapes with good mechanical and lamination behaviors, and decreased the additional organic additives such as plasticizer or defoamer. The glass transition temperature (T_g) is an important parameter for latex, and the T_g of this latex was below room temperature, about 273 K.

The dispersant used for stabilizing the alumina suspensions was PAA-NH_4, which is an ammonium polyacrylate salt solution with low molecular weight. The optimal amount of PAA-NH_4 was 1.0 wt% with respect to the alumina powder.

Three different concentrations of latex based on the gross volume (alumina powders + binder + water + additives) were 10 vol%, 20 vol%, and 30 vol%, respectively. The Al_2O_3 volume fraction was maintained at a constant level (43 vol%) and adjusted by deionized water.

(2) Tape Casting

Tape casting experiments were carried out on a laboratory tape caster (manufactured by Tsinghua University). This device enabled the production of tapes of 120 mm in width and 500 mm in length by fixing the container with two blades. Micrometer screws allowed setting the gap of blades from 0.05 mm up to 2 mm.

In the process, the support film is very important for stripping easily and getting continuous coiling tapes. Thus, we adopted the polyester film by silicon treatment whose thickness was 0.1 mm.

After some previous tests, a gap of 500 μm under the blade and a casting speed of 10 mm/s was selected for all experiments. According to the empirical equation (2.10), the shear rate generated under the blade was estimated to be 20 s^{-1}.

$$\gamma = v / h \tag{2.10}$$

where γ is the shear rate (s^{-1}), v the tape casting speed (m/s), h the blade gap height (m).

The drying process was tested in a container on the electronic balance at room temperature. As shown in Fig. 2.32, the container was fabricated by plastic sheets that were bonded on the support film, whose dimensions were: the length = 40 mm, the width = 40 mm, and the height = 1 mm. After casting the suspension in the container, the relationship between the weight loss of the samples and time was recorded every 2 min.

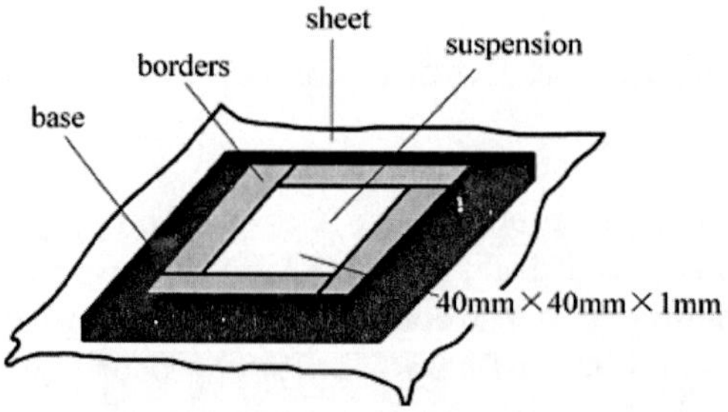

Figure 2.32 The mold used in the drying experiment

The bulk density of the green tapes was calculated by measuring the thickness, diameter and weight of the disk-shaped specimens. The average density of the unlaminated green sheets was estimated on the base of 10 samples disks that were punched along the dried coiling tapes.

2.3.2 The Forming Film Mechanism of Latex Binder

Latex can be defined as a stable colloidal dispersion of a polymeric substance in an aqueous solvent. The polymer particles are usually spherical in the size range of 30 – 500 nm. The styrene-acrylic latex acts as a binder that is absorbed on the surface of alumina particles in slurries. After the slurries were cast on the support film, the water continuously evaporates from the slurries all along. Finally, the latex particles could coalesce and form a polymeric film on the alumina particles when the water was lost. The alumina particles would be accumulative by bonding action of the polymer film, and then form the alumina tape. The latex binder forms a film on the surface of alumina particles as shown in Fig. 2.33.

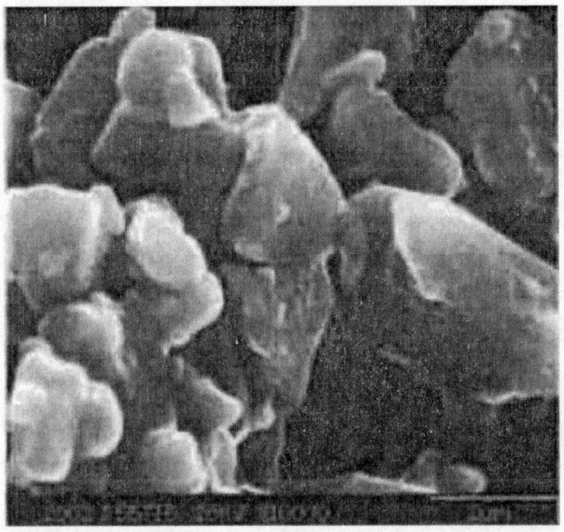

Figure 2.33 The SEM image of latex film on the surface of alumina particles

Figure 2.34 shows a drying curve of the suspension, which contains different fractions of latex binder under the same conditions. As shown in Fig. 2.34, the gross drying time decreases with the increasing of binder fraction in suspensions. The suspension with 30 vol% binder is dried quickly and the water loss remains invariable after 160 min. But the suspension with 10 vol% binder dried slowly and the water loss was kept at a fixed rate after 200 min. All weight-loss curves follow a linear relationship with time during the initial drying process before 80 min. Therefore, it may be concluded that the latex binder can not influence the first drying process of the suspensions because this binder is controlled by the evaporation process.

The drying curves deviate from the straight line after about 80 min. The drying curve of the suspension with 30 vol% binder is the earliest deviation curve, and the water loss rate is the slowest at the second stage among the three types of slurries. This is because the second drying process is controlled by the transportation of the water from the inner to the surface of suspensions after the surface of suspensions dried. The high fraction of latex binder should increase the resistance to diffusion at the second stage. Because the Al_2O_3 volume fraction is maintained at a constant level (43 vol%) and adjusted by deionized water, the suspension with 30 vol% would contain much less water than other suspensions, the gross drying time is the shortest.

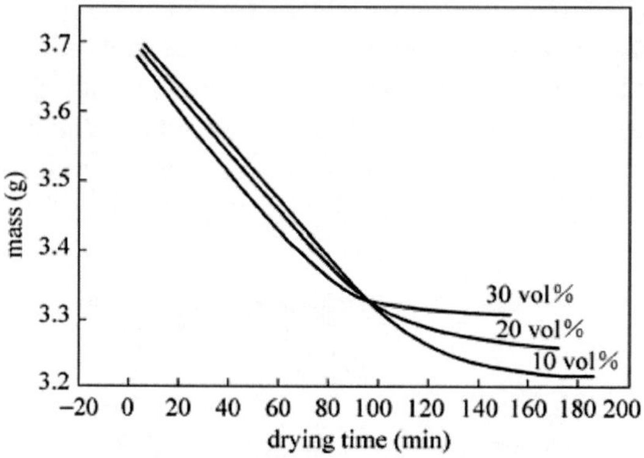

Figure 2.34 The drying curve of suspensions containing different fractions of latex binder

In summary, the solvent water removal from suspensions involves a two-stage process. The first stage is controlled by evaporation of the water, while the second stage is controlled by diffusion process of the water through the solidified part of the film with a decreasing drying rate because of the diffusion resistance. The surface of tapes dries when the first stage has been finished fast; therefore, the second stage is slower than the first.

During the first drying stage, the solvent water starts to evaporate from the top surface of the tape, allowing particles to approach each other. As a consequence, the curvature of the liquid/vapor interface in the pore channels among particles progressively increases. For a liquid that completely wets the solid surface, the difference in pressure across these curved surfaces, ΔP is given by:

$$\Delta P = -\gamma_{LV}(1/r_1 + 1/r_2) \tag{2.11}$$

where γ_{LV} is the liquid/vapor interfacial energy, r_1 and r_2 are the radius of curvature of the inside meniscus surface and outside meniscus surface, respectively. When the center of the curvature is in the vapor phase, the radius of curvature is negative and the liquid is in tension ($\Delta P > 0$). This tension of the liquid corresponds to a compressive force between ceramic particles, which is responsible for shrinkage of the casting slurries. Hence, the ΔP is a driving force for suspension forming film.

The glass transition temperature T_g of styrene-acrylic latex is 5℃, while the working temperature T is about 25℃. The difference of temperature ($T - T_g$) is enough to form film by drying. Since the binder ensures the high strength of the green tapes, and the plasticizer ensures a good toughness of the green tapes, the tapes can be coiled continuously by a roller. The role of the plasticizer is to controll the rate of evaporation of water in the drying process and to retain some water in the green tapes after drying.

Hence, the driving force ΔP and T_g are two key factors for forming a film in the tape casting processing.

2.3.3 Rheological Properties of the Alumina Slurries with Binder

The rheological behaviors of the slurries prepared with latex binders always show a typical shear-thinning behavior with different volume fractions of binders (see Fig. 2.35(a)). This behavior agrees well with the requirements for tape casting process. At a shear rate of 21.7 s^{-1}, the viscosity of suspension remains approximately 0.2 – 0.4 Pa·s with binder volume fraction of 20% and 30%. Hence, all the two slurries are suitable for tape casting. Figure 2.35(b) shows that the viscosity of suspensions increases with styrene-acrylic binder content by keeping constant solid concentration (constant solid fraction 43 vol% and constant shear rate). The increase of viscosity results from the increment of latex, thereby increasing the flowing resistance of suspensions.

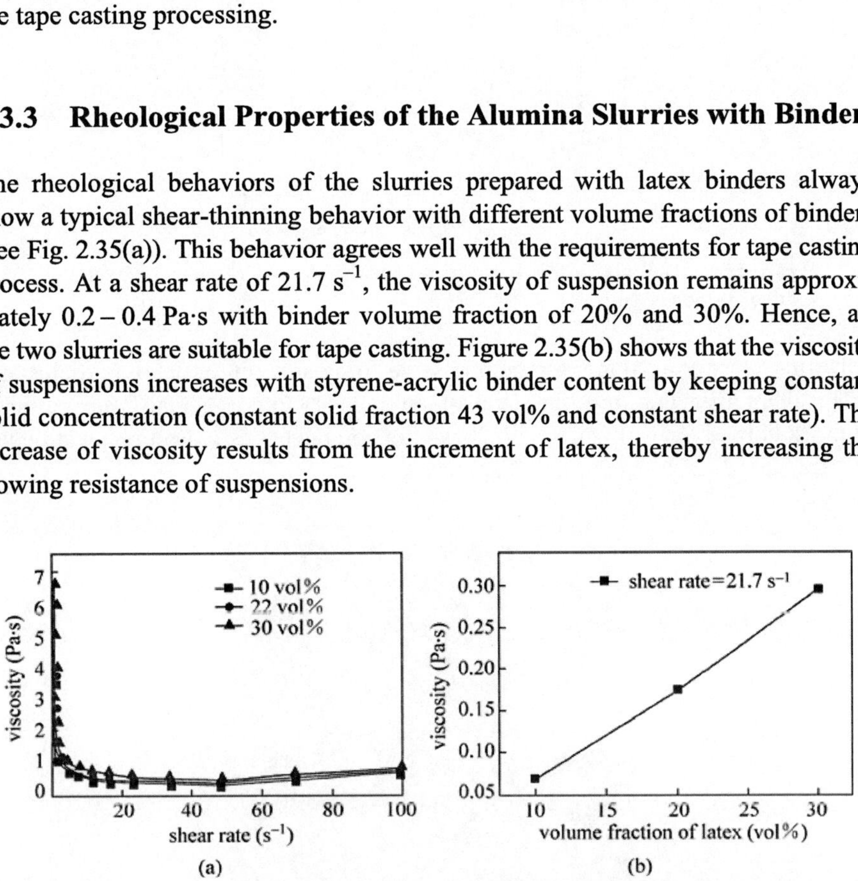

Figure 2.35 (a) The relationship between viscosity and shear rate with different latex fractions; (b) The relationship between viscosity and different latex fraction at constant shear rate

2.3.4 The Physical Properties and Microstructure of Green Tapes with Latex Binder

The bulk density of alumina green tape increases with the increase of latex volume fraction from 10 vol% to 30 vol% of the entire system. As shown in Fig. 2.36, it results from the increase of the gross solid volume (alumina + solid particles of latex), and the compressive force for latex particles becomes stronger because the drying driving force increases with the increase of latex volume.

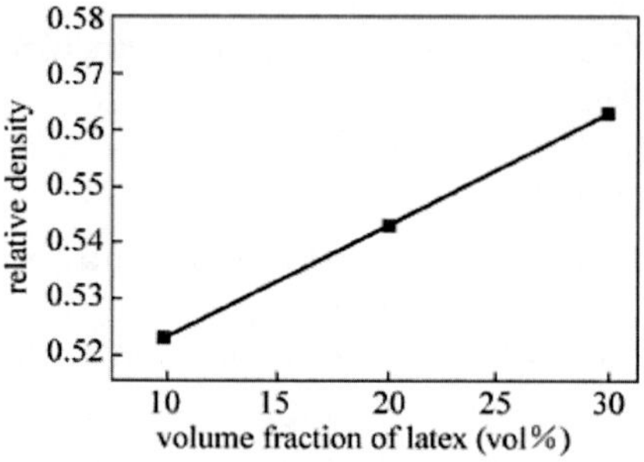

Figure 2.36 The relationship between the relative density of green body and the volume fraction of latex

During the drying stage, the compressive force acts on the latex particles and makes them distorting, resulting in cross-linking structure in contact latex particles of the solidified tapes. Figure 2.37 shows the homogeneous cross-linking structure in a fracture SEM image of the green tapes.

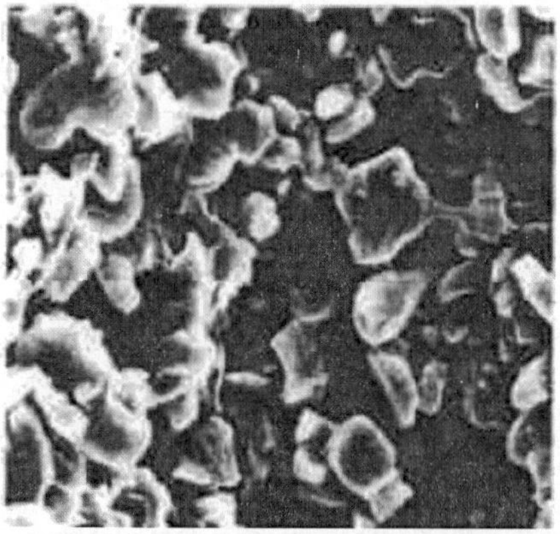

Figure 2.37 The SEM image of Al_2O_3 green tapes with latex binder

In this experiment, a smooth-surface and high-strength green tape was fabricated by aqueous tape casting with a latex binder system, as shown in Fig. 2.38.

It is clear in experiment that the well-dispersed aqueous slurries suitable for tape casting are prepared, and the slurries prepared with latex binders exhibit an obvious shear-thinning behavior.

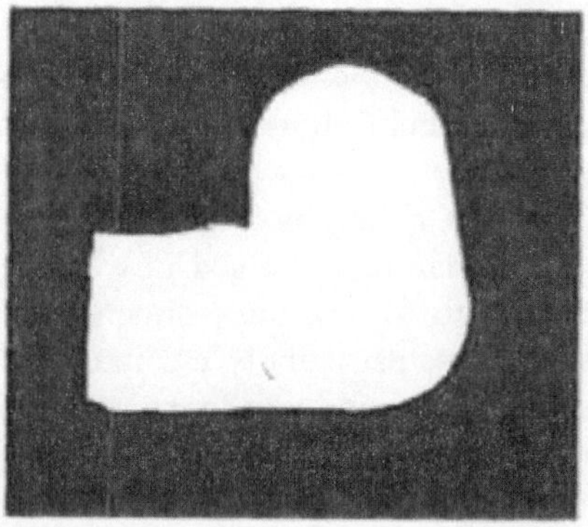

Figure 2.38 The coiling alumina green tapes with latex binder

The water-based tape casting has been used successfully to fabricate a smooth-surface and high-strength green tape with styrene-acrylic latex binder. The film drying mechanism involves two necessary conditions: the compressive force for latex particles resulted from solvent evaporation, and the glass transition temperature of latex that is lower than the work temperature. The driving force ΔP and T_g are the two key factors for forming a film in the tape casting processing.

2.4 A Gel-Tape-Casting Process Based on Gelation of Sodium Alginate

2.4.1 Why Study on Tape Casting of Sodium Alginate

1) Research Background

The tape casting process can be divided into non-aqueous and aqueous systems according to the solvents used (Hotez D., et al., 1995). Non-aqueous solvents typically have lower boiling points and can prevent the hydration of some ceramic powders, but require special precaution concerning toxicity and inflammability. In recent years, the environmental and health aspects of the tape casting process have drawn more attention. Therefore, aqueous solvents, which have the advantages of incombustibility, non-toxicity, and low cost, are considered as suitable substitutes for non-aqueous solvents. However, the aqueous suspension has a smaller tolerance to minor changes in drying conditions. Moreover, the drying rate of the aqueous suspension is invariably much slower than that of the non-aqueous systems (Hotez D., et al., 1995; Briscoe B. J., et al., 1998). Therefore, non-homogeneous microstructure and cracks usually occur in green tapes. To overcome these drawbacks, Dr. Xiang developed a gel-tape-casting process as an alternative route to the process, in which a monomer solution was used as a substitute for organic solvents and binders to coagulate ceramic suspensions (Xiang J. H., 2002). However, the monomer used is acrylamide with neural toxicity. In this paper, a natural innoxious polymer, sodium alginate, was used to consolidate ceramic suspensions.

Sodium alginate is a type of gelling polysaccharide commonly isolated from brown kelp. Sodium alginate can be dissolved in water at room temperature. It undergoes chemical gelation to form a three-dimensional (3D) network in the presence of multivalent cations, particularly calcium (Jin J., et al., 1993). The alginates are usually used as processing aids in many traditional ceramic-forming processes, including dry molding, extrusion, and slip casting, where they impart plasticity, workability, suspension stability, and suitable wet and dry strength. In fact, the chemical gelation of sodium alginate has also been used in gel-casting and solid free form fabrication techniques to produce ceramics (Suresh B., et al., 1998; Xie Z.P., et al., 2001). In this paper, sodium alginate was used to fabricate green tapes by using a sequestrant to control gelation behavior.

2) Experimental Procedure

(1) Starting Materials

The starting ceramic powder used here was α-Al_2O_3 (>99.7 wt% purity, Henan Xinyuan Aluminum Corporation, China) with a mean particle size of 2.9 μm, and the specific surface area 0.434 m^2/g. All additives of the suspensions used in this work are listed in Table 2.8.

Table 2.8 The additives of gel-tape-casting suspensions

binders	sodium alginate
	styrene-acrylic latex[a]
dispersant	ammonium citrate tribasic, $C_6H_{17}N_3O_7$
calcium salt	calcium phosphate tribasic, $Ca_3(PO_4)_2$
sequestrant	sodium hexameta phosphate, $(NaPO_3)_6$
acid	adipic acid, $C_6H_{10}O_4$
plasticizer	glycerol

[a] 50 wt% polymer, glass transition temperature (T_g), 5℃.

(2) Suspension Preparation and Tape Casting

The process flowchart is presented in Fig. 2.39. Sodium alginate was dissolved in the deionized water to form a solution, and then the styrene-acrylic latex was added into the solution and stirred for 30 min. After alumina powders, dispersant, plasticizer, sequestrant and calcium salts were added into the mixture of binders, the suspension was ball milled for 24 h. Before casting, the resulting suspension was deaired under vacuum (0.02 atm) for 10 – 15 min and then adipic acid was added. The prepared suspension was cast at room temperature, and was subsequently consolidated to form a green tape.

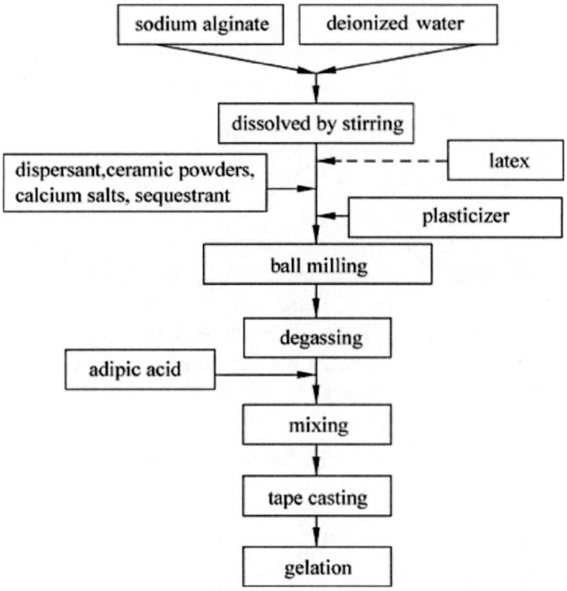

Figure 2.39 The flowchart of gel-tape-casting process

2.4.2 The Preparation of Aqueous Alumina Suspensions with Sodium Alginate and Calcium Phosphere Tribasic

Suspensions with 52 vol% solid loading (50 vol% alumina and 2 vol% calcium phosphate tribasic) and 1 wt% sodium alginate (based on water) were prepared. The effect of the amount of dispersant (based on ceramic powders) on the viscosity of the suspensions is shown in Fig. 2.40. It is obvious that the viscosity of the suspensions decreases with the increases of dispersant amount. The minimum of viscosity, corresponding to the best dispersion of alumina particles in the suspensions, is observed when about 0.3 wt% dispersant is added. When the amount of dispersant increases beyond 0.3 wt%, the viscosity increases again, which implies that there exists an optimum concentration at which just enough dispersant is present to provide maximized coverage of the ceramic particles and any excess dispersant may be harmful in decreasing viscosity. This optimal dispersant content was used for the preparation of all the suspensions in the following experiments.

In the range of shear rate tested, the suspension with 52 vol% solid loading and 1 wt% sodium alginate and different latex contents shows an initial shear thinning behavior at low shear rates followed by a shear thickening behavior (shown in Fig. 2.41). Such behavior is unfavorable for tape casting because of the high shear stress generated under the blade (about 50 s^{-1} in our conditions). The shear thickening behavior usually occurs in a high concentrated suspension.

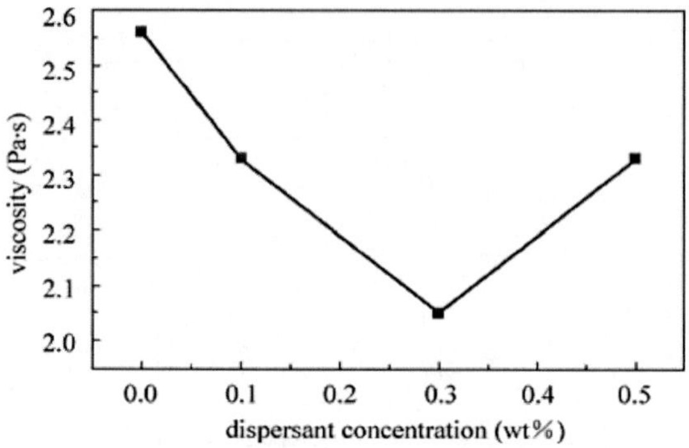

Figure 2.40 The effects of dispersant content on the viscosity of 50 vol% Al_2O_3 suspensions (shear rate: 50 s^{-1}; dispersant: ammonium citrate tribasic)

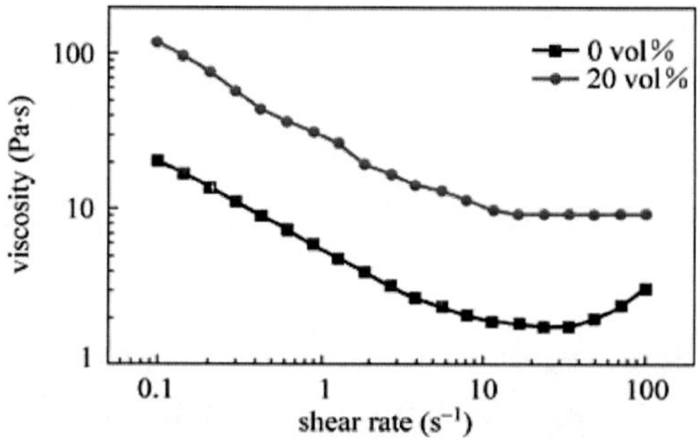

Figure 2.41 Viscosity curves of Al_2O_3 suspensions with different latex concentration

Indeed, the structure of the suspension is determined by the interaction forces between particles (Bender J., et al., 1996; Granja M. F. L., et al., 1997). At low shear rates, the agitation action can gradually break down the structure in suspension by overcoming the attractive energy and the viscosity of the suspension decrease. At high shear rates, the hydrodynamic forces between particles become dominant. This will lead to the formation of clusters in suspensions and the viscosity of the suspension will increase.

The green tapes obtained using the above mentioned suspensions were easily removed from the carrier and no obvious cracks were observed but tapes were not flexible enough for handling. The plasticizer (glycerol) could not effectively improve the flexibility of green tapes. In our previous study, alumina suspensions with 20 vol% styrene-acrylic latex and 1 vol% glycerol were found to be suitable for tape casting (Cui X. M., et al., 2002). Therefore, 20 vol% styrene-acrylic latex (based on sodium alginate solution) was simultaneously added into suspensions to improve the flexibility of the tapes here. As shown in Fig. 2.41, the addition of latex has a great effect on the viscosity of the suspensions and the type of their rheological behavior. The shear thickening behavior observed earlier disappears, and the suspensions containing latex exhibit a shear thinning behavior more suitable for tape casting now.

2.4.3 Control of the Gelation of Sodium Alginate

In order to complete casting processing under control, a controllable reaction of calcium salts with sodium alginate must be considered. Generally, it is difficult to control the gelation rate between calcium salts (such as $CaCl_2$, CaC_2O_4) and alginate, which makes it impossible to complete casting processing in a suitable period. In this study, the reaction rate was adjusted by sequestrant ($(NaPO_3)_6$) and adipic acid ($C_6H_{10}O_4$). Generally, adding $Ca_3(PO_4)_2$ and $(NaPO_3)_6$ into the ceramic suspensions simultaneously, the gelation reaction between calcium salts and sodium alginate will be prevented because a stable complex is formed from the reaction between $(NaPO_3)_6$ and $Ca_3(PO_4)_2$. After $C_6H_{10}O_4$ is added into the ceramic suspensions, the complex would be decomposed and calcium ions are released slowly, thus leading to gelation.

The gelation rate of suspension with 52 vol% solid loading was investigated. The beginning of gelation in ceramic suspensions was determined by changes in suspensions temperature, since the gelation is an exothermic reaction. The process was monitored in terms of induced time, $t_{induced}$, the time interval between the addition of the adipic acid and the commencement of gelation, which is equivalent to the time available for casting the suspensions during processing. Figures 2.42 and 2.43 show the variation of induced time with the amount of adipic acid and with sequestrant, respectively. These data clearly show that the induced time is reduced by an increase of acid content and increased by an increase of sequestrant content. The induced time, i.e., the release rate of calcium ions, can be easily adjusted by the concentration of the reagents, and thus the process can be easily completed under control.

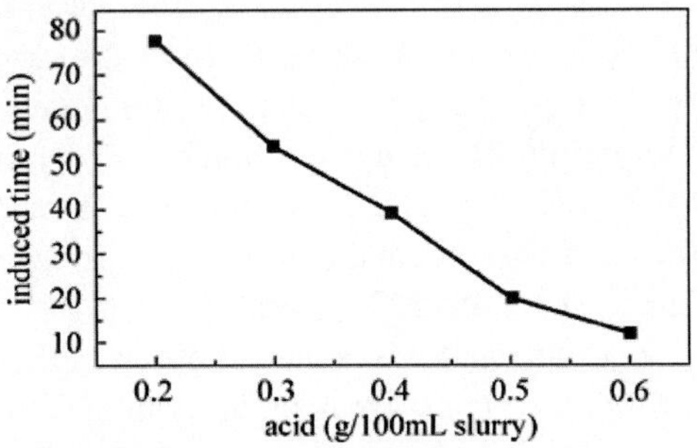

Figure 2.42 The relationship between amount of adipic acid and induced time

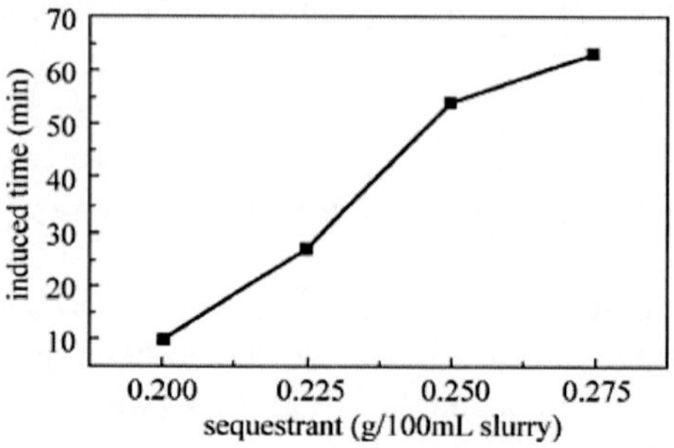

Figure 2.43 The relationship between amount of sequestrant and induced time

2.4.4 Characterization of Green Tapes

The green densities and the flexibility of green tapes using three formulations of binders are summarized in Table 2.9. Although the tapes with only sodium alginate showed poor flexibility, the tapes with alginate and latex showed good flexibility, which is similar to the tapes with only latex. The relative density of the green tapes with latex is higher than with only alginate. However, the addition of alginate to the suspensions with latex showed no obvious influence on the density of the green tapes.

Table 2.9 The relative green densities and flexibility of green tapes

binders	relative green density (% TD)	observation
alginate	~ 56	poor flexibility
latex	~ 58	good flexibility
alginate + latex	~ 58	good flexibility

In the conventional aqueous tape casting process, the drying procedure is a critical stage. Usually, the drying of green tapes is inhomogenous. During drying, both the anisotropic migration of the solvent and the differential shrinkage will generate stresses in the green tapes and usually lead to cracking. An obvious difference in microstructures of upper and bottom layers is usually observed (Zeng Y. P., et al., 2000; Cui X. M., et al., 2002; Bernd B., et al., 2002). In this process, the ceramic powders and organic additives are solidified in the 3D network due to the *in-situ* gelation of sodium alginate simultaneously in the whole tape. Therefore, there is no need for a special drying procedure in which the temperature and humidity are to be controlled carefully in order to avoid crack and inhomogeneity due to migration of solvent carrying binders, plasticizers and fine powders. Figure 2.44 shows the microstructures of upper surface, bottom surface of a green tape containing 20 vol% latex and 1 wt% sodium alginate. No obvious difference was observed between the upper surface and bottom surface. This shows that a more homogenous microstructure could be obtained by the addition of sodium alginate.

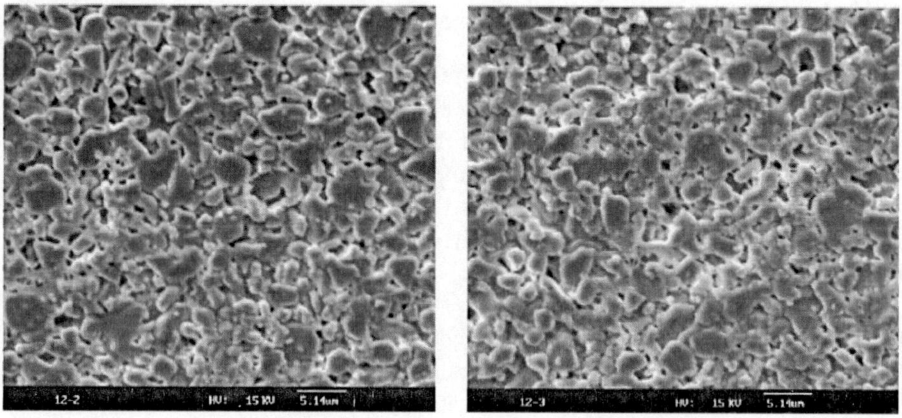

(a) (b)

Figure 2.44 SEM micrograph of Al_2O_3 green tapes
(a) upper surface; (b) bottom surface

A novel gel-tape-casting is described as following. In this method, sodium alginate is used to consolidate alumina tapes from aqueous suspensions. However, because of low flexibility of resulting tapes, styrene-acrylic latex was also added to enhance the flexibility. A special drying procedure, which is a critical stage in conventional tape casting, is avoided.

The use of a dispersant can effectively improve the fluidity of the suspension. The optimal amount of dispersant is about 0.3 wt%. The styrene-acrylic latex strongly affects the rheological behavior of the suspensions. Suspensions containing styrene-acrylic latex exhibit a typical shear thinning behavior which is suitable for this process.

The time available for casting (induced time) can be controlled by varying the concentration of adipic acid and sequestrant. The induced time is reduced by an increase of acid content or increased by an increase of sequestrant content. The processing can be easily controlled by changing the quantities of adipic acid and sequestrant.

The green tapes with homogeneous microstructure were fabricated by this novel forming process.

2.5 The Spray Trigger Fast-Curing for the Gel-Tape-Casting Process

2.5.1 The Idea of the Spray Trigger Fast-Curing

The gel-tape-casting process includes the following steps: adding the initiator and catalyst into the prepared slurry → placing the slurry into the slurry reservoir → flowing into the tape → heat curing.

The acrylamide polymerization rate is related to the temperature. Polymerization is slow at room temperature, and quickly accelerate when the temperature exceeds 40 – 50℃. In casting process, the use of this feature, the curing rate of the slurry can be controlled by adjusting the temperature, the slurry maintain a lower temperature (room temperature or even lower) before the gel-casting, avoid curing in the slurry reservoir; and then the slurry will be fast solidified into the tape or sheet after heated to 60 – 70℃.

Since the slurry can still be polymerized slowly at low temperatures, the casting operation of the slurry must be completed in a certain period of time and have a certain limitations. Otherwise, the slurry solidification will not be taken place simultaneously and lead to cracking or inner stress of the green bodies and in further breakage in the ensuing process. If the slurry comes into contact with the initiator in the course of flow extension, the operation of time constraint can be eliminated. To meet this requirement and to improve and control the slurry polymerization, the spray trigger fast-curing cast process was proposed (Xiang Junhui, 2001; Huang Yong, et al., 2000a, b).

The spray trigger fast-curing tape casting process is still based on gel-tape-casting. The slurry without initiator flows extension on the baseband first and then initiator is sprayed on the flow extension tape and triggers slurry polymerization, completed curing. Therefore, in this manner, the beginning of slurry solidification can be controlled and operating hours can be an unlimited extension in theory.

According to the technological features of trigger fast-curing spray-forming extension, an assist device for the spray initiator is installed on the gel-tape-casting machine (as shown in Fig. 2.45), so that it can meet the needs of the spray trigger fast-curing process.

2.5.2 Outline of the New Process

Figure 2.46 shows the process flow of the Al_2O_3 slurry fast-curing forming by the usage of a spray initiator trigger. In comparison with gel-tape-casting, this process has no the operation time limit and thus is more flexible. Table 2.10 lists two process comparisons.

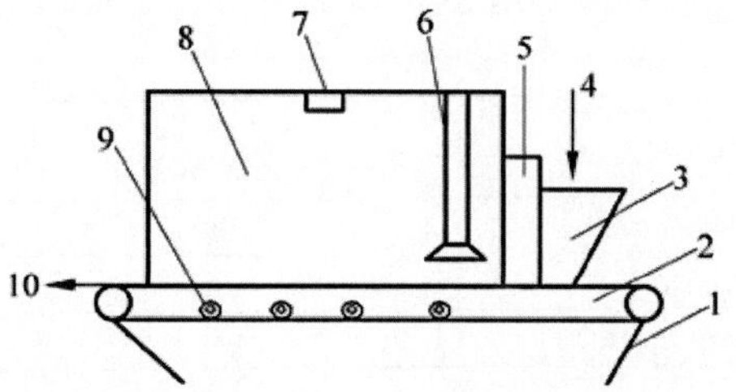

Figure 2.45 The schematic diagram of the tape casting machine for the spray trigger fast-curing
1–baseband; 2–baseplate; 3–slurry reservoir; 4–slurry; 5–doctor blade; 6–nozzle; 7–gas inlet; 8–the protective shield of the atmosphere; 9–heating wire; 10–green tape

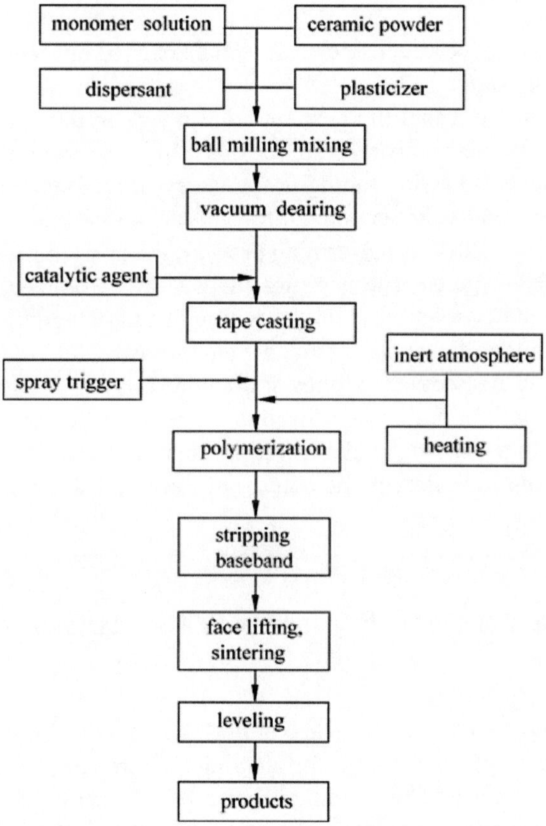

Figure 2.46 The process flow for the fast-curing forming process by using the spray initiator trigger

Table 2.10 The comparison between the spray trigger cast gel-forming and the gel-tape-casting

parameter	spray trigger fast-curing cast extension molding	gel-tape-casting
solid loading of slurry (vol%)	55	55
initiator (vol% of slurry)	>1	0.5 – 1
thickness of green tape (mm)	<0.3	0.1 – 1

In the spray trigger fast-curing forming process, the initiator is sprayed on the moving baseband in the course of slurry flow. Because only the surface layer of slurry is in contact with the initiator, the thickness of forming tape must be controlled to ensure the slurry cured completely. The results show that when the forming thickness is less than 0.3 mm, a smooth paste can be cured completely to form a green tape.

Green tape forming for larger thickness, can be carried out repeatedly by using the same way many times.

It must be pointed out that the concentration of the initiator for spray trigger couring should decrease relatively in order to avoid local polymerization, in fact equivalent to reduce the solid loading in the slurry; therefore, the density of the green tape is lower relatively than one of the gel-tape-casting.

In the spray trigger fast-curing forming process, the spray nozzle capability for shaping the quality of green tape is critical. If it is less atomized, larger droplets will exist in the initiator solution, so that resulting in a partial dilution phenomenon of the slurry, leading to formation of an uneven microstructure of the green tape.

This process is characterized as follows: usage of less organic matter, low level of environmental pollution, fast film-forming of the slurry, to achieve continuous production; shorten the drying cycle; lower production costs. At the same time, it is necessary to avoid limit time in the process of forming and get more convenient operation.

2.6 The Features and Prospects of the Aqueous Tape-Casting

In this chapter, the traditional tape casting and the aqueous gel-tape-casting (or novel gel-tape-casting) processing for ceramic green sheet production were introduced systematically. The ceramic sheet products, with good quality, high productivity, significantly reducing pollution, lower costs, significant social and economic benefits, were fabricated by this new gel-tape-casting process. The researches have demonstrated that this new aqueous tape casting technology is universally applicable for a variety of ceramic powders.

The main features of the contents in this chapter are as follows:

① The solidification mechanism of the gel-tape-casting is analyzed. The polymerization rate of AM depends on the properties of the slurry and the reaction conditions. The concentration of the monomer solution should be kept in the range of 30 – 40 wt% in the premixed solution to obtain a suitable polymerization rate. A suitable range of pH of the Al_2O_3 slurry and temperature are 7 – 9 and 40 – 80℃, respectively. The AM sensitivity to temperature can be used to control the starting time of polymerization accurately. For the studied Al_2O_3 slurry, the suitable amounts of initiator and catalyst are 0.5 – 1 vol% and about 0.5 vol% of the Al_2O_3 slurry, respectively.

② Ceramic thin sheets of 0.1 – 1 mm have been successfully formed by the gel-tape-casting process. In this process, polymerization of organic monomers is used to form the network in which a ceramic green body is solidified. Special drying and burnout procedures, which are critical stages in conventional tape casting, are omitted, and the production efficiency is highly improved.

AC is used as a dispersant in Al_2O_3 slurry. It can efficiently improve the dispensability of Al_2O_3 powder and the fluidity of the slurry. Accordingly, high solid loading slurry with 55 vol% can be obtained. The recommended amount of AC is 1 wt%. G is used as a plasticizer to ensure that the cast tape is soft enough; however, it increases the viscosity of the slurry. The amount of G added in the slurry should be kept less than 5 wt% in order to achieve the balance between the flexibility and the strength of the cast tape.

Gel-tape-casting was performed under the protection of inert atmosphere in order to overcome the effect of oxygen anti-polymerization. Al_2O_3 slurry is polymerized successfully in a few minutes, the temperature being around 50℃ with the protection of nitrogen gas. The relative density of the Al_2O_3 green tape is about 61%.

The Al_2O_3 sample is sintered at 1550℃ for 2 h and its relative density is 97.7%. A homogeneous microstructure without exaggerated grain growth is observed. The variation of solid loading of slurry (53 – 55 vol%) shows no influence on the density of sintered sample.

③ Alumina green tapes have been prepared through gel-tape-casting, which combines the merits of gel-casting and tape casting. A slurry with adequate fluidity can sustain a solid loading of more than 55 vol% when 1 wt% dispersant is used. A plasticizer is used to improve the flexibility of green tapes. As the content of plasticizer in slurry increases, the flexibility of green tape is enhanced, but its strength decreases to a certain extent. A surfactant can decrease the surface tension of aqueous slurry obviously due to its molecular structure, so that the slurry can easily wet the carrier film. Surfactant D can reduce the interface tension between the green tape and organic carrier film dramatically, thus making it possible for the green tape to be released from the carrier film. Foams are produced during mixing, which disrupt the microstructure of the green tape. Foam removal processes, including addition of defoamers in slurry and vacuum deairing, are taken to eliminate the foams. A green tape with a high solid loading, high strength, flexibility, and good releasing property is thereby successfully fabricated.

④ The efficiency of styrene-acrylic binder is obvious for water-based tape casting. The rheological behaviors of the slurries prepared with latex binders always showed a typical shear-thinning behavior with different solid loading at first, and then showed shear-thickening from a critical point. The PAA-NH4 was found to be an effective dispersant for a slurry of latex binder, and the main experimental factors that influenced the stability of the slurry were the properties of dispersant and pH. The research results showed that the glass transition temperature of latex (T_g) was also important for film formation. The drying speed of slurries was largely decided by the drying temperature and the slurry coating thickness.

⑤ The well-dispersed aqueous slurries suitable for tape casting are prepared. The slurries prepared with latex binders exhibit an obvious shear-thinning behavior.

Water-based tape casting has been used successfully to fabricate a smooth-surface and high-strength green tape with styrene-acrylic latex binder. The film drying mechanism is studied, which involves two necessary conditions: the compressive force for latex particles results from solvent evaporation, and the glass transition temperature of latex is lower than the work temperature. The driving force ΔP and T_g are two key factors for forming a filming in a tape casting processing.

⑥ A novel gel-tape-casting has been described. Sodium alginate has been used to consolidate alumina tapes from aqueous suspensions. However, because of low flexibility of resulting tapes, styrene-acrylic latex was also added to enhance the flexibility. Special drying procedure, which is critical stages in conventional tape casting, is avoided.

The use of dispersant can effectively improve fluidity of the suspension. The optimal amount of dispersant is about 0.3 wt%. The styrene-acrylic latex strongly affects the rheological behavior of the suspensions. Suspensions containing styrene-acrylic latex exhibit a typical shear thinning behavior suitable for this process.

The time for casting (induced time) can be controlled by varying the concentration of adipic acid and sequestrant. The induced time is reduced by an increase of acid content or increased by an increase of sequestrant content. The processing can be easily controlled by changing the addition amounts of adipic acid and sequestrant.

Green tapes with homogeneous microstructure were fabricated by this novel forming process.

⑦ On the base of gel-tape-casting, another novel process, spray inducing instant solidification of gel-tape-casting, was developed to take off the limitation of the operating time. This process is fit for the formation of ceramic sheets which thickness less than 0.3 mm.

References

Bender J., Norman J., 1996. Reversible shear thickening in monodisperse and bidisperse colloidal dispersions. J. Rheol., 40: 899 – 916.

Bernd B., Jurge G. H., 2002. Aqueous tape casting of silicon nitride. J. Eur. Ceram. Soc., 22: 2427 – 2434.

Briscoe B. J., Biundo G. Lo, Ozkan N., 1997. Drying of aqueous ceramic suspensions. Key Eng. Mater., 132 – 136: 354 – 357.

Briscoe B. J., Biundo G. Lo, Ozkan N., 1998. Drying kinetics of water-based ceramic suspensions for tape casting. Inter. Ceramic, 24: 347 – 357.

Burnfield K. E., Peterson B. C., 1992. Cellulose ethers in tape casting formulation. In: M. J. Cima (ed.). Ceramic Transactions, Vol. 26, Forming and Technology for Ceramics, American Ceramic Society, Westerville OH., 191 – 196.

Carisey T., Laugier-Wert A. H., Brandon D. G., 1995. J. Eur. Ceram. Soc., 15: 1 – 8.

Chartier T., Bruneau A., 1993. Aqueous tape casting of alumina substrates. J. Eur. Ceram. Soc., 12: 243 – 247.

Chartier T., Penarroya R, Pagnoux C., et al., 1997. Tape casting using UV curable binders. J. Eur. Ceram. Soc., 17: 765 – 771.

Cui X. M., Ouyang S. X., Yu Z. Y., et al., 2002. The study of water-based tape casting with latex binder system. In: Proceedings of the 12th China National Conference on High Performance Ceramics, Foshan, China, 133 – 136.

Cui X. M., Ouyang S. X., Huang Y., et al., 2003. A study on LOM with water-based tape casting. Mater. Lett., 57: 1300 – 1304.

Doreau F., Tari G., Pagnoux C., et al., 1998. Processing of aqueous tape casting of alumina with acrylic emulsion binders. J. Euro. Ceram. Soc., 18: 311 – 321.

Doreau F., Tari G., Guedes M., et al., 1999. Mechanical and lamination properties of alumina green tapes obtained by aqueous tape casting. J.Euro. Ceram. Soc., 19: 2867 – 2873.

Egashira M., Shimizu Y., Takatsuki S., 1991. Chemical surface treatments of aluminium nitride powder suppressing its reactivity with water. Journal of Materials Science Letters, 10: 994 – 996.

Fiori C., Portu G., 1986. Tape casting: a technique for preparing and studying new materials. In: R. W. Davidge (Ed.). Novel Ceramic Processes and Applications. Br. Ceram. Proc, vol. 38. The Institute of Ceramics, Stoke-on-Trent, UK, 213 – 225.

Ford R. W., 1986. Ceramic Drying. Oxford: Pergamon Press.

George W. Scherer., 1990. Theory of drying. Journal of the American Ceramic Society, 73(1): 3 – 14.

Grader G., 1993. J. Am. Ceram. Soc., 76: 1809 – 1814.

Granja M. F. L., Doreau F., Ferreira J. M. F., 1997. Aqueous tape casting of silicon carbide. Key Eng. Mater., 132 – 136: 362 – 365.

Graule T., Grauckler L. J., 1993. Electrostatic stabilization of aqueous alumina suspension by substituted phenols. In: Processing of Ceramics, Third Euro Ceramics. Edited by P. Duranand, J. F. Fernandez. Faenza Editrice Iberica. 1.1: 491 – 500.

Greil P., Kulig M., Hotza D., et al., 1994. Aluminium nitride ceramics with high thermal conductivity from gas-phase synthesized powders. Journal of the European Ceramic Society, 13(3): 229 – 237.

Gutierrez C. A., Moreno R., 2001. Influence of slip preparation and casting conditions on aqueous tape casting of Al_2O_3. Mater. Res. Bull., 36: 2059 – 2072.

Hong Xiaoyin, Feng Hanbao, 2001. Dope Chemistry. SPCC, 9220 (in Chinese).
Hotza D., Greil P., 1995a. Review: Aqueous tape casting of ceramic powders. Mater. Sci. Eng. A, 202: 206 – 217.
Hotza D., Sahling O., Greil P., 1995b. Hydrophobing of aluminium nitride powders. Journal of Materials Science, 30: 127 – 132.
Howatt G. N., 1952. Method of producing high-dielectric high-insulation ceramic plates. U. S. Pat., 2582993.
Howatt G. N., Breckenridge R. G., Brownlow J. M., 1945. Report No.540, Div.14, N. D. R. C. at MIT.
Huang Yong, Xiang Junhui, Xie Zhipeng, et al., 2000a. The thin ceramic green device body of rapid curing and forming *in-situ* through spray initiator. China Patent, ZL00107495.4.
Huang Yong, Xiang Junhui, Xie Zhipeng, et al., 2000b. China Patent, ZL00102922.3.
Hyatt E. P., 1986. Am. Ceram. Soc. Bull., 65: 637 – 638.
Jin J., Lin M. J., 1993. The Application and Processing of Marine Alga. Beijing: Science Press.
Kita K., Fukuda J., Ohmura H., et al., 1982. Green ceramic tapes and method of producing them. U. S. A. Patent, 4353958.
Kristoffersson A., Carlstrom E., 1997. Tape casting of alumina in water with an acrylic latex binder. J. Euro. Ceram. Soc., 17: 289 – 297.
Landham R. R., Nahass P., Leung D. K., et al., 1987. Am. Ceram. Soc. Bull., 66: 1513 – 1516.
Lee H. D., Pober R. L., Calvert P. D., et al., 1993. J. Mater. Sci. Lett., 5: 81 – 83.
Medowski G. O., Sutch R. D., 1976. cited in Williams J. C., Doctor-Blade process. In: F. F. Y. Wang (ed). Treatise on Materials Science and Technology. Vol. 9. Ceramic Fabrication Processes. New York: Academic Press, 173 – 198.
Michel Descamps, Monique Mascart, Bernard Thierry, et al., 1995. How to control cracking of tape-cast sheets. American Ceramic Society Bulletin, 74(3): 89 – 92.
Mister R. E., 1990. Tape casting: The basic process for meeting the needs of the electronics industry. Am. Ceram. Soc. Bull., 69(6): 1022 – 1026.
Mister R. E., 1998. Tape casting: Past, present, pontential. Am. Ceram. Soc. Bull., 77(10): 82 – 86.
Mister R.E., Shanefield D.J., Runk R.B., 1978. Tape casting of ceramics. In: G.
Moreno R., 1990. Am. Ceram. Soc. Bull., 69: 1022 – 1026.
Moreno R., 1992a. The role of slip additives in tape casting technology: part Ⅰ—Solvents and dispersants. Am. Ceram. Soc. Bull., 71(10): 1521 – 1531.
Moreno R., 1992b. The role of slip additives in tape casting technology: part Ⅱ—Sovents and plasticizers. Am. Ceram. Soc. Bull., 71(11): 1647 – 1657.
Morris J. Jr., 1986. Organic component interactions in tape casting slips of barium titanate (Ph. D. thesis). State University of New Jersey.
Myers D., 1992. Surfactant Science and Technology (2nd Ed.). New York: VCH, 255 – 272.
Nagata K., 1993. Rheological behavior of suspension and properties of green sheet-effect of compatiblity between dispersant and binder. Journal of the Ceramic Society of Japan, 101(10): 1271 – 1275.
Nahass P., Rhine W. E., Pober R. L., et al., 1990. A comparison of aqueous slurries for tape casting. In: Nair K. M., Pohanka R., Buchanan R. C. (Eds.). Ceramic Transactions. Vol. 15, Materials and Processes in Microelectronic Systems. American Ceramic Society, Westerville OH, 355 – 364.

Nahass P., Pober R. L., Rhine W. E., et al., 1992. J. Am. Ceram. Soc., 75: 2373 – 2378.

Onoda Jr., L. L. Hench (Eds.). Ceramic Processing before Firing. New York: John Wiley & Sons Inc., 411 – 448.

Pagnoux C., Chartier T., Doreau F., et al., 1998. Aqueous suspensions for tape casting based on acrylic binders. J. Eur. Ceram. Soc., 18: 241 – 247.

Qin Zhengqi, Qian Xingnan, He Panfa, 1996. New Type of Ceramic Material Manual. Nanjing: Jiangsu Science and Technology Publishing House, 10

Rechard E., Mistler R. E., 1998. Tape casting: Past, present, potential. American Ceramic Society Bulletin, 77 (10): 82 – 86.

Schuetz J. E., Khoury I. A., DiChiara R. A., 1987. Water-based binder for tape casting. Ceramic Industry, 66(10): 42 – 44.

Schwartz B., Wilcox D. L., 1967. Laminated Ceramics—Proceedings of Electronic and Computer Conference, 17 – 26.

Smith D. J., Newnham R. E., Yoshikawa S., 1990. UV curable system for ceramic tape casting. IEEE Seventh International Symposium on Applications of Ferroelectrics. Urbana-Champain, IL, USA.

Stetson H. N., Gyurk W. J., 1967. Development of two micoinch (CLA) as-fired alumina substrates. American Ceramic Society Bulletin, 46(4): 387 – 389.

Suresh B., Manupin G. D., Graff G. L., 1998. Free form fabrication of ceramics. Am. Ceram. Soc. Bull., 7: 53 – 58.

Tadros T. F., 1984. Surfactants. New York: Academic Press, 287 – 322.

Tan Mengkwang, 1996. The role of coalescent of film forming mechanism in latex pain. Doping Industry, (4): 32 – 35 (in Chinese).

Theodore P., 1995. Tape casting and roll compaction. Ceramic Engineering Science Processing, 16(3): 71 – 75.

Waack R., Venkataswamy K., Novich B. E., et al., 1988. Polymerizable binder solutions for low viscosity, highly loaded particulate suspensions, and methods for making green articles therefrom by tape casting or injection molding. U. S. A. Patent, PCT/US88/01232.

Wayne N. J., 1969. Chemistry of Acrylamide. Parsippany, N. J.: American Cyanamid Co.

Xiang Junhui, 2001. Study of a novel process-instant solidification of aqueous gel-tape-casting for ceramic sheet formation. In: Chinese Dissertation of Tsinghua University, Beijing.

Xiang Junhui, 2002. Processing of Al_2O_3 sheets by the gel-tape-casting process. Ceram. Int., 28: 17 – 22.

Xie Z. P., Huang Y., Chen Y. L., et al., 2001. A new gel-casting of ceramics by reaction of sodium alginate and calcium iodate at increased temperatures. J. Mater. Sci. Lett., 20: 1255 – 1257.

Yoshikawa S., Heartling C., Smith D. J., et al., 1988. Patterned ceramic green films using UV curable binders. In: G. L. Messing (Ed.). Ceramic Powder Science Ⅲ. American Ceramic Society, Westerville, OH, 533 – 560.

Zeng Y. P., Jiang D. L., Greil P., 2000. Tape casting of aqueous Al_2O_3 slurries. J. Eur. Ceram. Soc., 20: 1691 – 1697.

Chapter 3 Gelation Forming Process for Low Toxicity System

Abstract In this chapter, the forming process of ceramics by gelling of agarose, gelatine and sodium algaecide has been presented, and another gel-casting system with low toxicity by using HEMA is also described. This study shows that gelled green bodies by the addition of agarose of 0.7 – 1.05 wt% and a solid loading of 53 – 55 vol% had a homogeneous density distribution and a green strength >3 MPa. Some simple and complex parts, such as a turbine rotor, were formed by this method. Gelatine can be dissolved in water above 40℃, and the solutions gel at 15 – 20℃. Wet green compacts can be formed from the alumina slurry containing 4 – 5 wt% gelatine (based on water) by gelation. A novel casting method by gelatine and enzyme catalysis is also described. As urease is added into the suspension, the gelatine molecules attract each other and change to a three-dimensional (3D) network structure through hydrogen bonding due to urea decomposition by urease. Then the slurry containing ceramic powder and gelatine will be consolidated *in-situ* by the gelation under ambient conditions. For the another gel-casting process by alginate, in which solidified agent and chelator are added into sodium alginate solution simultaneously; thus, the gelation between calcium ions and sodium alginate is avoided in this stage before casting. Free calcium ions are released when hexanedioic acid is added into the suspension. They react with sodium alginate and form a 3D network. Therefore, ceramic particles are fixed in this 3D network. The slurry is consolidated and near-net-shape green parts are formed. In addition, a new low toxic gel-casting system has been successfully developed by using HEMA or HEMA-PVP, by which dense and homogeneous complex-shaped alumina and silicon carbide parts have been successfully fabricated.

Keywords ceramic slurry, gelation, agarose, gelatine, sodium algaecide, HEMA

Gel-casting has been widely studied during the last decade (Alford N. M., et al., 1987; Lange F. F., 1989; Reed J. S., 1994; Zhang J. G., et al., 1989; Xie Z. P., et al., 1995). In this process, the slurry made from ceramic powder and a water-based monomer solution is poured into a mold, and then polymerized *in-situ*. The gelled part is removed from the mold while still wet, and then it is dried

Y.Huang et al., *Novel Colloidal Forming of Ceramics*
© Tsinghua University Press, Beijing and Springer-Verlag Berlin Heidelberg 2010

and fired. The dried green body is strong enough to be machined. However, the process is not perfect in that the polymerization of monomers is difficult to control in the ceramic suspension. In addition, acrylamide has higher toxicity. Therefore, a new gel-casting process that is toxicity-free or has reduced toxicity has been investigated in this study.

In fact, many polymer solutions, such as agar, agarose, gelatine, carrageenan, and sodium algaecide, can gelate in suitable conditions (Whistler R. L., 1973; Amott S., et al., 1974). Some of them have been employed in food industry. When the polymer is dissolved in water, the molecular chains attract each other and form a 3D network by hydrogen bonds or van der Walls forces. The gelling property of agar has been used in the water-based injection molding (Hoffman R. L., 1972).

Millan et al. (Millan A. J., et al., 2002a, b) studied the influence of aqueous solutions of agar, agarose and carrageenan on the rheological properties of alumina slurries, and the mechanical and microstructural characterization of green alumina bodies. Gel-casting performance is a function of slurry viscosity, gelling behavior and time, and body deformation during drying. In general, carrageenan leads to higher shape distortion in the body. Better shape retention is achieved with precursor solutions with $\geqslant 3$ wt% additive and total content of gel of $\geqslant 0.5$ wt%. The bend strength up to 4 MPa is obtained, which is significantly higher than those corresponding to slip cast alumina without gelling additives, and they increase with the final concentration of polysasccharide, while Young's modulus values are mainly influenced by the concentration of additive in the precursor solution.

Santacruz et al. (2002) studied the rheological characterization of synergistic mixtures of carrageenan and locust bean gum for aqueous gel-casting of alumina. The rheological behaviors of 2 wt% solutions of the polysaccharides and their mixtures are measured under mixing conditions (60℃) and by recording the viscosity and elastic modulus on cooling. The effect of the addition of these solutions to 50 vol% alumina slurries up to a concentration of 0.5 wt% has been studied. Although the gelling time increased, the resulting gels were stronger than that for carrageen alone. Gelcast alumina bodies with green and sintered densities of 57% and 97.6% of theoretical values have been obtained.

In this study, the forming process of ceramics by the gelling of agarose, gelatine and sodium algaecide is presented, and another gel-casting system with low-toxicity by using HEMA is also described.

3.1 Gelation Forming of Ceramic Suspension with Agarose

3.1.1 Characteristics of Agarose

Agarose is soluble in hot water and forms a gel on cooling. The characteristic strengths of the gels produced by these gelling agents vary greatly, and agarose is the strongest. Fanelli et al. (1989) have examined aqueous injection molding of mixes of agar, agarose, and alumina powder (33 – 39 vol%) with an apparent viscosity of 300 Pa·s or more. The purpose of the present work was to investigate gelation casting of ceramic suspensions of low apparent viscosity with agarose as a binder for producing homogeneous complex shaped green bodies.

The alumina powder used was a commercial grade An-05 (from Henan Xinyuan Alumina Industry Co. Ltd., China), prepared by calcining γ-Al_2O_3 in a high temperature revolving furnace. The chemical composition is given in Table 3.1. The average particle size was about 1.8 μm. PMAA-NH_4 of molecular weight about 1×10^4 was used as a dispersant.

Table 3.1 Chemical compositions of alumina powder

composition	Al_2O_3	α-Al_2O_3	SiO_2	Fe_2O_3	Na_2O	B_2O_3	weight loss
content (wt%)	99.7	⩾95	0.05	0.03	0.05	0.04	0.10

Agarose, a purified derivative from agar, was used as a gelation agent. Analysis of the agarose revealed 0.2 wt% ionic impurities in the form of SO_4^{2-} and an ash content <1 wt%. In general, agarose molecules are stable and do not decompose over the pH range of 4 – 10. It absorbs water and swells in water at ambient temperature, and dissolves in hot water at about 80℃. For the agarose used, the gelation strength of a solution with 1.5 wt% agarose reached more than 600 g·cm^{-2}.

The rheological properties of the agarose solution and ceramic suspensions, prepared at pH = 7, were measured with a rotary rheometer (NSX-11 Chengdu). The gel strength of the agarose solution was measured using a technique commonly employed in the industrial gum industry, the details of which are described in literature (Whistler R. L., 1973). Green body strength was measured by three-point bending tests on sample bars of 6 mm × 5 mm × 30 mm. For comparison, the green samples without agarose were prepared by centrifugal casting and had a density of 67.5% of the theoretical value.

Agarose is composed of alternate units of two types of form of the sugar molecule. The polymer chains form rigid bundles of twisted helical chains, and by hydrogen bonding, the hydroxyl groups along the polymer backbone interact strongly with water molecules in the gel network (Amott S., 1974). The apparent viscosity of an aqueous agarose solution is shown in Fig. 3.1. It may be seen that the viscosity increased with a decrease in temperature, and reached 10 Pa·s near the gel point. At the gel point, gelation resulted in a tough mass. The gel strength increased linearly with agarose concentration. For example, the gel strengths with 1.5 and 3 wt% agarose were 650 and 1300 g·cm^{-2} respectively. Figure 3.2 illustrates the rheological behavior of a 3 wt% agarose solution at temperatures above the gelation point. Shear thickening and slight viscosity increasing with shear rate were observed. This is similar to results on colloidal stable PVC suspensions reported earlier by Hoffman (1972). In general, shear thickening often occurs in concentrated suspension and monodisperse systems. Several researchers (Boersma W. H., et al., 1991) have suggested that shear thickening can be explained in terms of the balance between the viscous forces and interparticle repulsive forces.

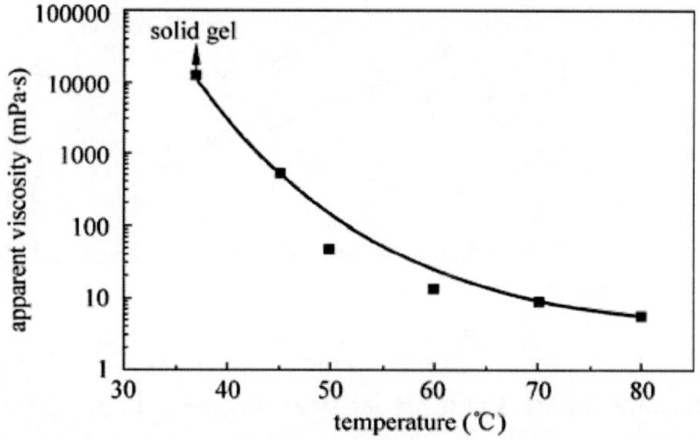

Figure 3.1 The apparent viscosity of a 3 wt% agarose solution

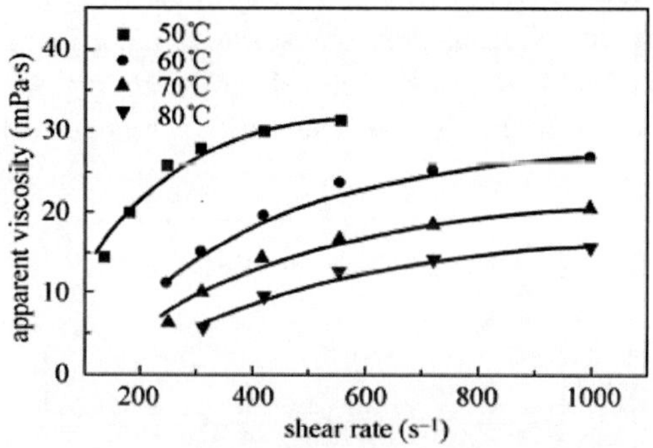

Figure 3.2 The rheological behavior of a 3 wt% agarose solution

3.1.2 The Effect of Agarose Contents on the Rheology of Aqueous Ceramic Suspensions

The variations of the zeta potential with the pH for as received alumina and the suspension containing PMAA-NH_4 as dispersant are shown in Fig. 3.3. The isoelectric point (IEP) of the alumina powder was matched with pH=5.3. After adding PMAA-NH_4, the IEP of the suspension changed to pH 3.1 due to adsorption of the dispersant on the alumina particles. In addition, the zeta potential increased exponentially in the range pH ≤ 5.3. This is very helpful for making slurries with low viscosity in deionized water where the pH is nearly 7.

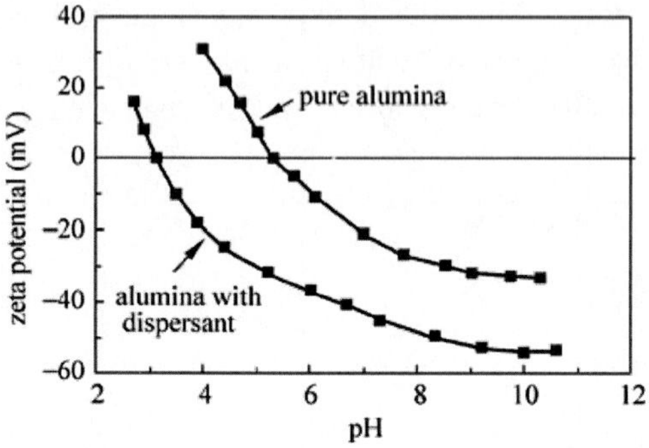

Figure 3.3 The variation of zeta potential with pH for as-received alumina and alumina with dispersant

The rheology of aqueous ceramic suspensions generally depends on solids content, particle characteristics and binder type. Figure 3.4 shows that the viscosity varied with shear rate for a solid loading of 55 vol% and agarose content of 0.7 wt% (Al_2O_3 basis) at pH = 7. For the solution with the agarose, the suspension showed shear thickening. This is due to a combination of the high alumina solid loading and the agarose solution. Previous studies (Yang J. L., et al., 1998) have also shown that alumina suspensions exhibit shear thickening at high concentrations >50 vol%, and shear thinning at low concentrations, respectively.

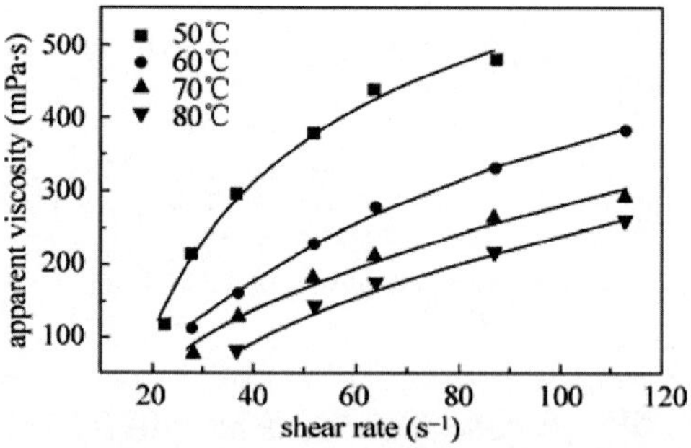

Figure 3.4 The relationship between viscosity and shear rate for a suspension with 55 vol% solid loading and 0.7 wt% agarose (Al_2O_3 basis)

It is important to have low viscosity and high solid loading for the suspension used in gelation forming and these are influenced by the type and amount of binder. Figure 3.5 illustrates the effect of agarose additions on the viscosity of a suspension containing 53 vol% solid contents. It may be seen that the viscosity was

<1 Pa·s above the gelation temperature for agarose additions from 0 to 0.7 wt% (Al_2O_3 basis). At 1.05 wt% agarose addition, the viscosity increased considerably and casting became difficult. The agarose addition for solid loading >50 vol% should therefore be <1.05 wt%. The influence of solid loading viscosity is shown in Fig. 3.6. Suspensions containing <53 vol% and 0.7 wt% agarose exhibited good fluidity above the gel point.

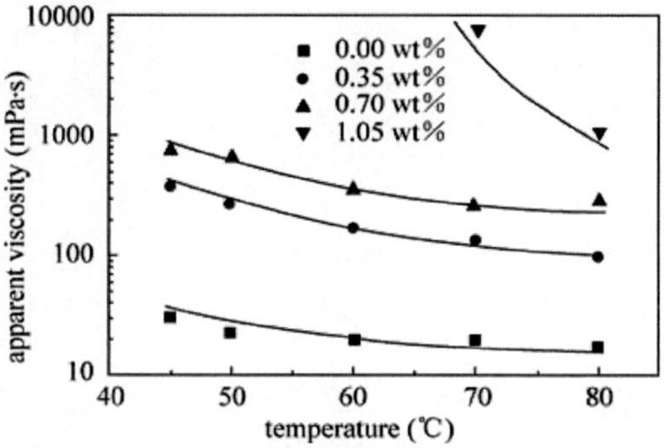

Figure 3.5 The effect of agarose content on viscosity of suspension with 53 vol% solid loading

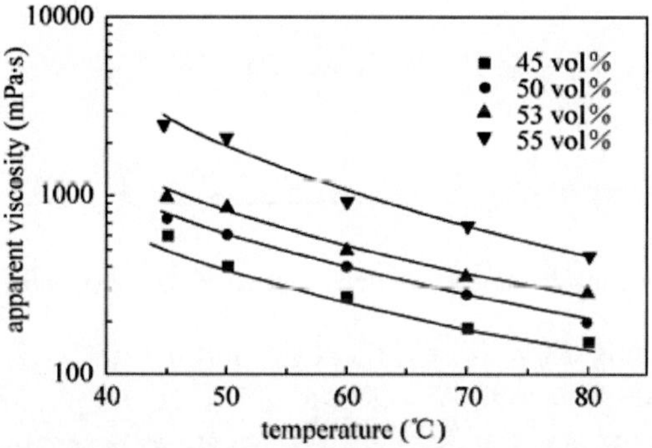

Figure 3.6 The influence of alumina solid loading on viscosity of suspension with 0.7 wt% agarose

3.1.3 The Forming Courses of the Aqueous Ceramic Suspensions with Agarose

1) Forming

The flow chart of the forming process is shown in Fig. 3.7 for alumina slurry with agarose. Agarose and dispersant (PMAA-NH_4) were added to deionized water, and then mixed with alumina powder. The resulting slurry was then ball milled for 24 h. After degassing in a rotary evaporator under vacuum, the slurry was heated to 80℃ in a water bath to dissolve the agarose completely and to prevent water from evaporating. This was done in a closed chamber. The warm slurry was then cast into a nonporous mould, which could be made of metal, glass or plastic. When cooling to the agarose gel point, the slurry gelled to the shape required, and then the wet green body was demoulded.

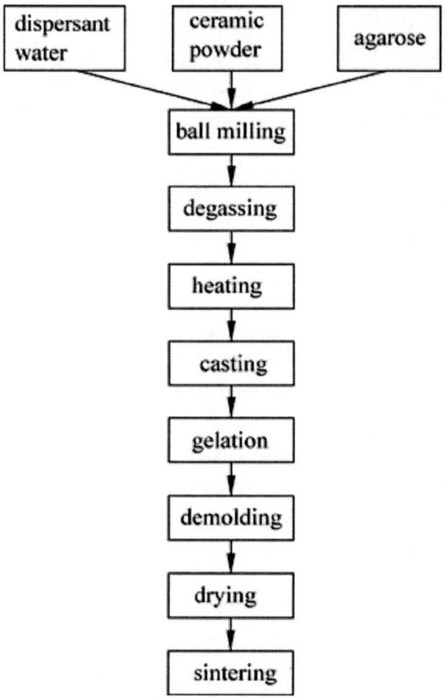

Figure 3.7 The flow chart of the forming process for alumina slurries with agarose

A system containing 53 vol% solid loading and 0.7 wt% agarose (Al_2O_3 basis) is suitable for gelation casting processing and some simple and complex parts, such as a turbine rotor (see Fig. 3.8), were formed by this system. For forming of these parts, it should be noted that the mould needs to reach a temperature above the gel point in order to prevent the slurry fluidity decreasing significantly during the mould filling due to gelation. In addition, the wet green parts should be demoulded after a tough gel is formed. In general, several hours (about 4 – 8 h) are needed to obtain a gel strength high enough for complex parts, such as turbine rotors and blades, to be removed from the mould. After demoulding, the wet green parts have accurate dimensions and a very smooth surface, similar to an injection moulded body made with thermoplastic vehicles.

2) Drying and Compact Characteristics

Linear shrinkage measurements were performed on gelled cake 40 mm in diameter and 20 mm thickness. Figure 3.9 shows the average linear shrinkage in

Figure 3.8 The turbine rotor (110 mm dia.) shaped by gelation forming

diameter and the resulting relative green density vs. the solid loading during drying. The results illustrate that the green bodies with more than 50 vol% solid loading had shrinkage lower than 1.7% which increased to 3.2% for 45 vol% solid loading. After drying, green bodies gelled from the suspensions with 50, 53, and 55 vol% solid loading, respectively, displayed 51, 54, and 56% theoretical densities. However, the green density for a solid loading of 45 vol% reached 50% theoretical density because of greater drying shrinkage. In general, a larger shrinkage of the green body tends to cause deformation or cracking especially in complex shaped parts. This implies that solid loading above 50 vol% is needed for successful processing.

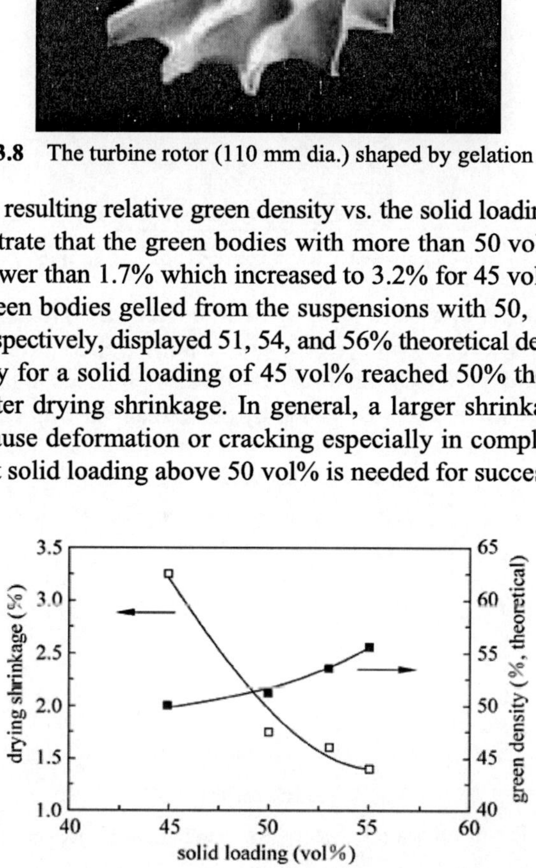

Figure 3.9 The drying shrinkage and relative density vs. solid loading

The demolded wet green parts can be dried at ambient temperature or in a drying chamber without cracking or damage. To control the starting alumina particle size and the resultant homogeneous pore size distribution in the green part are responsible for the rapid drying. Analysis by autoscan mercury porosimeter (Quantachrome Corp., USA) showed that a narrow, monodispersed pore size distribution was obtained in the green body, whereas the same alumina powder compacted by isostatic pressing exhibited a wide pore size distribution with double peaks (Yang J. L., et al., 1998).

A uniform density distribution was also observed for the complex shaped green compacts. From density measurements of the turbine rotor shown in Fig. 3.10, for example, the density of 2.26 g/cm^3 was the same in different parts of the rotor. This implies that gelation forming from the slurry to the green compact was completed homogeneously.

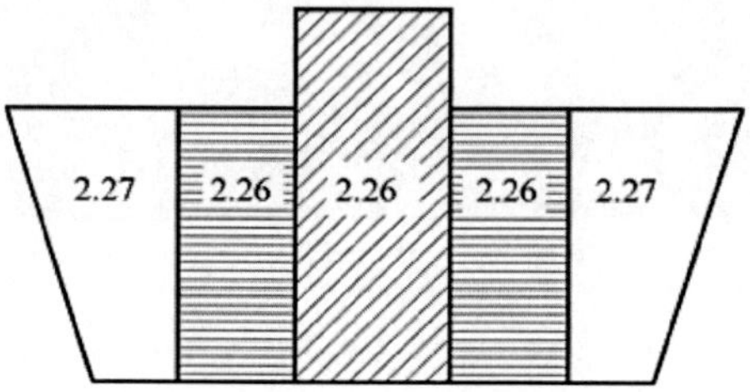

Figure 3.10 The density distribution in green body of turbine rotor (g/cm^3)

The green strength increased with agarose content as illustrated in Fig. 3.11. For example, the green strength, for a green body gelled from a concentrated suspension containing 0.7 wt% agarose (Al_2O_3 basis), reached 3.2 MPa, comparing with 1.5 MPa for a green sample formed by centrifugal casting without agarose.

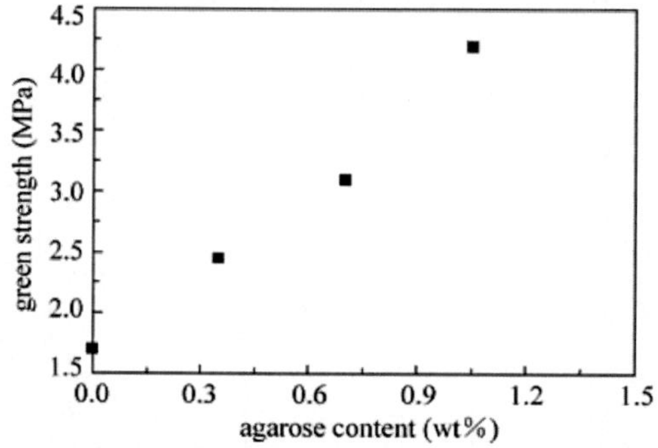

Figure 3.11 The relationship between green strength and agarose content (Al_2O_3 basis)

3) Sintering

Binder burnout is normally carried out before sintering for some forming methods such as injection molding, tape casting, and gel-casting where the binder content ranges from 4 wt% to 18 wt%. In the case of the present forming method, the gelled green body could be sintered directly because the organic binder addition was <1.05 wt%. Thermogravimetric analysis for a green body containing 53 vol% solid loading and 0.7 wt% agarose showed that weight loss started at 250℃ and completed at 500℃. For the green turbine rotor, sintering was carried out at a heating rate of 60 K/h and held for 2 h at 1600℃. The sintered sample had a 97.8% theoretical density and a 16% linear shrinkage in diameter, which is comparable with that for samples made by die pressing. According to many reports, macro defects, such as cracks, often occur in the center of turbine rotors prepared by injection molding by using organic based vehicles due to nonuniformity in both density distribution and shrinkage. However, no cracks were observed in sections of the sintered rotor shaped by the gelation forming method. This is attributed to its uniform distribution of composition and density in the green state.

The gelation forming of ceramic compacts by agarose gelation has very good near net shape capability for complex shaped parts requiring precise dimensions and very smooth surface finish. For the casting of alumina, an agarose addition of 0.7 – 1.05 wt% and a solid loading of 53 – 55 vol% were found to be successful. The gelled green bodies had a homogeneous density distribution, and a green strength >3 MPa. Green parts, such as turbine rotors, can be sintered directly without special binder burnout procedures, and without deformation or cracking in the finished component.

3.2 Alumina Casting Based on Gelation of Gelatine

3.2.1 Characteristics of Gelatine

1) Why Study Gelatine

The gelation property of agarose has been used in the water-based injection molding (Fanelli A. J., et al., 1989; Zhang T., et al., 1994) and gel-casting (Xie Z. P., et al., 1999) of alumina. In these cases, a well-dispersed suspension of alumina and agarose powder was heated at 85 – 90℃ to dissolve the agarose completely and then it is cast into a nonporous mould. At about 35℃, the agarose gelled, enclosing ceramic powder and water in its gel network to form a green body with the desired shape.

The problem of agarose is that the dissolving temperature is fairly high (about 85℃). During the process of heating, water vaporized heavily, resulting in agglomerates and flocculation, which increased the viscosity of the slurry.

Compared with agarose, gelatine (protein polymer composed of various amino acids) has some excellent properties (Ward A. G., et al., 1977). The most attractive is that gelatine can be dissolved quickly at low temperature (about 40℃). This work investigated the gelling property of gelatine and the influence of gelatine on an alumina ceramic suspension. Rigid alumina green bodies were also prepared by *in-situ* coagulation of slurries containing gelatine.

2) Properties of the Gelatine

The gelatine powder with molecular weights ranging from 15000 to 90000 was provided by Shanghai Chemical Reagent Works. The analysis of the gelatine is shown in Table 3.2. A 1 wt% gelatine solution changes into strong and stable gel over a wide pH range of 4 – 10 when cooled.

The alumina powder used here has been shown in Section 3.1.1 in this chapter.

Table 3.2 Impurities in gelatine powder

composition	chloride (Cl^-)	heavy metal ash (Pb)	ash	sulfur dioxide (SO_2)	arsenide (As)	iron (Fe)
content (<wt%)	0.2	0.008	2	0.015	0.0003	0.007

The rheological properties of the gelatine solution and mixed ceramic slurry were measured by a rotary rheometer (NXS-11, Chengdu, China). The gel strength was examined by a simple device usually used in the gums industry (Whistler R. L., 1973; shown in Fig. 3.12). Supposing the total weight recorded on the load cell was W when the push pin broke the surface of gel sample, and the cross-sectional area of the push pin was A, then the gel strength could be calculated using the formula $G = W/A$, where the unit of G is g/cm^2.

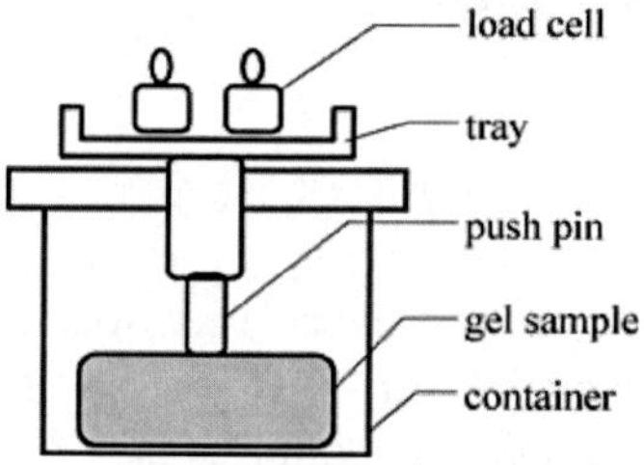

Figure 3.12 The device for measuring gel strength

3) The Effect of Temperature Change on Viscosity of Gelatine Solution

The change in the apparent viscosity of a 6 wt% gelatine solution with temperature is shown in Fig. 3.13. Above 25℃, the apparent viscosity changes only slightly, ranging from 2 to 8 mPa·s. Above this temperature, the shape of polymer molecule changed from random clew ball to single-helix chain. When the temperature dropped below 25℃, single-helixes entangled with each other, resulting in an obvious rise of apparent viscosity. With the apparent viscosity rising sharply to more than 20 Pa·s, a 3D network was formed by the attraction between molecular chains. At the joints of the gel network, the molecular chains were ordered into microcrystalline zones (Zhou Z. K., et al., 1996; Jin J., et al., 1993; Ward A. G., et al., 1977).

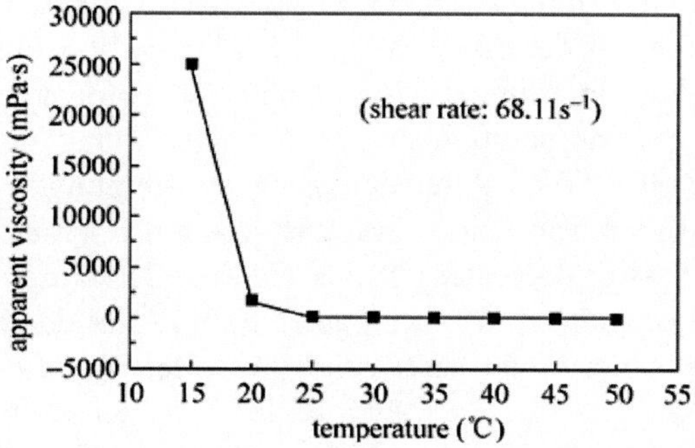

Figure 3.13 The change of apparent viscosity for a gelatine solution (6 wt%) with temperature

4) The Effect of the Gelatine Contents on Gel Strength

The gel strength increased with the increase of the gelatine content (see Fig. 3.14) As more microcrystalline zones were formed by gelatine polymer chains, the structure of the gel network became tighter and the gel strength increased. Compared with agarose, the strength of the gelatine gel is relatively low (Fanelli A. J., et al., 1989; Zhang T., et al., 1994). With the content of 3 wt%, the strength of agarose gel is greater than 1000 g/cm^2, while the strength of gelatine gel is only 100 g/cm^2. Because the strength of the gel, along with the solid loading, ultimately determines the strength of the wet and dry green body, adding more gelatine into the slurry to increase the strength of the green body is more effective. However, in order to realize the forming process of gel-casting, the slurry should have an appropriate viscosity and fluidity. Therefore, the content of gelatine must also be determined from the rheology of the slurry besides the strength of green body.

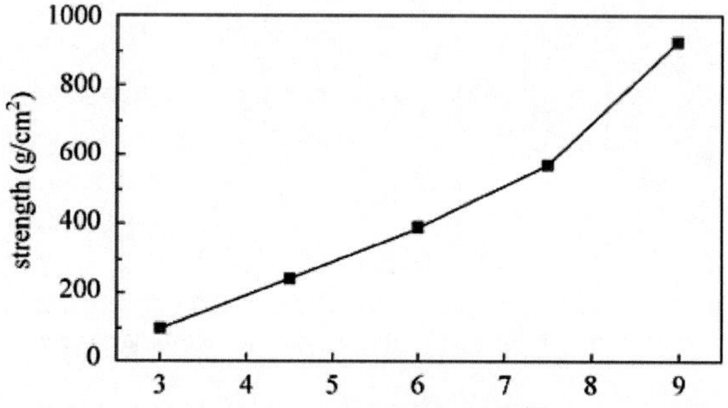

Figure 3.14 The change of gel strength vs. gelatine concentration

3.2.2 The Gelation Process of the Ceramic Slurry with Gelatine Solution

1) The Forming Process of the Alumina Slurry with Gelatine Solution

The forming process is described in Fig. 3.15. Firstly, the alumina powder and dispersant were added to deionized water, ball-milled for more than 12 h, and then mixed with gelatine solution prepared by dissolving gelatine powder in deionized water at 40 – 50℃. After degassing by vibrating or stirring under a vacuum (10 mmHg), the warm slurry was cast into a nonporous mould that could be made of steel, glass or polyester plastics. Cooling down to the gel point of the gelatine, the slurry coagulated to form a green body. After demolding and drying under 80℃ for 24 h, the green parts could be sintered without special binder removal processing.

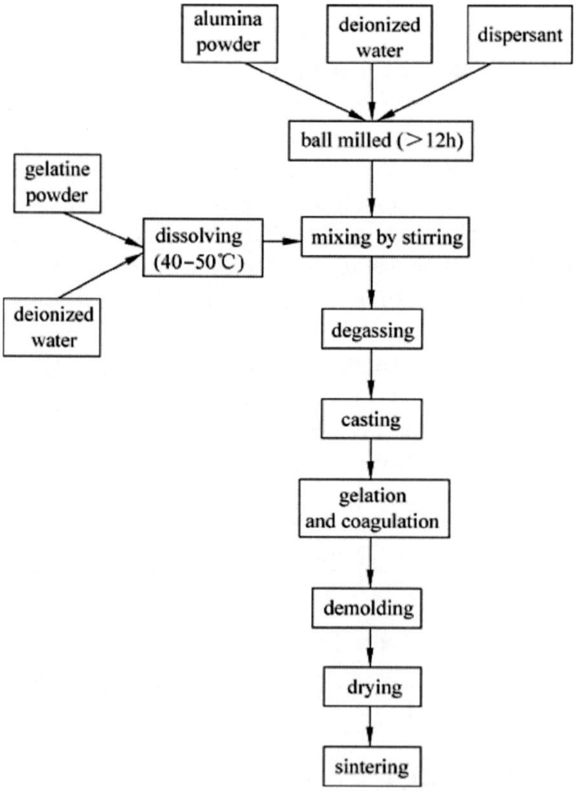

Figure 3.15 The flow chart of the forming process for alumina slurries with gelatine

2) The Rheology of the Alumina Slurry with Gelatine Solution

Figure 3.16 shows the flow curves of the Al_2O_3 suspensions with different gelatine concentrations at 30℃. The pure alumina slurry with solid loading of 53 vol% had an average viscosity of 50 mPa·s. On adding gelatine, the viscosity increased greatly. When the content of gelatine was less than 4.5 wt% (based on water), the apparent viscosity was lower than 1 Pa·s, which allowed the casting process to take place. Slurries with gelatine contents greater than 6 wt% were thick and cannot flow into the mould easily, as their apparent viscosity was higher than 3 Pa·s. Under such conditions, agglomerates or flocculation may occur and the homogeneous microstructure of ceramic body will be destroyed.

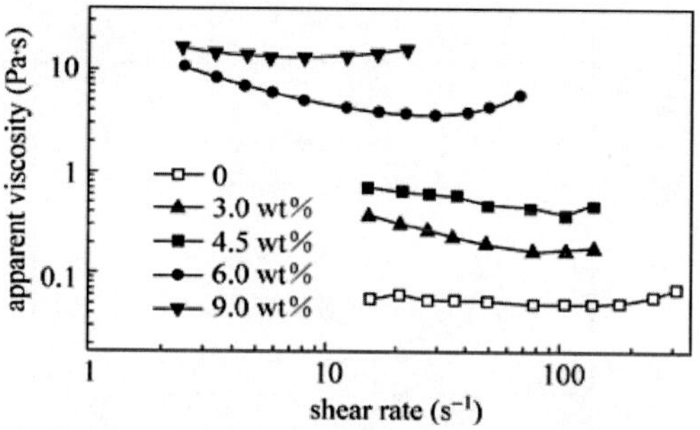

Figure 3.16 The effect of gelatine concentration on viscosity of the suspension containing 53 vol% solid loading at 30℃

The rheological properties of a slurry with 53 vol% Al_2O_3 and 3 wt% gelatine at different temperatures are shown in Fig. 3.17. Before gelation, the slurry is a pseudo plasticity fluid, because gelatine polymer chains could turn and transform to decrease the shearing resistance. Since the process of slurry degassing and casting are usually performed under dynamic conditions rather than static, pseudo plasticity is helpful.

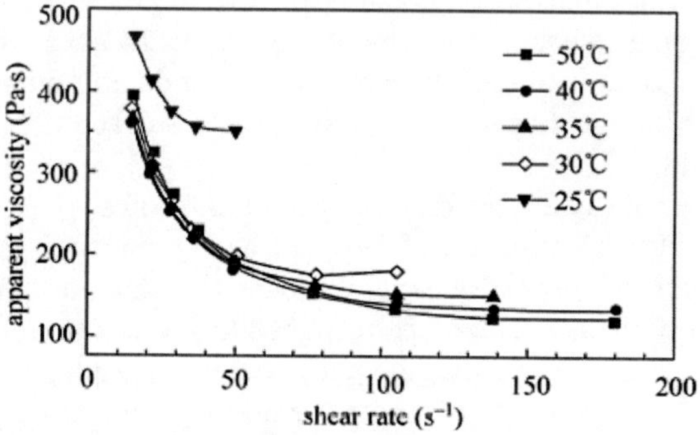

Figure 3.17 The rheological behavior of the slurry with 53 vol% alumina and 3 wt% gelatine at different temperatures

3.2.3 The Preparation of Green Body Using Slurry with Gelatine Solution

1) The Forming of a Wet Green Body

On casting the warm slurry of alumina and gelatine into a nonporous mold, it would coagulate at the gel-point of gelatine. The change in apparent viscosity of the slurry (see Fig. 3.18) is similar to that of a pure gelatine solution (see Fig. 3.13). Above the gel-point, the slurry had a relatively low viscosity (less than 200 mPa·s). At the temperature between 15 – 20℃, the viscosity rose abruptly to more than 30 Pa·s.

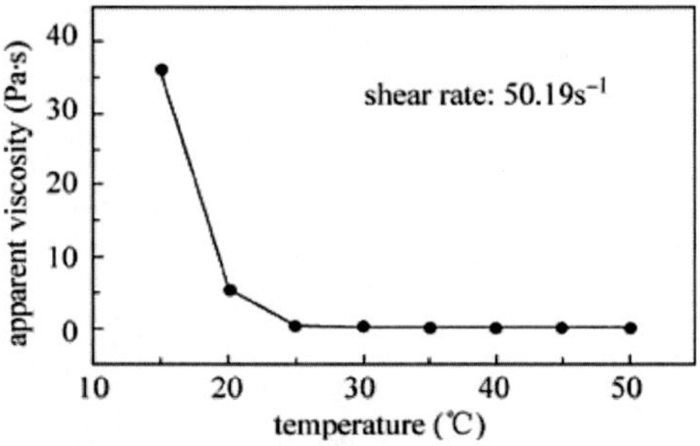

Figure 3.18 The apparent viscosity change for the slurry with 53 vol% alumina and 3 wt% gelatine

It has been reported that the network of gelatine gel contains many microcrystalline zones, which act as the joints of gel. After gelation, the polymer chains continue to rearrange in the network to form more microcrystalline zones, and thus increase the rigidity of the gel.

The wet green body was very weak and had low strength when just formed. It could not be demoulded until the structure of gelatine gel developed through the arrangement of polymer chains. A simple ceramic part, such as a disk with 150 mm in diameter and 20 mm in thickness could be demoulded 6 – 8 h after slurry casting. For a complex compact, such as a turbine blade, nearly 24 h was required in order to avoid deforming or cracking of the wet body when demoulding. In general, the demoulded bodies had precise size and very smooth surface if the surface of the moulds was smooth.

2) The Properties of the Dry Bodies and Sintered Bodies

The green body has a shrinkage of 2 – 3% (linear) during drying. Figure 3.19 shows the SEM photocopy of a dry body. It can be seen that powders are connected by slender polymer chains, which are responsible for the strength of the green body. The influence of the gelatine content on the strength of the dry body is shown in Fig. 3.20. The strength is greater than 8 MPa when the gelatine content is more than 4.5 wt%. Although the body is stronger with more gelatine (with gelatine content higher than 4.5 wt%, strength is greater than 8 MPa), the error bar is also enlarged, indicating that homogeneity is reduced. This may be attributed

Figure 3.19 The SEM micrograph of a dry green body

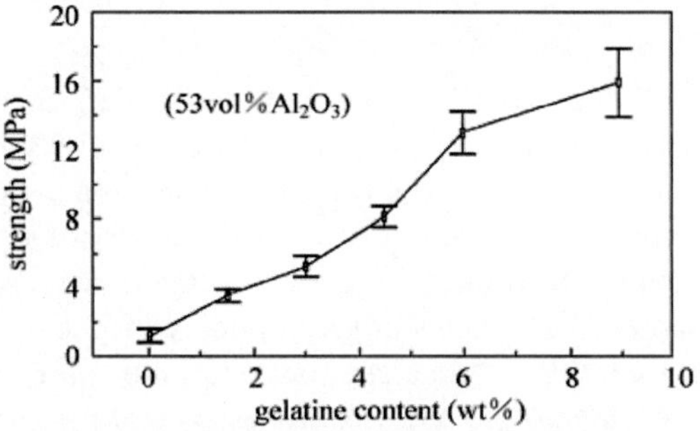

Figure 3.20 The influence of the gelatine content on the strength of a dry green body

to the incomplete degassing, inhomogeneous slurry mixing, or the presence of agglomerates resulting from the higher viscosity.

Regarding the fluidity of the slurry and the strength of the green body, 4 – 5 wt% gelatine (based on water) is suitable for the suspension containing 50 – 55 vol% alumina powder used in this research. Various ceramic green bodies were prepared by this forming process (shown in Fig. 3.21). Green parts can be sintered directly without special binder removal procedures because the organic binder content

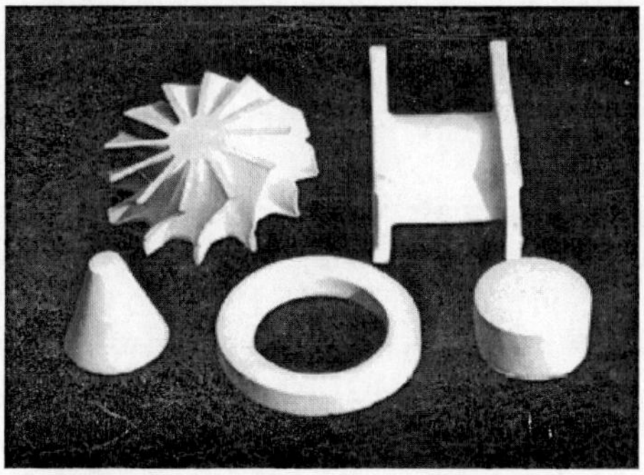

Figure 3.21 The examples of ceramic green parts made by the forming method

is 1 wt% based on the powder weight. Ceramics consisting of 95 wt% alumina and 5 wt% clay made by this method have linear shrinkage of 15– 16% during firing. The relative density of the final product is 96.5% and the bending strength reaches 304.6 MPa.

3.3 A Casting Forming for Ceramics by Gelatine and Enzyme Catalysis

3.3.1 Research Background

1) Academic Idea

We have reported that a ceramic suspension was gelled to a green body using agarose and gelatine in previous study (Xie Z. P., et al., 1998; Chen Y. L., et al., 1999). In these cases, a well-dispersed ceramic suspension containing agarose or gelatine was ball milled and heated at 80 or 40℃ to dissolve them completely, and then cast into a non-porous mould. During cooling down to room temperature, the suspension gelled and formed a wet green body with the desired shape. Some ceramic parts with a complex shape, such as a turbine rotor, were produced by this method. However, the problem of the process is that the suspension has to be heated to a certain temperature before casting, which might increase the viscosity of the slurry due to water vaporization during the heating process.

To overcome this drawback of the forming process, a new kind of gelation process using gelatine and enzyme catalysis for forming a ceramic green body is developed in this study. The forming method is based on the process in which gelatine, urea and urease are used. Urea can prohibit a hydrogen bond from forming between gelatine molecules when the hot gelatine solution cools down. As urease is added into the suspension, gelatine molecules attract each other and become a 3D network structure through hydrogen bonding due to urea decomposition by urease. Then the suspension containing ceramic powder and gelatine will be consolidated *in-situ* by the gelation under ambient conditions. This process can prevent water vaporization and increase of viscosity before casting. Also, the process is more easily controlled to obtain a homogeneous microstructure in the green body.

2) Materials Used

The gelatine polymer with molecular weights ranging from 15000 to 90000 was provided by Shanghai Chemical Reagent Works. The analysis of the gelatine has been reported in previous study (Chen Y. L., et al., 1999). The gelatine powder can be dissolved in water in less than 20 min at 40℃, and shows pH=5.5–7.0 for 1 wt% solution. Urea with a purity of 99% is a commercial grand (Beijing Chemical Company). The alumina powder used here was a commercial grade An-05, and the details about the powder were described (Chen Y. L., et al., 1999).

3) Process

The forming process is described in Fig. 3.22. First, the urea and gelatine were dissolved in deionized water at 40℃. As the solution cooled down to room temperature, alumina powder and dispersant were added to the solution. Then mixture was ball milled for more than 12 h to obtain a well-dispersed suspension with good fluidity. After degassing, urease solution was added and mixed at the lower temperature of 2 – 5℃. Then the slurry was cast into the mould under ambient conditions and coagulated to a green body through gelation due to enzyme-catalyzed decomposition of urea.

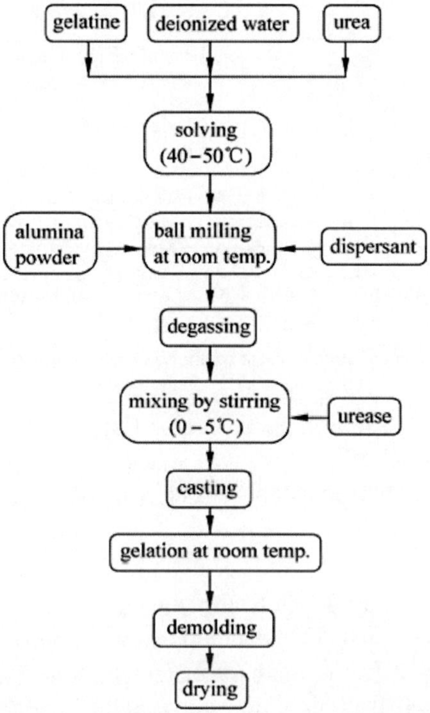

Figure 3.22 The flowchart of the forming process for alumina slurries with gelatine and urea

3.3.2 The Gelation Mechanism of Gelatine Solution with Urea under Enzyme Catalysis

1) Prevention of Urea on Attraction of Gelatine Molecule

Gelatine, a kind of protein polymer, is mainly composed of various amino acids. Urea can react with amino acids with hydroxide radical and restrain the formation of hydrogen bonds between polymer molecules or polymer and water. Figure 3.23 shows that the viscosity of 2 wt% gelatine solution changes with urea content under the cooling stage from 40 to 20℃. The ratio of urea to gelatine is expressed as α. When $\alpha < 3$, the solution viscosity was high due to gelation of the gelatine. When $\alpha = 4$, the solution showed some fluidity and still appeared in a viscous state, in which flocculation was observed. However, the viscosity decreased to 10 mPa·s or lower when $\alpha > 5$. In this case, sufficient urea prevented gelatine from gelling completely and a clear gelatine solution with perfect fluidity could be prepared. For the system with 3 wt% gelatine solution which also was studied, the results, almost similar to those explained above were observed.

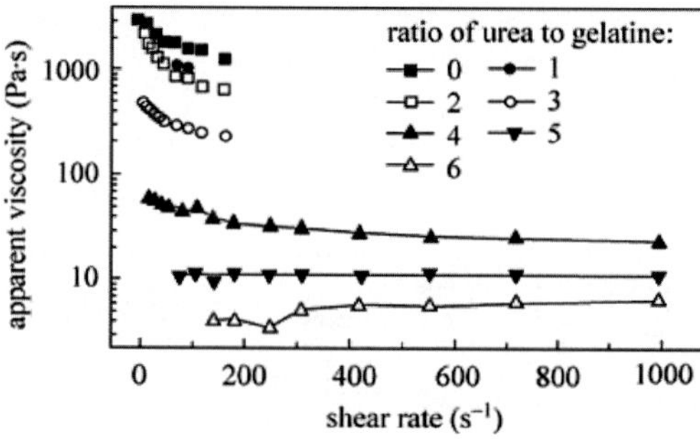

Figure 3.23 The effect of urea content on the viscosity of gelatine solution (2 wt%)

2) Gelling of Gelatine Solution under Urease Catalysis

When urease, as a kind of catalyst, was added into the gelatine solution, urea could be decomposed by urease catalysis. The reaction is as follows:

$$NH_2\text{—}CO\text{—}NH_2 + H_2O + \text{urease} \longrightarrow NH_4^+ + HCO_3^- \tag{3.1}$$

The reaction speed mainly depends on temperature and the amount of urease. At a temperature lower than 5℃, decomposition of urea is very slow, but it would increase greatly with the increase of temperature. For example, one unit will liberate 1 μmol NH_3 from urea per min at pH = 7 and 25℃ (Omatete O. O., et al., 1991). During the decomposition of urea, the gelling of gelatine solution will occur through hydrogen bonding among gelatine molecules, and the gelation can take place at room temperature as the gelling point ranges from 15 to 25℃. Figure 3.24 shows the viscosity change with decomposition time after adding urease at 15℃ for the solution containing 2 wt% gelatine and 10 wt% urea. The viscosity increased slowly at the initial stage due to less significant urea decomposition and weak hydrogen bonding among molecules. However, the solution lost fluidity and gelled after about 3 h. This resulted from a 3D network formed by attraction among molecular chains.

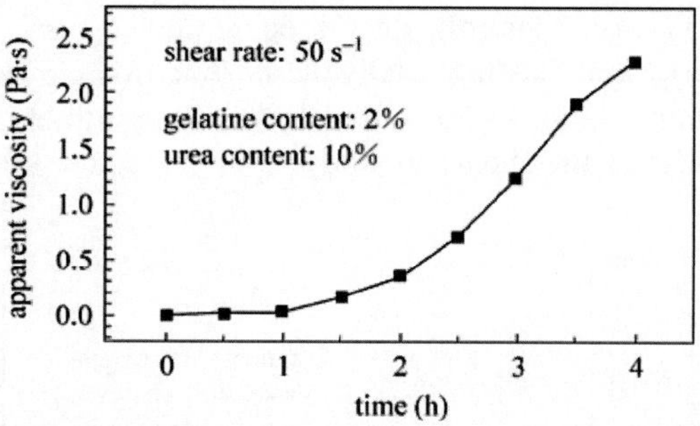

Figure 3.24 The relationship between viscosity of gelatine solution and time after addition of urease

The time to reach consolidation was clearly influenced by the amount of urease. From Fig. 3.25, it can be seen that about 10 h were required for a urease concentration of less than 5 units/mmol urea, and only 3 h for a urease concentration of 10 units/mmol urea. There was almost no influence of urease content on the time as the urease concentration reached 16 units/mmol urea.

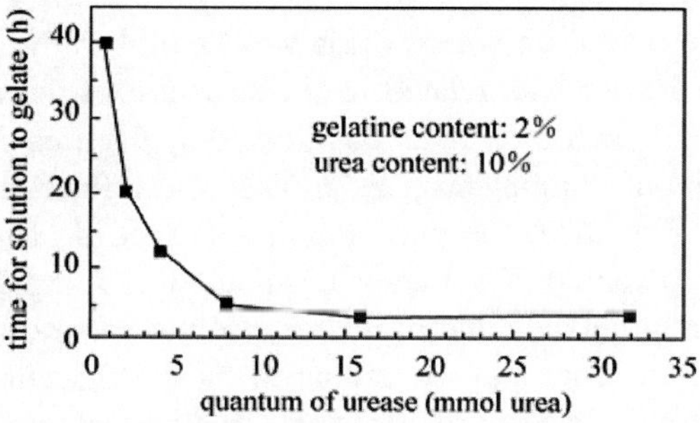

Figure 3.25 The effect of the urease concentration on the gelation time

3.3.3 The Rheology and Zeta Potential of Alumina Suspension Containing Gelatine and Urea

A suspension with 50 vol% alumina and 2 wt% gelatine (based on water) was prepared according to the flow chart in Fig. 3.22. The suspension was ball milled for 24 h. The influence of the amount of urea on rheological properties was illustrated in Fig. 3.26. When the ratio of urea to gelatine (α) was equal to 4, the system showed a relatively large viscosity. Also, a dilatant flow model was observed in which viscosity increased with shear rate. It was implied that there still was attraction among polymer molecules that resulted in a large viscous force during shearing. Under those conditions, agglomerates or flocculation may occur and the homogeneous microstructure of the ceramic body will be destroyed.

When $\alpha \geqslant 5$, the viscosity sharply decreased to lower than 100 mPa·s, which facilitated degassing and casting easily carry out. Moreover, the suspension showed a Newtonian model. In this case, the viscosity almost did not vary with shear rate, especially for the shear rate above 50 s^{-1}.

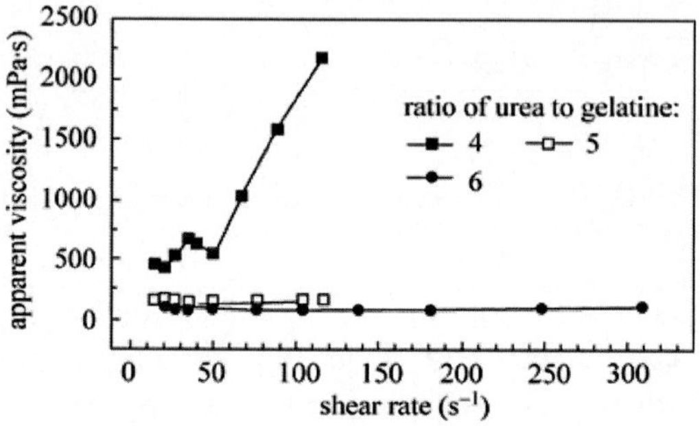

Figure 3.26 The effect of the urea content on the fluidity of slurry with gelatine and alumina

Similar results were also observed for the suspension with 50 vol% alumina and 3 wt% gelatine (based on water). This was helpful to the colloidal process. This rheological behavior was related to the structure of the system where the attraction between polymer molecules was limited by the urea content.

The zeta potential of alumina suspension was also influenced by gelatine and urea. As shown in Fig. 3.27, the pure alumina suspension had IEP = 6.0 and a relatively high zeta potential (absolute value about 40 mV at pH = 8 – 10). For the suspension containing gelatine and urea, the absolute value of the zeta potential was clearly decreased. This may be explained by the fact that the electrostatic double layer was compressed by ions at high gelatine content. However, it was noted that the suspension still showed a good fluidity when the ratio of urea to gelatine (α) was higher than 5.

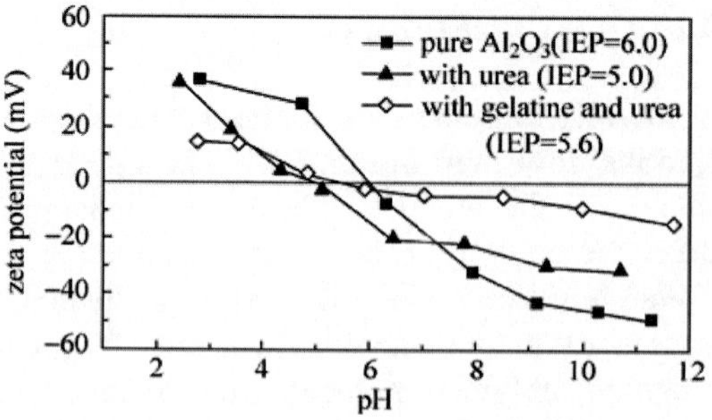

Figure 3.27 The change of zeta potential with pH

3.3.4 The Coagulation Forming and Microstructure of Green Body

The relationship between apparent viscosity and time was measured at 15℃, when the urease (16 units per mmol urea) was added into 50 vol% alumina suspension with 2 wt% gelatine (based on water) and 10 wt% urea (based on water). Figure 3.28 shows that the change in apparent viscosity of the slurry was similar to that of a gelatine solution (see Fig. 3.24). The viscosity increased relatively slowly at the initial stage and rose greatly after 3 h. This is due to the fact that the coagulation of the slurry resulting from gel network formed by gelatine molecules strongly depended on the speed of the enzyme-catalyzed decomposition of urea. The viscosity reached more than 25 Pa·s after 7 h and then remained constant thereafter. At this point, a 3D gel network was formed which resulted in complete coagulation of the slurry, and a rigid wet green body was obtained.

The microstructure and pore distribution of the dried green body were examined by an SEM and a mercury porosimeter. As shown in Fig. 3.29, a homogenous structure could be observed, in which large agglomerates and bubble were removed. Also, the pore distribution was characterized by a narrow single peak while two peaks were observed for dry pressed green body. These results suggested that a

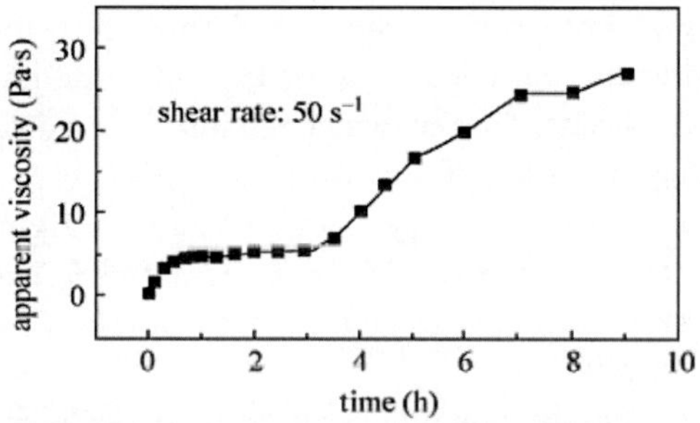

Figure 3.28 The slurry viscosity change with time after adding urease

(a) (b)

Figure 3.29 The SEM micrograph of dry green body

well-dispersed slurry, good degassing and casting could be performed by the colloidal forming process.

3.4 The Alumina Forming Based on Gelation of Sodium Alginate

3.4.1 Research Background

Alginate is a type of gelling polysaccharide, which can be dissolved in water at room temperature and then begin gelling by cross-linking with divalent metal ions at increased temperature after casting. The mechanism of cross-linking in alginate gels can be considered in terms of an "egg-box" model involving cooperative bonding of calcium ions between aligned polyguluronate ribbons (Braccini I., 1999). Calcium iodate has been applied in this process. It has low solubility at room temperature and high solubility at increased temperature of 60℃. Thus, the gel-casting process can be controlled by the heating rate and the final temperature (Xie Z. P., et al., 2001).

A novel gel-casting process by alginate is discussed in this chapter. In this process, a solidified agent and chelator are added into sodium alginate solution simultaneously. Thus, the gelation between calcium ions and sodium alginate is avoided in this stage before casting. Free calcium ions are released when hexanedioic acid is added into the suspension. They react with sodium alginate and form a 3D network. Therefore, the ceramic particles are fixed in this 3D network. The slurry is consolidated and near-net-shape green parts are formed.

1) Materials and Chemicals

The composition of the alumina powder used here has been shown in Section 3.1.1 of this chapter. Controllable casting and solidification are possible through the use of sodium alginate, the solidified agent, the chelator and the hexanedioic acid ($C_6H_{10}O_4$). Sodium alginate (Na-alginate) is used as a gelation reagent, which can dissolve in water at room temperature. Three kinds of solidified agents: namely, calcium carbonate ($CaCO_3$), calcium phosphate secondary ($CaHPO_4$), and calcium phosphate ($Ca_3(PO_4)_2$), and two chelators: sodium hexametaphosphate ($(NaPO_3)_6$) and tri-sodium citrate ($Na_3C_6H_5O_7$), are employed in the process.

2) Procedure

The forming process is shown in Fig. 3.30. Sodium alginate was added into the deionized water to form a solution, which had been decomposed at 70℃ for 48 h. Then, the alumina powder, solidified agent and chelator were added into the solution. The resulting slurry was then ball milled for 24 h. After degassing in a rotary evaporator under vacuum, hexanedioic acid was added into this slurry before it was cast into a nonporous mold. The slurry consolidated to the shape required, and then the wet green body was demolded and dried at room temperature.

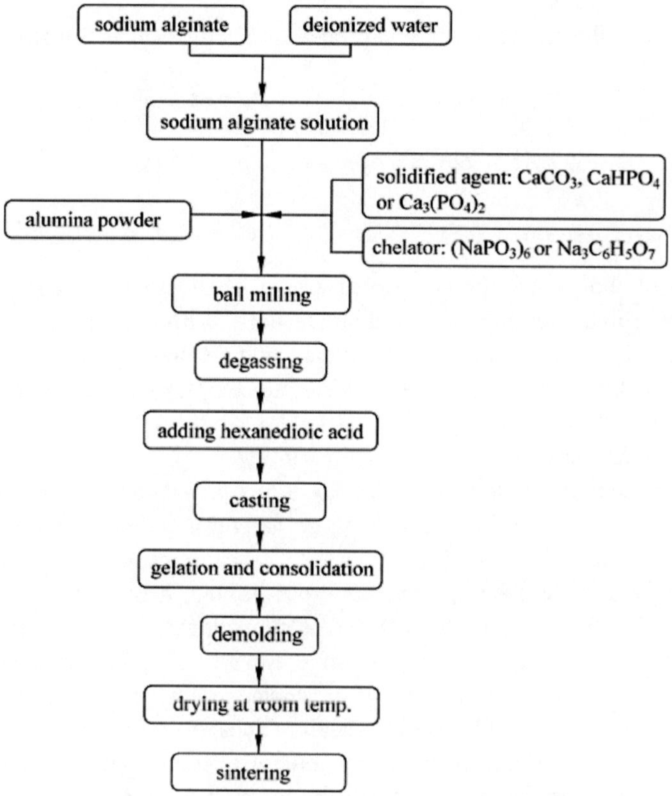

Figure 3.30 The flow chart of the forming process for alumina slurries with sodium alginate

3.4.2 The Gelation Principle of Sodium Alginate

When calcium salts are added into sodium alginate solution, they will form a gel very quickly owing to the reaction between calcium ions and sodium alginate, as shown in Eq. (3.2). Gelation occurs during casting due to the fast gelation rate, therefore, results in varying degrees of cross-linking density and heterogeneity within the gel. So, how to control the reaction rate by a chemical process is important. In this case, a chelator is used in the process. It forms a chelate complex with calcium salts (shown in Eq. (3.3)). Therefore, free calcium ions are chelated in the chelate complex, which allows them to be uniformly distributed in the alginate solution and does not induce gelation immediately. Then just before casting, hexanedioic acid is added into the slurry and the calcium ions are released slowly from the chelate complex (shown in Eq. (3.4)). Gelation takes place at this stage, which is shown in Eq. (3.2).

The reactions that occur during the process are expressed as following:

$$2\text{Na-alginate} + Ca^{2+} = \text{Ca-alginate} + 2Na^{+} \quad (3.2)$$

$$2C_6H_5O_7^{3-} + 3Ca^{2+} = Ca_3(C_6H_5O_7)_2 \quad (3.3)$$

$$Ca_3(C_6H_5O_7)_2 + 3C_6H_{10}O_4 = 2C_6H_8O_7 + 3Ca^{2+} + 3C_6H_8O_4^{2-} \quad (3.4)$$

It has been shown that the slow-release acid is effective in producing homogeneous and relatively slow setting alginate gels. With the addition of chelator and hexanedioic acid, the time-delayed release of cross-linking calcium ions allows the calcium-alginate suspension to be molded into complex geometry before gelation occurs (Kuo C. K., 2001).

1) Effect of Different Chelators on Gelation

Comparing samples listed in Tables 3.3, 3.4, 3.5. All the results showed that the gelation start time prolonged and the gel strength decreased with increasing amount of the chelator. This might indicate that the conformational rearrangement of the polymer (alginate) is important for cross-linking. A higher chelate complex concentration at the constant condition would result in slower rearrangement of the polymer conformations to form effective cross-linking. Higher content of the chelator yielded higher concentration of the chelate complex, and naturally needed more hexanedioic acid to liberate the calcium ions from the chelate complex to initiate gelation. Therefore, for the same content of the hexanedioic acid and the solidified agent, the calcium ions would be released incompletely with elevated content of chelator; and the gel with lower gel strength was achieved due to this factor.

Table 3.3 The effect of solidified agent ($CaCO_3$) on gelation

sample	Na-alginate (mL)	$Na_3C_6H_5O_7$ (g)	$CaCO_3$ (g/mL)	$C_6H_{10}O_4$ (g)	gelation start time (min)	gel strength (g/mm^2)
1	20	0.19	0.01	0.08	7	9.09
2	20	0.20	0.01	0.08	12	8.92
3	20	0.20	0.02	0.08	3	17.17
4	20	0.23	0.02	0.08	8	19.17

Note: The concentration of the Na-alginate solution is 1 wt%.

Table 3.4 The effect of solidified agent ($CaHPO_3$) on gelation

sample	Na-alginate (mL)	$Na_3C_6H_5O_7$ (g)	$CaHPO_3$ (g)	$C_6H_{10}O_4$ (g)	gelation start time (min)	gel strength (g/mm^2)
1	20	0.19	0.1	0.08	5	11.72
2	20	0.21	0.1	0.08	10	8.99
3	20	0.23	0.1	0.08	12	7.16
4	20	0.23	0.1	0.15	4	26.16
5	20	0.26	0.1	0.15	4	19.51
6	20	0.35	0.2	0.15	8	17.29
7	20	0.40	0.2	0.15	9	10.24

Note: The concentration of Na-alginate is 1 wt%.

Table 3.5 The effect of solidified agent ($Ca_3(PO_4)_2$) on gelation

sample	Na-alginate (mL)	$Na_3C_6H_5O_7$ (g)	$(NaPO_3)_6$ (μL)	$Ca_3(PO_4)_2$ (g)	$C_6H_{10}O_4$ (g)	gelation start time (min)	gel strength (g/mm^2)
1	20	0.15	—	0.1	0.08	5	12.86
2	20	0.18	—	0.1	0.08	8	8.43
3	20	0.20	—	0.1	0.08	>180	—
4	20	0.20	—	0.15	0.15	2	29.18
5	20	—	60	0.1	0.08	2.5	43.9
6	20	—	120	0.1	0.08	5	37.3
7	20	—	120	0.3	0.15	1	103.7
8	20	—	180	0.3	0.15	2	99.33
9	20	—	240	0.3	0.15	3	70.51

Note: ① The concentration of the Na-alginate is 1 wt%. ② The concentration of $(NaPO_3)_6$ is 20 wt%.

It could also be seen that the gelation of alginate with various chelators differed considerably. Samples 1 and 6 in Table 3.5 were prepared with the case of the same amount of hexanedioic acid and solidified agent. It was observed that the gel strength with the chelator of $(NaPO_3)_6$ increased much greater than $Na_3C_6H_5O_7$. In addition, the concentration of $(NaPO_3)_6$ solutions was lower than that of $Na_3C_6H_5O_7$, which brought less ions. Therefore, it would have less effect on the suspension and the sintering of green body. Thus, the chelator of $(NaPO_3)_6$ was suitable for gel-casting.

2) Rheological Property Contrast

As shown in Fig. 3.31, there was no obvious effect on the viscosity when $CaCO_3$,

$Ca_3(PO_4)_2$ were mixed with aqueous alumina suspensions. However, the slurry viscosity increased sharply with the addition of $CaHPO_4$ after ball-milling for 20 h. $CaHPO_4$ (solubility product constant $K_{sp}=1\times10^{-7}$; Dean J. A., 1985) was found to be slightly soluble in water at room temperature and produced a number of calcium ions after a long stirring time, which initiated calcium gelation and increased the viscosity of the slurry. While $CaCO_3$ ($K_{sp}=2.8\times10^{-9}$), $Ca_3(PO_4)_2$ ($K_{sp}=2\times10^{-29}$) had very low solubility in pure water, allowing its uniform distribution in alginate solution before gelation occurred. Therefore, in order to ensure that the slurry has sufficient fluidity during casting, $CaHPO_4$ should be added into the slurry at the later stage during ball-milling.

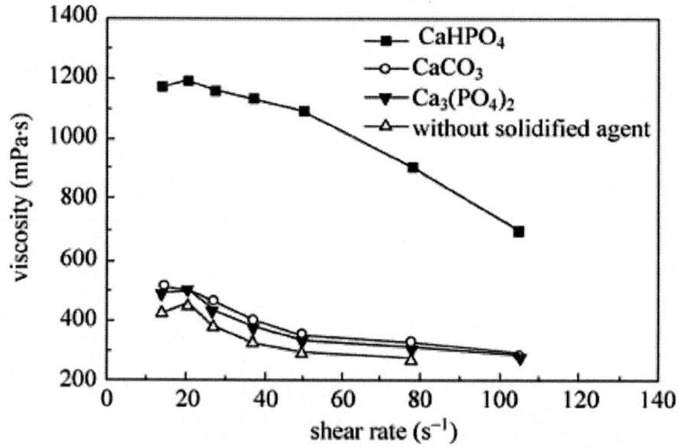

Figure 3.31 The rheological property of the suspensions with different solidified agent

3.4.3 The Preparation Process of Alumina Green Bodies and Samples by Sodium Alginate

1) Property of the Green Bodies and Sintered Samples

From samples 2 and 3 in Table 3.3, it was found that the gelation rate and the gel strength increased with increasing the amount of the solidified agent. This implied that the availability of calcium ions played a critical role in the gelation, and a high concentration of calcium ions led to faster gelation and higher gel strength.

It was found that the microscopic structures of the green body varied greatly with different solidified agents, as shown in Figs. 3.32 and 3.33. $CaCO_3$ could react with acid and release some CO_2 gas. And CO_2 could not be degassed completely and formed pores in the green part. These pores functioned as defects and resulted in poor mechanical behavior of the ceramic parts after consolidation since they could not be eliminated during sintering. $CaHPO_4$ had to be added into the slurry during ball-milling, which resulted in heterogeneity in the green body. However, $Ca_3(PO_4)_2$ had no such problems and appeared to be a kind of proper

Figure 3.32 The SEM micrograph of the cross-section of the green body with solidified agents of (a) $CaCO_3$, (b) $CaHPO_4$, and (c) $Ca_3(PO_4)_2$

solidified agent. To further investigate the difference among the solidified agents, sintered samples with $Ca_3(PO_4)_2$ and $CaHPO_4$ as solidified agents were also prepared. Figures 3.32 and 3.33 showed that the microscopic structure of these samples. It illustrated that the sintered body made from $Ca_3(PO_4)_2$ had a dense and homogeneous structure; while there were several holes and agglomerates in the structure of that made from $CaHPO_4$. This suggested that $Ca_3(PO_4)_2$ was more suitable for the process than the other agents.

2) Effect of the Hexanedioic Acid on Gelation

Samples 3 and 4 in Table 3.4 showed that the gelation rate and gel strength increased with increasing the amount of the hexanedioic acid. Similar results were observed in the case of samples 3, 4 and 6, 7 in Table 3.5. The time to form gel

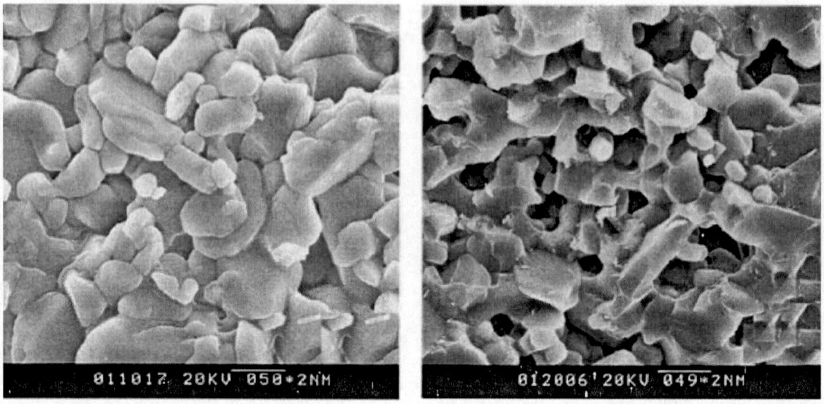

(a) (b)

Figure 3.33 The SEM micrograph of the sintered body with the solidified agents of (a) $Ca_3(PO_4)_2$, (b) $CaHPO_4$

strongly depended on the reaction between calcium ions and alginate. As mentioned, hexanedioic acid was employed to release the calcium ions from the chelate complex and to realize the time-delayed gelation. Therefore, the content of the free calcium ions in the solution was mainly determined by adding the hexanedioic acid. Higher content of hexanedioic acid could result in larger concentration of free calcium ions, and correspondingly faster gelation.

3) Consolidation

The solidified agent of $Ca_3(PO_4)_2$ and chelator of $(NaPO_3)_6$ were employed here. It was clear that the addition of alumina powder decreased the gelation rate, comparing samples 2 and 4 in Table 3.6. The suspension exhibited greater viscosity due to the addition of alumina, which led to slower rearrangement of the polymer chain conformations to form effective cross-linking. Similar results from the gelation process could also be observed in the consolidation process, for example, the consolidation rate increased with decreasing content of the chelator.

Table 3.6 The effect of alumina powder on gelation

samples	Na-alginate (mL)	$(NaPO_3)_6$ (μL)	$Ca_3(PO_4)_2$ (g)	$C_6H_{10}O_4$ (g)	Al_2O_3 (g)	gelation time (min)
1	20	300	0.3	0.15	72	10
2	20	400	0.3	0.15	72	13
3	20	1000	2.3	0.6	72	28
4	20	400	0.3	0.15	—	7
5	20	400	0.3	0.15	—	10
6	20	1000	2.3	0.6	—	26

Note: ① The concentration of the Na-alginate is 1 wt%. ② The concentration of $(NaPO_3)_6$ is 20 wt%.

3.5 The Gel-Casting of SiC Based on Gelation of Sodium Alginate

3.5.1 Introduction of the Research

1) Research Background

SiC ceramics have attracted great interest for their use in the field of refractories as well as in the field of high-temperature structural applications for a long time. SiC ceramics are known to possess a variety of virtues, such as high mechanical strength, high chemical stability, good thermal conductivity, and semi-conductivity and so on. They possess specific applications in the petroleum, mechanical, and aerospace industries (Zhou L. J., et al., 2000; Rao R. R., et al., 1999a, b; Zhou L. J., et al., 1998).

In this section, gel-casting of SiC ceramics by using an alginate gelation is discussed. For this process, a solidified agent and a chelator are added into sodium alginate solution simultaneously. Thus the gelation between calcium ions and sodium alginate is avoided in this stage due to the formation of a chelate complex from chelator and sodium alginate. When hexanedioic acid is added into the suspension, free calcium ions are released from the chelate complexes and they react with sodium alginate, forming a 3D network. Therefore, the ceramic particles are fixed in this 3D network and the slurry is consolidated.

The starting ceramic powder consisted of coarse SiC and fine SiC powders in order to achieve a high packing density. The mass ratio of coarse powder to fine powder was 70:30. Main impurities in the powders included: the content of Fe is 520×10^{-6}, O 1.17 wt%, Si 0.26 wt% and free carbon 0.35 wt%. The particle size of mixture containing coarse and fine SiC powders ranged from 0.2 to 350 μm (Yi Z. Z., et al., 2001).

The controllable casting and consolidation are possible through the use of sodium alginate (Na-alginate), sodium hexametaphosphate ($(NaPO_3)_6$), tertiary calcium phosphate ($Ca_3(PO_4)_2$), and hexanedioic acid ($C_6H_{10}O_4$). Sodium alginate is used as a gelation reagent, which dissolves in deionized water at room temperature. When the sodium alginate and calcium ions react, the molecular chains attract each other to form a 3D network. Sodium hexametaphosphate, a vitriform solid, is used here as a chelator while tertiary calcium phosphate is used as a solidified agent. Three dispersants, namely, TMAH, DOLARPIX PC33 (PC33), and tri-ammonium citrate are employed.

2) Research Program

The flowchart of forming process is shown in Fig. 3.34. Sodium alginate was dissolved in the deionized water to form a solution, which had been decomposed at 70℃ for 48 h to reduce solution viscosity. And then SiC powder, dispersant, solidified agent and chelator were added into the solution. The resulting slurry was then ball-milled for 24 h. After degassing in a rotary evaporator under vacuum, hexanedioic acid was added into this slurry before it was cast into a nonporous mould, after which the consolidation occurred. After several hours, the wet green body was demolded and dried at room temperature.

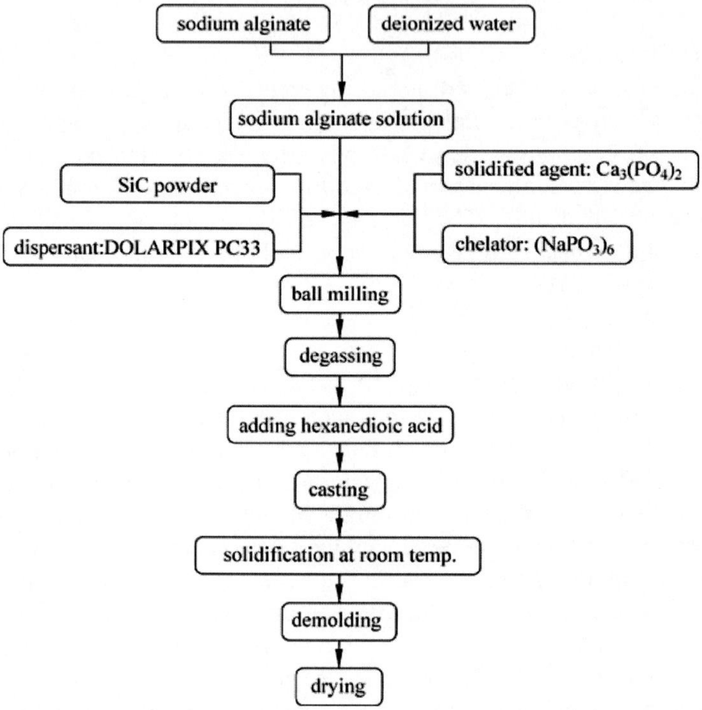

Figure 3.34 The flowchart of the forming process for SiC slurries with sodium alginate

3.5.2 The Effect of Dispersant on the Colloidal Behaviors of the SiC Suspension

The variations of zeta potential with pH for silicon carbide and the suspension containing TMAH, DOLARPIX PC33 and tri-ammonium citrate as dispersants respectively are shown in Fig. 3.35. The addition of sodium alginate produced a pronounced effect on electrokinetic behaviors of SiC, wherein the IEP of the SiC suspension changed from pH = 3.02 to 2.36. This is caused by the absorption of carboxyl negative charge on the alginate chains onto the surface of SiC particles. At pH = 7 – 9, a negative zeta potential about –(10 – 15) mV was observed. This may be attributed to the fact that the electrostatic double layer was compressed by ions with a high value in sodium alginate (Pierre A. C., et al., 1999; Teng W. D., et al., 1997).

TMAH is an organic alkali, PC33 is a kind of polyelectrolyte, and tri-ammonium citrate is an inorganic salt. TMAH and PC33 are alkaline, and result in better dispersion of the SiC powder since SiC powder can be suspended in a basic solution, larger repulsion between ceramic particles and a more stable suspension are achieved with increase of the zeta potential, as shown in Fig. 3.35. In contrast, tri-ammonium citrate is an acid and does not provide good dispersion (Zhou L. J., et al., 2000; Yi Z. Z., et al., 2001). DOLARPIX PC33 is more effective among the three kinds of dispersants due to its high zeta potential absolute value (about 40 mV at pH = 7 – 9) and poly-electrolyte stabilization, which is helpful for the stability of suspensions (Cesarano Ⅲ J., et al., 1988). In this case, DOLARPIX PC33 is a suitable dispersant for the sodium alginate solution compared to the other two.

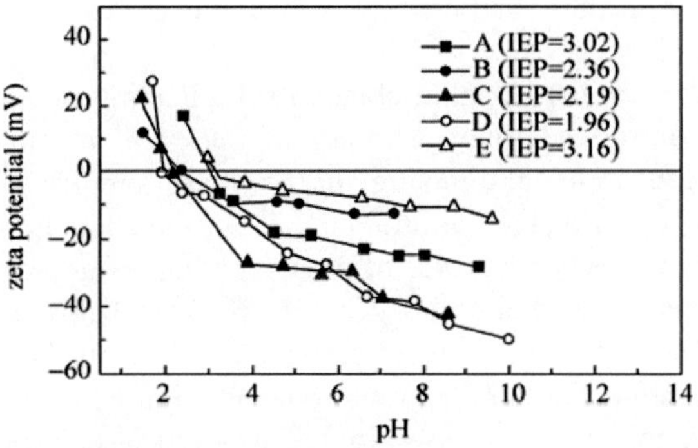

Figure 3.35 The change of zeta potential vs. pH of SiC suspension in deionized water (A), Na-alginate solution (B), Na-alginate solution with dispersant TMAH (C), PC 33 (D) and tri-ammoniumcitrate (E) respectively

3.5.3 The Rheological Property of SiC Suspension

A suspension with 50 vol% SiC and 1.5 wt% sodium alginate (based on water) was prepared, as shown in Fig. 3.34. The influence of the amount of dispersant (DOLARPIX PC33) on the rheological property of SiC suspension is illustrated in Fig. 3.36, which indicates that the dispersant PC33 (curves B, C, D) increases the fluidity of the suspensions compared to a suspension without the dispersant (curve A). When the shear rate is 100 s^{-1}, the apparent viscosity of the slurry with volume ratio of 10 μL : 1 mL (PC 33 to the sodium alginate solution) system (B) was lower than 1 Pa·s, which is the empirical value below which casting process was allowed. The viscosity of the slurry with a volume ratio of 20 μL : 1 mL system was about 1 Pa·s and was much higher than 1 Pa·s for a slurry with a volume ratio of

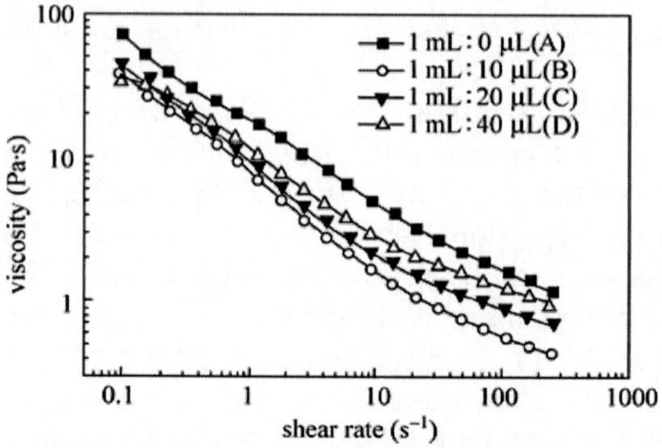

Figure 3.36 The effect of the dispersant (DOLARPIX PC33) on the viscosity of the slurry of a 50 vol% SiC suspension (volume ratio of PC33 to 1.5 wt% Na-alginate solution)

40 mL: 1 mL, which implies that the excessive polyelectrolyte (DOLARPIX PC33) is not absorbed on the SiC particles, therefore, the viscosity of the suspension increases.

On the one hand it is beneficial to obtain a solid loading as high as possible in order to better control the final geometry of the part, and to achieve higher dimensional tolerances by minimizing the possible complex distortion during drying and sintering shrinkage. On the other hand, the shaping of intricate parts requires the slurry to easily fill molds of possible complex geometry. This in turn means that the viscosity of the slurry has to be kept below some acceptable level. The two-boundary conditions necessitate a well-controlled tailoring of the suspension that can meet both of the requirements. Figure 3.37 demonstrates that the suspension has relatively low viscosity and pseudo plastic property when solid loading is below 55 vol%. When the solid loading increased up to 60 vol%, the

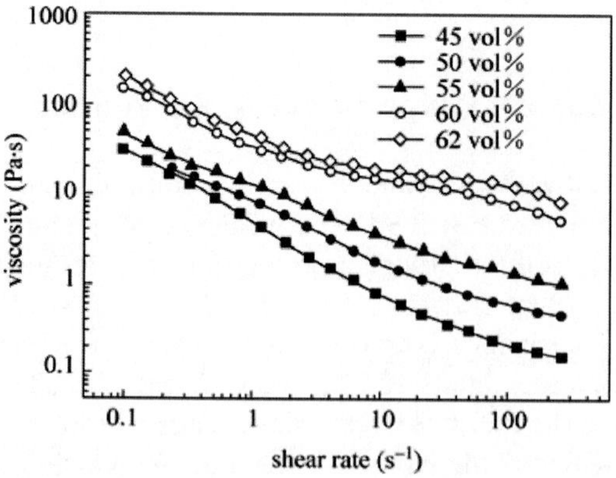

Figure 3.37 The rheological property of the slurry with various solid loading (based on 1.5 wt% Na-alginate)

viscosity was 7.79 Pa·s under the shear rate of 100 s^{-1}. For the slurry of 62 vol%, the viscosity reached 12.46 Pa·s, which does not render it suitable for gel-casting. Therefore, the suspension with solid loading of 45 – 55 vol% is suitable for this process.

3.5.4 The Sedimentation Behavior of the SiC Suspension

When ceramic powder is dispersed in a liquid medium, the particles are influenced by a variety of forces such as force of gravity, drag force from the medium and Brownian motion of particles, etc. Assuming that the other forces are negligible or of non-varying type, the ceramic particles tend to settle under the force of gravity. The state of dispersion and its stability is measured by the particle settling rate (manifested in terms of an increase of sediment height or sediment volume) vs. time (Pierre A. C., et al., 1999; Rao R. R., et al., 1999a, b). The particle settling rate manifested in terms of an increase of sediment height or sediment volume as a function of time gives a measure of the state of dispersion and its stability. Relative sedimentation height here refers to the ratio of the mixed suspension height (including the stable suspension and the sedimentation layer) to the whole height of the suspension. The sedimentation height is inversely related to the dispersion of the powder; the more dispersed the powder, the smaller the sediment height.

The sedimentation behavior was conducted in test tubes that had a capacity of 50 mL. Control samples with solid loading of 20, 30, 45, 50, 55 vol% were included to provide a basis for contrast. Figure 3.38 shows that the sedimentation of suspensions with low solid loading, such as 20 and 30 vol%, is obvious; while that at high solid loading, such as 45, 50 and 55 vol% cannot even be observed (Rao R. R., et al., 1999a, b).

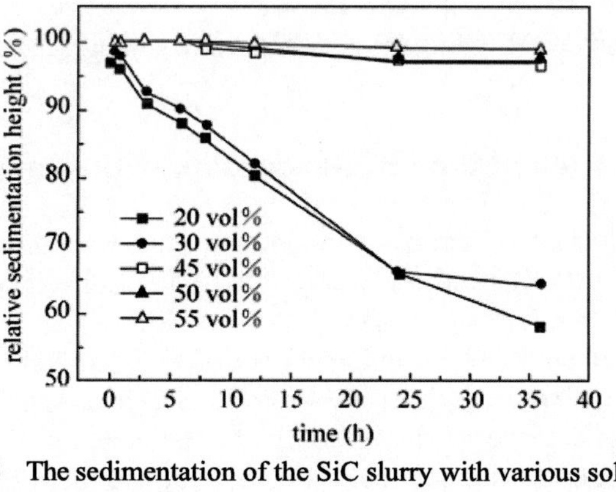

Figure 3.38 The sedimentation of the SiC slurry with various solid loading

3.5.5 The Gelation Principle and Process of the Alginate Solution

The mechanism of cross-linking in alginate gels can be considered in terms of an "egg-box" model involving cooperative bonding of calcium ions between aligned polyguluronate ribbons (Wber K., et al., 1997; Cesarano Ⅲ J., et al., 1988). However, it is generally difficult to control the reaction rate, which makes it impossible to complete the casting process at a certain period. Therefore, the calcium salts at a controlled reaction rate with alginate are considered in this study. The concentration of the calcium ions released from the salt can be adjusted by the formation and decomposition of the chelate complexes. The principle sketch of the process is shown in Fig. 3.39. Tertiary calcium phosphate (the solidified agent) and sodium hexametaphosphate (the chelator) were added into the alginate solution, and calcium ions were chelated in the chelate complexes. Therefore, calcium ions could not react with alginate and the gelation between calcium ions and alginate was avoided in this stage before ceramic slurry casting. Then hexanedioic acid was added into the solution, the chelate complexes decomposed and released the calcium ions slowly. The time-delayed release of cross-linking calcium ions allowed the Ca-alginate suspension to be molded into complex geometry before gelation occurred. Thus, gelation rate control was realized.

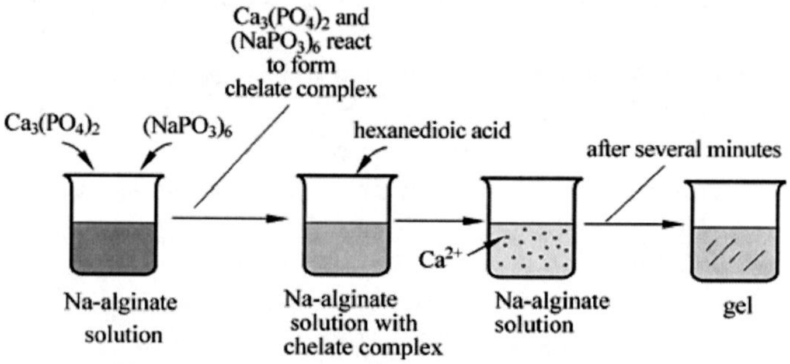

Figure 3.39 The principle sketch of the gelation process

3.5.6 The Gelation of the SiC Suspension with Alginate

To study the influence of various Na-alginate concentrations, the amount of solidified agent ($Ca_3(PO_4)_2$) and chelator ($(NaPO_3)_6$) on consolidation and property of the green body, several groups of samples were fabricated, as shown in Table 3.7.

For the same content of solidified agent ($Ca_3(PO_4)_2$) and chelator ($(NaPO_3)_6$), samples 1, 2, and 3 were made from 45 vol% SiC and 1.0, 1.5 and 2.0 wt% Na-alginate solutions, respectively. Accordingly, the consolidation time increased from 2 to 5 h, as the Na-alginate concentration increased from 1.0 to 2.0 wt%.

Table 3.7 The properties and gelation of the suspension with 45 vol% SiC (400 μL DOLARPIX PC33 in 20 mL Na-alginate solution) under various conditions

Sample	Na-alginate concentration (wt%)	$Ca_3(PO_4)_2$ (g)	$(NaPO_3)_6$ (mL) ①	pH	Consolidation time ② (h)	Bending strength (MPa)	Relative green density ③ (%)
1	1.0	4	1.0	8.18	2	2.49	47.86
2	1.5	4	1.0	8.36	3	2.85	52.41
3	2.0	4	1.0	8.66	5	3.09	50.57
4	1.5	6	1.0	8.71	1	3.47	52.16
5	1.5	1.2	1.0	8.04	4	2.54	51.57
6	1.5	4	2.0	8.30	5	3.26	53.99
7	1.5	4	0.4	8.95	0.5	2.67	51.74

Note: ① Concentration of $(NaPO_3)_6$ is 20 wt% (based on deionized water).
② Consolidation time refers to the time of change from slurry to wet green body.
③ The theoretical density of SiC is 3.217 g/cm^3.

These results exhibited the decrease in the consolidation rate with the increase of the polymer (sodium alginate) concentration, which may be an indication that the conformational rearrangement of the polymer chains is important for cross-linking formation. A higher polymer concentration likely resulted in higher viscosity, leading to slower rearrangement of the polymer chain conformations to form the effective cross-linked structure. On the other hand, the resulting bending strength and relative green density increased with the increase of the alginate concentration, which indicated that the strength of the green body was mainly determined by the 3D network. It could be concluded that the stronger network and higher strength of the green body was achieved with the increase of the alginate concentration. However, the consolidation rate also increased with the increase of alginate concentration, and some agglomerates were observed in the consolidation process of the slurry dispersed in 2.0 wt% Na-alginate solution, which probably results in nonuniform structure. In this case, 1.5 wt% Na-alginate solution was employed in the following experiment.

Further characterization of mechanical properties of the green bodies of samples 4, 2 and 5 were studied, which were fabricated from suspensions with decreasing content of 6, 4 and 1.2 g $Ca_3(PO_4)_2$ (with 45 vol% SiC and 20 mL 1.5 wt% Na-alginate solution). On comparing samples 5(2.54 MPa) with 2(2.85 MPa) and 4(3.47 MPa), it was found that the strength of the green body increased with the calcium content, presumably due to the increase of cross-linking density. The sodium ions on the Na-alginate molecules would be further replaced when more calcium salt was added into the slurry. In this case, the overall contribution of the effective ionic interactions between calcium ions and Na-alginate became greater at higher calcium contents, leading to better mechanical properties of the green body.

Therefore, at low content of $Ca_3(PO_4)_2$ (1.2 g), the sodium ions on the Na-alginate molecules would inevitably be replaced incompletely. Thus, the resulting network was not strong, and the density and strength of the green body were reduced. It should also be noted that the suspension with high content of $Ca_3(PO_4)_2$ (6 g), on the other hand, consolidated too quickly, which might have resulted in structural heterogeneity, leading to lower relative density. Therefore, the appropriate content of $Ca_3(PO_4)_2$ was 4 g for this study.

Samples 6, 2 and 7 were produced from 45 vol% SiC slurries and 1.5 wt% Na-alginate solution with different content of 2, 1, 0.4 mL $(NaPO_3)_6$ (20 wt%) and the corresponding green bars were also made. When the chelator ($(NaPO_3)_6$) addition is not enough, the calcium ions will be chelated incompletely and the redundant calcium ions will react with sodium alginate to form partially gelation, which is detrimental to the fluidity of the slurry. Furthermore, it has a negative effect on the uniformity and density of the green body, as shown in sample 7. Therefore, a high content of $(NaPO_3)_6$ is helpful to the fluidity of the slurry and the ultimate strength of the green body. With increasing the content of $(NaPO_3)_6$, however, the consolidation rates decrease promptly due to the time-delayed calcium releasing period. Considering the effect of the chelator addition on both the consolidation rate and the strength of the green body, 1 mL $(NaPO_3)_6$ was found to be appropriate in this study.

Addition of hexanedioic acid plays a critical role in this process, as it can release the calcium ions from the chelate complex and initiate the consolidation. To study the effect of hexanedioic acid on the consolidation, a solid loading 45 vol% slurry was prepared, which was composed of 52.4 g SiC powder, 20 mL 1.5 wt% Na-alginate solution, 1 mL 20 wt% $(NaPO_3)_6$, 1.2 g $Ca_3(PO_4)_2$ and 1 mL PC33. Disk shaped green bodies with different hexanedioic acid addition were fabricated. Calcium ions will be released incompletely because of the small amount of hexanedioic acid (<0.3 g), and therefore, consolidation will not be available. It was observed that a threshold amount of hexanedioic acid is needed to form a fix-shaped green body; below this threshold amount, consolidation is impossible. Table 3.8 demonstrates that the consolidation rate increases with increasing the amount of hexanedioic acid and only when the amount of hexanedioic acid reaches 0.6 g/20 mL complete consolidation of Na-alginate solution will occur.

Table 3.8 The influence of hexanedioic acid on consolidation

hexanedioic acid (g)	<0.3	0.5	0.6	0.8	1.0	1.2	>1.5
consolidation time	>24 h	50 min	50 min	35 min	35 min	20 min	18 min
consolidating degree	not consolidated	fair	good	good	good	good	good
the morphology of the green body	poor	fair	good	good	good	good	good

Green parts with the strength of 3 MPa and perfect surface are obtained in this process. The microstructure of the cross-section of the green body (from the slurry composition of sample 2) was studied by the SEM, which shows that the structure of the green body is homogeneous and dense.

3.6 The Alumina Gel-Casting with a Low-Toxicity System of HEMA

3.6.1 The Academic Idea and Research Program

HEMA is a low toxic reagent, which is commonly used for soft lens material and as an ingredient in some adhesive resins. HEMA is prone to polymerization due to the vinyl bond while the hydroxy group makes it possible to form a water-compatible polymer. Here HEMA is selected as a new monomer and it seems to be a successful one. The rheological properties of the ceramic slurries, the microstructure of the green body obtained from the new system and the pyrolysis property of the system during sintering are carefully investigated.

The alumina powder used here has been shown in Section 3.1.1 in this chapter.

HEMA produced by Beijing Easten Acrylic Chemical Co. was used as the monomer and MBAM as the cross-linker. HEMA and MBAM were dissolved in deionized water to prepare the premix solution. TEMED and $(NH_4)_2S_2O_8$ were used as catalyst and initiator respectively. A commercial product JN281 and TAC were selected as the dispersants. Before casting the ceramic powder, dispersants were added into the premix solution and thoroughly ball-milled for 24 h to prepare a well dispersed suspension. After being deaired for 15 min the polymerization was initiated by adding the initiator and catalyst. Then the slurry was cast into the molds and heated to 60 – 80℃ to form the ceramic green bodies of the desired shapes.

3.6.2 The Colloidal Chemistry and Rheological Property

In order to obtain high quality ceramic parts by gel-casting, it is important to obtain homogeneously dispersed suspensions with high solid loading. In ceramic suspensions an intimate relation exists between the rheological property and the suspension structure (i.e. the spatial particle distribution in the liquid), which depends on the zeta potential of the powders. The higher this potential with the same polarity, the more important the electrostatic repulsion between the particles. The zeta potentials of different Al_2O_3 suspensions are shown in Fig. 3.40. The zeta potential of Al_2O_3 suspensions changes from 7.98 mV at pH=2.51 to –16.85 mV at pH = 11.02, with an IEP at pH = 4.9. Addition of a monomer does not shift the IEP but it slightly decreases the absolute value of the zeta potential. This suggests that the uncharged HEMA molecule may screen the charge developed by the powders in the solution. With the addition of 0.25 wt% TAC and 1 wt% JN281 the IEP of the Al_2O_3 suspensions are shifted to pH = 3.9 and pH = 2.6 respectively, and zeta potential become more negative. JN281 can provide a more negative zeta potential, and is likely to be more effective for a well dispersed suspension.

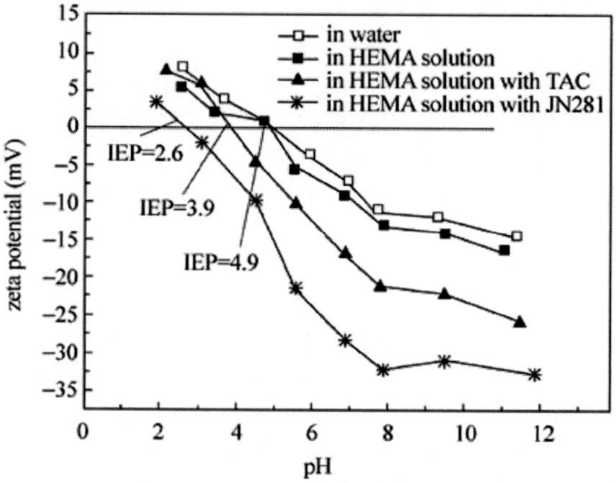

Figure 3.40 The zeta potential of different Al_2O_3 suspensions

Figure 3.41 shows the viscosity of the alumina (50 vol%) slurries in the premix solution containing 25 wt% of HEMA. The viscosity curve of the slurry in AM premix solution is also illustrated for comparison. The high viscosities at the beginning of the curves indicate a "Bingham" behavior of all the suspensions. Compared with the suspension in AM solution, which has almost the same viscosity as the suspension in pure water, addition of HEMA slightly increases the viscosity. The slurry in AM solution shows a shear-thinning behavior while the slurries in HEMA premix solution show a shear-thinning behavior at low shear rate and a shear-thickening behavior at high shear rate. The shear-thinning behavior reveals that the flow results in a more favorable 2D structure arrangement rather than a three one of the particles. The 2D arrangement is a layered structure and results in a low resistance of the particle movement between different layers. Thus, a low viscosity is obtained (Barnes H. A., et al., 1989). Due to the damage of the unstable 2D arrangement at very high shear rate, over the critical shearing rate, viscosities increase again and a shear-thickening behavior is observed in the HEMA system. Although the TAC is a very effective dispersant for AM system, JN281 seems more appropriate than TAC for HEMA system. JN281 can provide a much lower viscosity and the critical shearing rate is moved to a higher level than that of the TAC system in HEMA solution. This suggests that JN281 is more effective than TAC in stabilizing the suspension and in decreasing the so-called phase solid loading

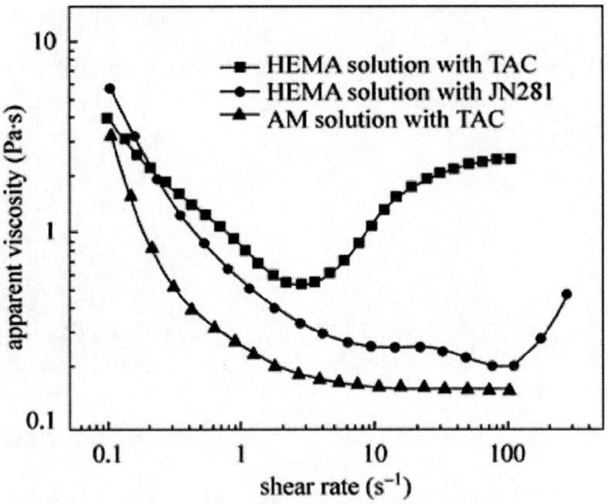

Figure 3.41 The change of the viscosity with shear rate for the Al_2O_3 suspensions

(Barnes H. A., 1989) in HEMA solution. These results agree well with the zeta potentials, indicating that the dispersants work by an electrostatic stabilization mechanism.

The strength of the green bodies derived from different HEMA premix solution with 50 vol% solid loading is shown in Table 3.9. As expected, green strength increases with the increase of the HEMA concentration. Considering both the mechanical strength and production cost, we should adopt a HEMA concentration of 25 wt%. The strength of the green bodies derived from AM or gelatine at the same Al_2O_3 loading is also listed. It can be seen that although the green strength derived from the HEAM system is lower than that from the AM system, it is much higher than that derived from the gelatine (Chen Y. L., et al., 1999).

Table 3.9 The strength of the green bodies for different gel-casting systems

	HEMA (20 wt%)	HEMA (25 wt%)	HEMA (30 wt%)	HEMA (40 wt%)	AM (20 wt%)	gelatine (4.5 wt%)
green strength (MPa)	14	18	18	20	28	8

Figure 3.42 shows the SEM photos of the green bodies (50 vol% Al_2O_3) obtained by the HEMA system or AM system respectively. The homogeneous structure can be obtained through both systems, suggesting that the HEMA system is suitable for gel-casting.

3.6.3 The Binder Burnout and Application of the New System

The TGA curves of the green bodies (50 vol% Al_2O_3) made from a 25 wt% HEMA premix solution and that from a 30 wt% AM solution are shown in Fig. 3.43. The

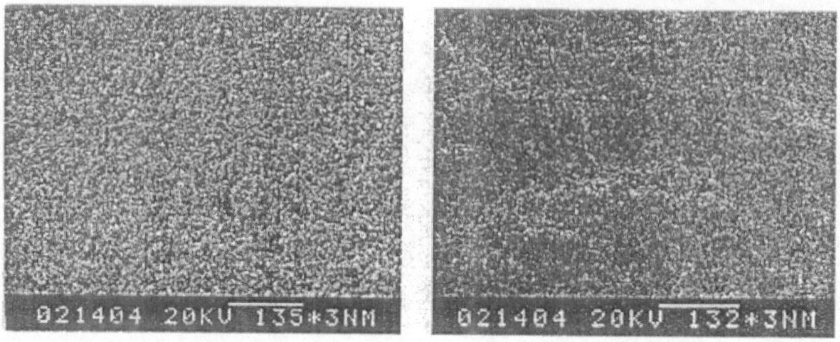

(a) (b)

Figure 3.42 The comparison of the microstructure of the green bodies with 50 vol% Al_2O_3

(a) HEMA system; (b) AM system

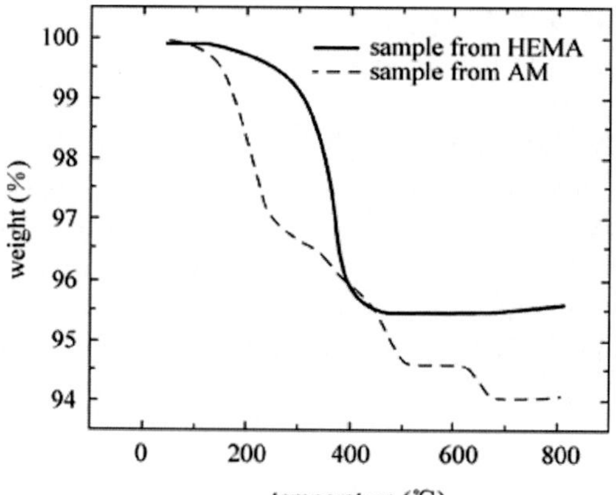

Figure 3.43 The pyrolysis of the green bodies with 50 vol% Al_2O_3 made from the HEMA and AM system

two gelcast bodies show similar pyrolysis behavior. Pyrolysis occurs mainly in the range of 120 – 680℃ and there are four weight loss peaks for the green body from the AM system. The weight loss in the temperature range of 160 – 320℃ may be mainly due to the removal of water while those in the range of 320 – 680℃ may be ascribed to the pyrolysis of PAM polymer. However, there is only one focused weight loss process in the range of 200 – 470℃ for the green body from HEMA system, suggesting that the polymerized HEMA can be removed more easily than PAM. Based on the above results, complex-shaped alumina parts (see Fig. 3.44) are successfully fabricated through a premix solution containing 25 wt% HEMA with JN281.

A new innoxious gel-casting system has been successfully developed by using HEMA. For the Al_2O_3 suspension in the new system, JN281 is more effective than the frequently used TAC. SEM and TGA results indicated that the green bodies

Figure 3.44 The ceramic parts fabricated by the new system

made from the new system have a homogeneous microstructure and the organic substance can be removed easily. Dense and homogeneous complex-shaped alumina parts were successfully fabricated.

3.7 The Synergistic Low-Toxicity Gel-Casting System by Using HEMA and PVP

3.7.1 The Academic Idea and Research Program

1) PVP Improving Strength Distribution in Green Bodies Formed by the HEMA System

We have reported that using HEMA (2-hydroxyethyl methacrylate) as a new monomer. It was exciting that a green strength as high as 18 MPa was obtained from a 25 wt% HEMA aqueous premix solution (Cai K., et al., 2005). However, a few percent of green bodies might show a much lower strength (about 8 MPa as minimum) than the average strength (18 MPa) because of the heterogeneity of the HEMA hydrogel network. Although the percentage is very low, it may constitute a potential problem in cases when narrow strength distribution and high reliability are required. Polyvinyl pyrrolidone (PVP) is a low-toxicity organic reagent widely used in various industries such as cosmetics, pharmaceuticals, and agricultural formulations as a covering, dispersing, viscosity-enhancement reagent and adhesive. Owing to the hydrophilic C=O group, it is water soluble and can form a hydrogel. This paper reports our results of alumina gel-casting with a PVP-modified low-toxicity system. Adoption of PVP as a synergistic gel-forming reagent resolved the problem of wide strength distribution in the green bodies formed by the HEMA system. Besides, the surface exfoliation phenomenon that seems inherent to the AM gel-casting system was also successfully eliminated in the

HEMA-PVP system. Analysis of the interaction between HEMA and PVP suggested that the improved microstructure and strength homogeneity, as well as the elimination of surface exfoliation in the new system, were mainly due to their ability to form an intermolecular hydrogen bonding, which greatly improved the homogeneity of the entire poly-HEMA (PHEMA)-PVP co-gel and prevented phase separation and oxygen diffusion.

HEMA produced by Eastern Acrylic Chemical Co. (Beijing, China) was used as the monomer. PVP (K-30 type) was an analytically pure reagent with an average molecular weight of 40000 provided by the Beijing Chemical Reagent Company (Beijing, China). Tri-ammonium citrate (TAC) provided by Beijing Chemical Reagent Company and a JN281-type polyelectrolyte (ammonium salt of poly (acrylic acid)) provided by Beijing Pinbao Co. were used as the dispersants to achieve a concentrated alumina suspension.

2) Sample Preparation and Characterization

The procedure for forming green bodies is similar to that of the AM system. HEMA and PVP were dissolved in deionized water, and then the alumina powder and dispersant were added. After ball-milling for about 24 h, the catalyst and initiator were added to the alumina suspension and heated at 80℃ for *in-situ* solidification.

3.7.2 Zeta Potentials and Rheological Properties

Development of stable high solid-loading ceramic suspension with a low viscosity is still a prerequisite for the successful application of gel-casting. An effective and necessary route to achieve this is to use a surfactant. TAC is an effective dispersant for alumina and we found that 0.3% TAC (by weight of alumina powder) would be the optimum level for a solution with AM monomer. Its dispersing effect may be mainly due to its electrostatic repulsion. Our results indicated that a polyelectrolyte dispersant, ammonium salt of poly (acrylic acid), had a better effect than TAC for the HEMA monomer solution. Therefore, the commercial polyelectrolyte was also used as the dispersant for the HEMA-PVP system.

The ζ of the particle is defined as the electrical potential between the shear plane surrounding the particle and the bulk solution in a suspension. It is the most important parameter of an electrical double layer and a good measure of the magnitude of the repulsion or attraction between particles. The ζ of different suspensions are shown in Fig. 3.45. The alumina particles show small ζ relative values from 8.37 to –14.9 mV in a wide pH range from 3.2 to 11.6 with an IEP at about pH=5.8. Addition of HEMA or PVP slightly decreases the absolute ζ of the particle, while they show no obvious effect on the IEP. These may indicate that the neutral HEMA or PVP molecules absorbed on the alumina particle have a charge screening effect. The dispersant obviously decreases the ζ under the same pH conditions and moves the IEP to pH=3.2, which can be due to the adsorption of ionized negative chains of the dispersant molecule on the particles. Therefore, it may be expected that the viscosity of the alumina suspension will be increased by adding HEMA and PVP and be decreased by adding JN281 as far as the electrostatic repulsive force is concerned.

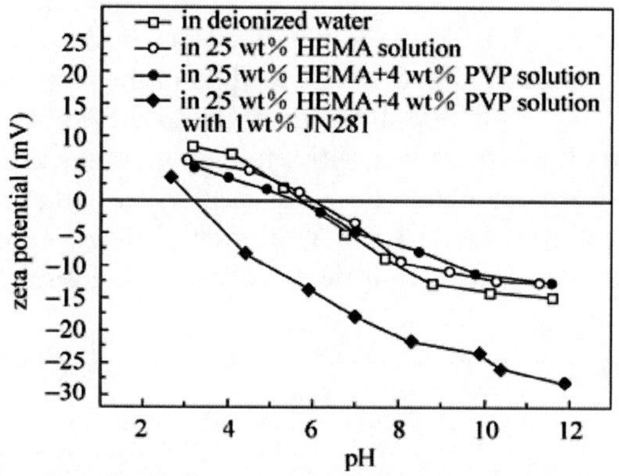

Figure 3.45 The comparison of pH dependence of zeta potentials for different suspensions

Consistent with the ζ, we can see from the shear dependence of viscosity in Fig. 3.46. The alumina suspensions with dispersants, especially JN281, show much lower viscosities compared with the suspension in pure water even with a much smaller solid loading of 40 vol%. Addition of 4 wt% PVP slightly increases the viscosities of the suspension containing 25 wt% HEMA. Here, suitable contents of HEMA and PVP were selected based on our preliminary tests and the dispersant contents are the optimum values with which the minimum viscosities could be obtained. The data were based on suspensions with pH=7 (without any pH controlling chemicals) because our previous results indicated that the pH value had a very small role in decreasing the viscosities of the concentrated ceramic slurries (Cai K., et al., 2005). The data agree with the zeta potential, indicating that PVP and the effect of the polyelectrolyte on viscosities can partly be attributed to their influence on the electrostatic repulsive forces. All three kinds of 50 vol% suspension (with dispersant JN281 or TAC) show shear-thinning behavior at low shear rates and shear-thickening behavior at high shear rates. Shear-thinning behavior generally indicates that the flow can induce a more favorable 2D layered structure arrangement of the particles because the flow-induced layered arrangement can decrease the resistance of particle movement between different layers; thus, a low viscosity can be obtained (Barnes H. A., 1989). For all concentrated suspensions of non-aggregating solid particles in a Newton fluid, shear-thickening will occur at a sufficiently high shear rate due to the instability and damage of the 2D arrangement at the high shear stress. The shear rate where viscosity starts increasing is called as the critical shear rate (Barnes H. A., et al., 1989; Barnes H. A., 1989). The critical shear rate is related to various conditions such as effective phase volume, particle size distribution, and continuous phase viscosity. Under the same condition, a lower critical shear rate is undesirable because it means decreased stability of the flow-induced layered structure. Therefore, both the lower viscosities and the higher critical shear rates of the suspension with JN281 compared with those of the solution with TAC indicate that JN281 is better than TAC in stabilizing the suspension of the HEMA system and in decreasing the so-called phase solid loading of the suspension. Besides, it is also apparent from Fig. 3.46 that although the addition of PVP in the HEMA suspension increases the viscosity slightly, it increases the critical shear rate to a higher level. In fact, we can obtain flowable suspensions with the same maximum solid loading of about 50 vol% in HEMA solution. This suggests that addition of PVP to the HEMA system has very little effect on viscosities.

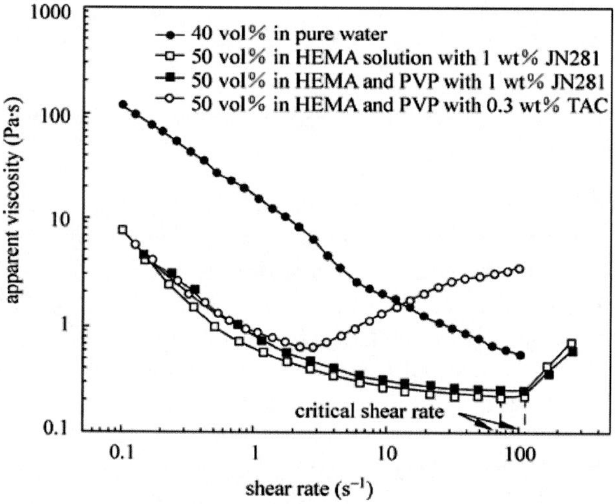

Figure 3.46 The shear dependence of viscosities for alumina suspensions with different solid loading

As for the polyelectrolyte's effect, we could not attribute its effect only to the increased electrostatic repulsive forces as indicated by the zeta potentials. It has been reported by Tari et al. (1998) that the increase of the ζ by adding a dispersant may not assure a good dispersion, because each particle may have an "interaction size" or "effective diameter" that is approximately its real diameter plus twice the range of total surface forces. The larger effective diameter of particles with higher ζ may cause a high viscosity in its suspension. But in our opinion, it is difficult to judge under what circumstance the increase of the "interaction size" can counteract the effect of electrical repulsion in preventing the particle aggregation, because the "effective diameter" may be only applicable for a suspension with high solid loading and small particle sizes. However, irrespective of whether the higher ζ of JN281 means a better electrostatic stabilization or not, it seems reasonable that its better dispersion should also be attributed to its steric stabilization. Generally, steric stabilization requires adsorption of a reasonably dense polymer layer on each particle, and in turn a sufficient supply of polymer in the solution (Horn R. G., 1990). This might account for the much higher optimum value of JN281 than that of TAC. An additional level of 1 wt% JN281 can yield a 50 vol% alumina suspension with a viscosity of 8.2 Pa·s at a shear rate of about 0.1 s^{-1}, which is generally indicative of a ceramic suspension with good flowability. A further increase of the dispersant content has an inverse effect while a little higher solid loading can greatly increase the viscosity as predicted by the Krieger-Dougherty equation (Barnes H. A., 1989; Guo D., et al., 2003).

On the other hand, because PVP is a macromolecule (the PVP used here has a molecular weight of 40000), it should also have a more or less steric effect that can prevent particles from aggregating. This may be the reason for the small influence of PVP on the suspension viscosity. In fact, PVP is also used as a steric dispersant for inorganic (e.g., CeO_2) and organic materials (e.g., PMMA. Cao K., et al., 2000; Yang X., et al., 1997). However, further increasing the content of PVP above 6 wt% will greatly increase the viscosity of the slurries. The influence of PVP concentration on the viscosities of a 50 vol% alumina suspension is illustrated in Fig. 3.47. Addition of a polymer to the suspension can have various and subtle effects on surface force of the particle depending on the chemical nature of the polymer, the solution, and the particle, as well as on the polymer molecular weight, particle concentration, etc. Bridging flocculation is unlikely to occur in this case for PVP because the polyelectrolyte dispersant is dominant on the particle surface. Therefore, another type of flocculation for soluble polymers, depletion flocculation, should be the reason for the increase of viscosity at a higher PVP concentration, since the osmotic pressure difference in the depletion flocculation mechanism will be enhanced at a higher polymer concentration (Horn R. G., 1990;

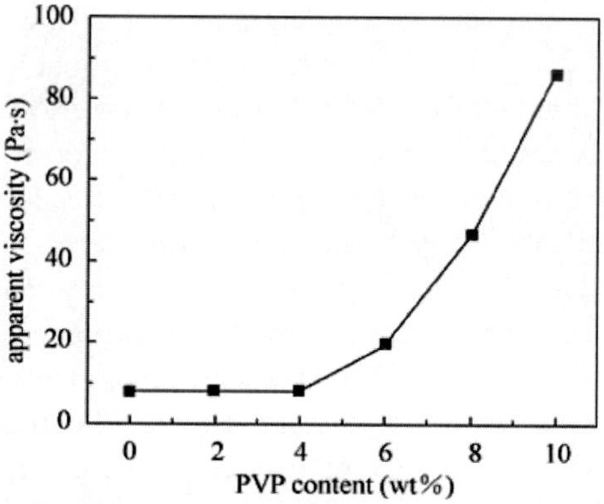

Figure 3.47 The influence of PVP concentration on the viscosities of 50 vol% alumina suspension at a shear rate of 0.1 s^{-1}

Kiratzis N. E., 1999). The effect of PVP on increasing the intrinsic viscosity of the liquid phase should also be accounted.

3.7.3 The Activation Energy and Solidification

Different monomers require different suitable cross-linker, initiator, and catalyst concentrations in order to obtain the suitable polymer network structure and strength. Based on our preliminary test, for a 25 wt% HEMA premix solution, the suitable cross-linker, initiator, and catalyst contents are about 8 wt%, 0.18 g/L and 0.3 vol%, respectively. The data are similar to those for the AM system. In order to compare the kinetic aspect of the solidification reaction of the HEMA system with that of the AM system, the activation energies E_a of the two systems with the same initiator and catalyst contents were compared. Since the HEMA slurry temperature showed no sharp increase with time after adding the initiator and catalyst, the solidification time t_s was tested as a measure of reaction rate deriving from the E_a of the gels with powders, namely, the time period from addition of the initiator to a solidified state where the slurry just lost its flowability in a flatly placed test tube (beginning of solidification). Due to the inverse proportional relationship of t_s with the reaction rate, its relationship with temperature T may be expressed by the Arrhenius-type equation:

$$t_s \propto \frac{1}{r} = A\exp\left(\frac{E_a}{RT}\right) \tag{3.5}$$

where r represents the reaction rate, R is the standard gas constant 8.3144 J/(K·mol), and E_a represents the activation energy of the total reaction. The logarithm plot of t_s with $1/T$ and the corresponding E_a data are shown in Fig. 3.48. We can see that with the same initiator and catalyst concentrations of 0.18 g/L and 0.3 vol%, the activation energies of the AM and HEMA system are about 40.1 and 56.4 kJ/mol, respectively. A small amount of PVP (4 wt%) showed no effect on the E_a of HEMA because there is no observable change in t_s. Although an error may exist for t_s due to the difficulty in precisely determination of the solidification point, the data can reveal that HEMA has obviously higher activation energy than AM. Therefore, the gel-forming reaction of HEMA is more sensitive to temperature than that of the AM monomer. The slurry in the AM premix solution can solidify (lost flow ability) in 4 – 5 min at 50℃, while it will take three times longer for the HEMA-PVP system to solidify at the same temperature. But at 50℃ the solidification time are 3 and 5 min for AM and HEMA, respectively. Therefore, we should heat the HEMA-PVP system to accelerate the solidification. Slight heating can substantially shorten the solidification time.

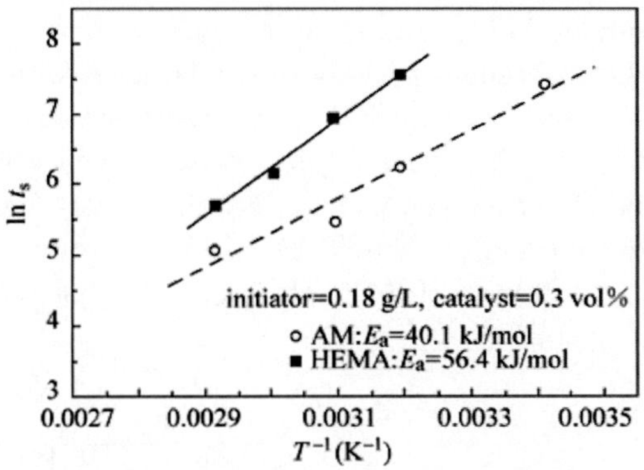

Figure 3.48 The comparison of activation energy (E_a) of acrylamide and 2-hydroxyethyl methacrylate gels with 50 vol% alumina powders

3.7.4 The Green Strengths and Microstructures

As mentioned above, a few percent of green bodies might show a much lower strength than the average strength value. As a result, these samples may fracture under small stress and hence result in large uncertainties in the survival probabilities. Although a wide strength distribution for green parts is not as detrimental as that for sintered ceramics because the sintering process can increase the density of the body and remove most pores, it is a serious problem since cracks may still remain in the final ceramics. The effect of PVP content on the alumina green strength derived from 50 vol% alumina slurries containing 25 wt% HEMA is shown in Fig. 3.49. It is evident that the green strength of the samples increases slightly with PVP content. It is also evident from the error bars in the figure that a small amount of PVP greatly decreases the standard deviation (where each value is the mean one for eight test data) of the green strength, indicating that the reagent can increase the strength (microstructural) homogeneity. Although a further increase of PVP content above 4 wt% can slightly increase the strength, it seems to have no obvious effect on the standard deviation of the data, and the slurry viscosity was found to increase obviously at a higher PVP level as shown in Fig. 3.49. Thus, the optimum PVP content was determined to be 4 wt%. A strength of 19 MPa can be achieved for the sample derived from solution containing 25 wt% HEMA and 4 wt% PVP. Although it is lower than that derived from AM with the same solid loading (28 MPa), it is much larger than that derived from gelatine (8 MPa) and chitosan (<1 MPa), etc. (Huzzard R. J., et al., 1997; Olhero S. M., et al., 2000; Chen Y. L., et al., 1999; Johnson S. B., et al., 1991). Such a strength is high enough for further handling and machining.

The SEM photographs of fractured green alumina bodies cast from the HEMA monomer with and without PVP are illustrated in Figs. 3.50(a) and (b), respectively.

The green body without PVP shows many grooves in the fractured surface that are indicative of a heterogeneous packing of the particles, while addition of PVP obviously improves the homogeneity of the green microstructure. The SEM pictures agree well with the strength data. PVP is a polymer, and a low PVP addition level of 4 wt% should have no significant effect on the polymerization of HEMA; therefore, its modification effect should be attributed to reasons other than its effect on the polymerization of HEMA. The reason for this will be discussed in detail in the following section.

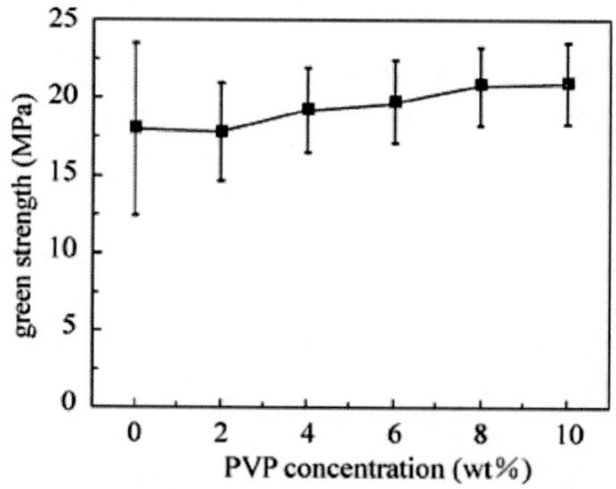

Figure 3.49 The effect of PVP content on the flexural strength of alumina green bodies (The error bars in the figure represent the standard deviation values of each group of data)

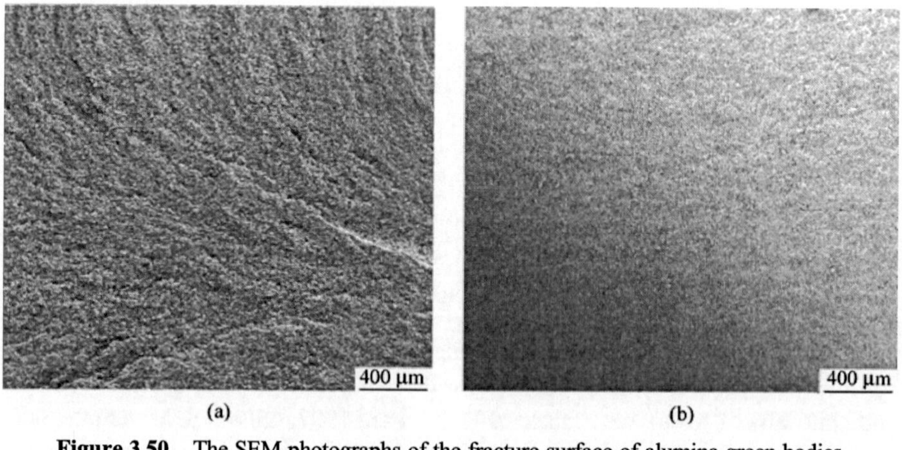

Figure 3.50 The SEM photographs of the fracture surface of alumina green bodies formed from the 2-hydroxyethyl methacrylate premix solution

(a) without polyvinyl pyrrolidone (PVP); (b) with PVP

3.7.5 The Exfoliation Elimination Effect and Analysis of the Interaction between PVP and HEMA Molecules

In the gelation process of gel-casting, the monomer undergoes a radical-type polymerization after the initiator is added. This chain reaction is strongly inhibited by oxygen, which scavenges very efficiently both the initiating and the polymer radicals to form relatively stable free radicals of peroxide. Consequently, when the polymerization occurred in ambient air, inhibition of gelation due to the presence of oxygen was encountered, and surface exfoliation occurred. A thin layer of about 1 mm will peel off and the geometry of the ceramics will be affected (see Fig. 3.51(a)), especially for small parts. The surface exfoliation also occurs for the HEMA system but not as severely as in the AM system because the peeled layer is thinner. Interestingly, another advantage of the PVP-modified HEMA system is that there is no exfoliation for the bodies that are gelcast from the modified system (see Fig. 3.51(b)). As noted above, the diffusion of oxygen from the surface to the inner part should be considered to estimate the exfoliation phenomenon during the polymerization. According to Stoke's Law, the relationship for the resistance force of spherical particle movement is given by:

$$F = 6\pi\eta r v \tag{3.6}$$

where η, r, and v are the medium viscosity, particle radius, and movement velocity, respectively. As a result, the diffusion coefficient of oxygen is inversely related to the medium viscosity. Although PVP has little effect in increasing the viscosity of the alumina suspension owing to its steric stabilization, it can increase the intrinsic viscosity of the water phase of the hydrogel. Therefore, the exfoliation suppression effect of the HEMA-PVP system may be partly ascribed to the relatively higher viscosity of the water phase during gelation, which in turn decreases the diffusion coefficient of oxygen and prevents it from diffusing into the system.

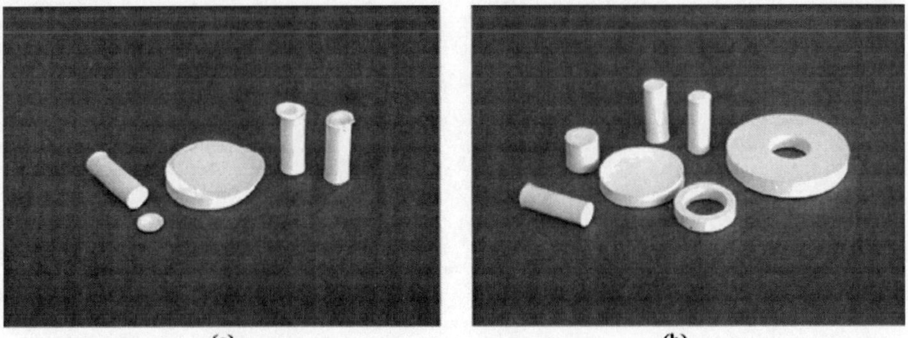

(a) (b)

Figure 3.51 The illustration of the alumina parts gelcast by different systems (a) formed by acrylamide; (b) formed by PVP improved 2-hydroxyethyl methacrylate

Dufaud's research on ceramic suspensions for stereolithography and Studer's study on curing of organic coating drew similar conclusions on oxygen diffusion (Studer K., et al., 2003; Dufaud O., et al., 2003).

Meanwhile, due to the polar hydroxyl group of poly-2-hydroxyethyl methacrylate (PHEMA), the negatively charged oxygen atom of the carbonyl group in PVP can form a hydrogen bond with the active hydrogen atom in PHEMA as shown in Fig. 3.52. The hydrogen-forming ability of the two chemicals is well known. Goh and Lee (1990) have proved that incorporation of HEMA into EMA (ethyl methacrylate) could increase the compatibility of the copolymer with PVP due to the hydrogen bonding between HEMA and PVP (Goh S. H., et al., 1990). The ability of PVP to form single-phase blends with various polymers containing hydroxyl groups and the existence of hydrogen bonding between hydroxyl and carbonyl groups has been confirmed by FTIR (Moskala E. J., et al., 1985; Thyagarajan G., et al., 1989; Costa R. O. R., et al., 2002). Polymers with hydrogen bonds are very good barriers against permeating gases because the bridging effect of the hydrogen bonds could slow down the diffusion of penetrant molecules (Karlsson G., E., et al., 2002). Besides, molecular dynamics simulations of the permeation processes of oxygen in PVA hydrogels showed that the processes are highly correlated with the orientational relaxation of side chains of polymers and the reorganization of the hydrogen bond networks (Tamai Y., et al., 1998). Because the oxygen molecule tends to move along the polymer chain, this movement is sensitive to the number of polymer-polymer hydrogen bonds. Therefore, the diffusivity depended uniquely on the number of the hydrogen bonds (Karlsson G. E., et al., 2004). It was also found that the life time of the hydrogen bonds between polymer chain segments was longer than those between water and polymer chain segments; therefore, water diffusion in the polymer was also constrained by the hydrogen bonds since the jump of the water molecule from one site to another requires breaking of hydrogen bonds (Karlsson G. E., et al., 2002; Tamai Y., et al., 1996). Based on these findings, we consider that another important reason or main reason for the exfoliation suppression effect is the hydrogen bonds formed between PVP and PHEMA.

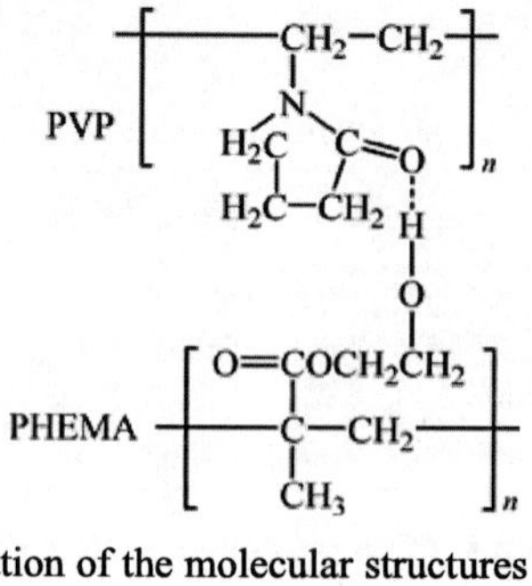

Figure 3.52 The illustration of the molecular structures of PVP and PHEMA and the hydrogen bonding

When PHEMA is obtained by polymerization in water in amounts exceeding the equilibrium water content (EWC 5 (swollen gel weight-dry gel weight)/swollen gel weight), phase separation will result. The EWC of the PHEMA hydrogel is very sensitive to the cross-linker and initiator concentration, etc., and phase separation can result in heterogeneity, cracks, and low strength in the hydrogel (Perera D. I., et al., 1996; Dalton P. D., et al., 2002; Liu Q., et al., 2000). Sometimes, we found that a very thin layer of moisture could be formed on the fractured surface of bodies that were gelcast through the HEMA solution. Therefore, the occasionally formed low-strength samples may have presumably ascribed to the phase separation caused by the decrease of EWC, which may be resulted from the difficulty in precisely controlling the initiator concentration due to its small quantity. It has been reported that incorporation of PVP can greatly enhance the EWC of the HEMA hydrogel and hydrogen bonding can bind the water more tightly and substantially prevent phase separation (Malak K. M., et al., 2003). This is another proof of the effect of PVP in improving the homogeneity through the formation of hydrogen bonding.

Combining the strength, SEM photos, and the exfoliation suppression effect, it seems reasonable to conclude that the hydrogen bonding between PVP and PHEMA is the basic reason for the increased gel homogeneity. Thus, we may call the PVP-modified system a synergistic gel-forming system. A combination of HEMA and PVP leads to an obvious improvement that cannot be obtained by either of the reagents. Furthermore, it may be reasonable to propose that the suppression of surface exfoliation and improvement of the gel homogeneity may also be achieved by other gel-forming reagents that can form strong hydrogen bonds with the PAM or PHEMA, provided that we can also achieve high solid loading required for gel-casting with the reagents.

References

Alford N. M., Brichalli J. D., Kedall K., 1987. High-strength ceramics through colloidal control to remove defects. Nature., 330: 51.

Amott S., Fulmer A., Scott W., 1974. The agarose double helix and its function in agarose gel structure. J. Mol. Biol., 90: 269 – 284.

Barnes H. A., Hutton J. F., Walers K., 1989. An Introduction to Rheology. Oxford: Elsevier Press, 1989.

Barnes H. A., 1989. Solid particles dispersed in Newtonian liquids. J. Rheol., 33(2): 329 – 366.

Boersma W. H., Baets P. J. M., Laven J., et al., 1991. Time-dependent behavior and wall slip in concentrated shear thickening dispersions. J. Rheol., 35: 1093 – 1119.

Braccini I., 1999. Conformational and configurational features of acidic polysaccharides and their interactions with calcium ions. Carbohydr. Res., 317: 119 – 130.

Cai K., Huang Y., Yang J. L., 2005. Alumina gel-casting by using HEMA system. J. Eu. Ceram. Soc., 25(7): 1089 – 1093.

Cao K., Yu J., Li B. G., et al., 2000. Micron-size uniform poly (methyl-methacrylate) particles by dispersion polymerization in polar media 1. particle size and particle size distribution. Chem. Eng. J., 78(2 – 3): 211 – 215.

Cesarano III J., Aksay I. A., 1988. Processing of highly concentrated aqueous alpha-alumina suspensions stabilized with poly-electrolytes. J. Am. Ceram. Soc., 71(12): 1062 – 1067.

Chen Y. L., Xie Z. P., Yang J. L., et al., 1999. Alumina casting based on gelation of gelatine. J. Eu. Ceram. Soc., 19(2): 271 – 275.

Costa R. O. R., Vasconcelos W. L., 2002. Structural modification of poly(2-hydroxyethyl methacrylate)-silica hybrids utilizing 3-methacry-loxypropyltrim-ethoxysilane. J. Non-Crystal. Sol., 304: 84 – 91.

Dalton P. D., Flynn L., Schoichet M. S., 2002. Manufacture of poly (2-hydroxyethyl methacrylate-co-methyl methacrylate) hydrogel tubes for use as nerve guidance channels. Biomaterials, 23(12): 3843 – 3451.

Dean J. A., 1985. Lange's Handbook of Chemistry (13th ed.). New York: McGraw-Hill.

Dufaud O., Corbel S., 2003. Oxygen diffusion in ceramic suspensions for stereolithography. Chem. Eng. J., 92: 55 – 62.

Fanelli A. J., Silvers R. D., Frei W. S., et al., 1989. New aqueous injection molding process for ceramic powders. J. Am. Ceram. Soc., 72: 1833 – 1836.

Goh S. H., Lee S. Y., 1990. Compatibility of poly (*N*-vinyl-2-pyrrolidone) with ethyl-methacrylate 2-hydroxyethyl methacrylate copolymers. Makromol. Chem., 191(7): 3081 – 3085.

Guo D., Cai K., Huang Y., et al., 2003. Water based gel-casting of lead zirconate titanate. Mater. Res. Bul., 38(5): 807 – 816.

Hoffman R. L., 1972. Discontinuous and dilatant viscosity behavior in concentrated suspensions I. observation of a flow instability. Trans. Soc. Rheol., 16: 155.

Horn R. G., 1990. Surface forces and their action in ceramic materials. J. Am. Ceram. Soc., 73(5): 1117 – 1135.

Huzzard R. J., Blackburn S., 1997. A water-based system for ceramic injection molding. J. Eu. Ceram. Soc., 17(2 – 3): 211 – 216.

Jin J., Lin M., 1993. Utilization and Fabrication of Algae. Beijing: Science Press.

Johnson S. B., Dunstan D. E., Franks G. V., 1991. Rheology of cross-linked chitosan-alumina suspensions used for a new gel-casting process. J. Am. Ceram. Soc., 85(7): 1699 – 1705.

Karlsson G. E., Johansson T. S., Gedde U. W., et al., 2002. Physical properties of dense amorphous poly(vinyl alcohol) as revealed by molecular dynamics simulation. J. Macromol. Sci.-Phys., B41(2): 185 – 206.

Karlsson G. E., Gedde U. W., Hedenqvist M. S., 2004. Molecular dynamics simulation of oxygen diffusion in dry and water-containing poly(vinyl alcohol). Polymer, 45: 3893 – 3900.

Kiratzis N. E., Luckham P. F., 1999. The rheology of aqueous alumina suspensions in the presence of hydroxyethylcellulose as binder. J. Eu. Ceram. Soc., 19(15): 2605 – 2612.

Kuo C. K., 2001. Ionically crosslinked alginate hydrogels as scaffolds for tissue engineering: part 1. structure, gelation rate and mechanical properties. Biomaterials, 22(6): 511 – 521.

Lange F. F., 1989. Powder processing science and technology for increasing reliability. J. Am. Ceram. Soc., 72(1): 3 – 21.

Liu Q., Hedberg E. L., Liu Z. W., et al., 2000. Preparation of macroporous poly(2-hydroxyethyl methacrylate) hydrogels by enhanced phase separation. Biomaterials, 21(21): 2163 – 2169.

Malak K. M., Hill D. J., Whittaker A. K., 2003. Water sorption into poly[(2-hydroxyethyl methacrylate)-co-(1-vinyl-2-pyrrolidone). Polym. Int., 52(11): 1740 – 1748.

Millan A. J., Nieto M. I., Baudin C., et al., 2002a. Thermogelling polysaccharides for aqueous gel-casting—part I: a comparative study of gelling additives. J. Eu. Ceram. Soc., 22(13): 2217 – 2222.

Millan A. J., Nieto M. I., Baudin C., et al., 2002b. Thermogelling polysaccharides for aqueous gel-casting—part Ⅱ: influence of gelling additives on rheological properties and gel-casting of alumina. J. Eu. Ceram. Soc., 22(13): 2223 – 2230.

Moskala E. J., Varnell D. F., Coleman M. M., 1985. Concerning the miscibility of poly (vinyl phenol) blends[J]. Polymer, 26: 228 – 234.

Olhero S. M., Tari G., Coimbra M. A., et al., 2000. Synergy of polysaccharide mixtures in gel-casting of alumina. J. Eu. Ceram. Soc., 20(4): 423 – 442.

Omatete O. O., Janny M. A., Strelow R. A., 1991. Gel-casting a new ceramic forming process. Ceramic Bulletin, 70(10): 1641 – 1649.

Perera D. I., Shanks R. A., 1996. Swelling and mechanical properties of cross-linked hydrogels containing *N*-vinyl pyrrolidone. Polym. Intern., 39(2): 121 – 127.

Pierre A. C., Ma K., 1999. DLVO theory and clay aggregate architectures formed with $AlCl_3$. J. Eu. Ceram. Soc., 19(8): 1615 – 1622.

Rao R. R., Roopa H. N., Kannan T. S., 1999a. Dispersion, slip casting and reaction nitridation of silicon-silicon carbide mixtures. J. Eu. Ceram. Soc., 19(12): 2145 – 2153.

Rao R. R., Roopa H. N., Kannan T. S., 1999b. Effect of pH on the dispersability of silicon carbide powders in aqueous media. Ceramics International, 25: 223 – 230.

Reed J. S., 1994. Principles of Ceramics Processing. New York: John Wiley & Sons Inc., 447 – 502.

Santacruz I., Nieto M. I., Moreno R., 2002. Rheological characterization of synergistic mixture of carrageenans and locust bean gum for aqueous gel-casting of alumina. J. Am. Ceram. Soc., 85(10): 2432 – 2436.

Studer K., Decker C., Beck E., et al., 2003. Overcoming oxygen inhibition in UV-curing of acrylate coatings by carbon dioxide inerting: part Ⅱ. Prog. Organ. Coat., 48: 101 – 111.

Tamai Y., Tanaka H., Nakanishi K., 1996. Molecular dynamics study of polymer-water interaction in hydrogels. 1. hydrogen-bond structure. Macromolecules, 29: 6750 – 6760.

Tamai Y., Tanaka H., 1998. Permeation of small penetrants in hydrogels. Fluid Phase Equilibrium, 144: 441 – 448.

Tari G., Ferreira J. M., Lyckfeldt O., 1998. Influence of the stabilising mechanism and solid loading on slip casting of alumina. J. Eu. Ceram. Soc., 18(5): 479 – 486.

Teng W. D., Edirisinghe M. J., Evans J. R. G., 1997. Optimization of dispersion and viscosity of a ceramic jet printing ink. J. Am. Ceram. Soc., 80(2): 486 – 494.

Thyagarajan G., Janarthanan V., 1989. FTIR and thermal analysis studies of poly(vinyl alcohol)-poly(vinyl pyrrolidone) blends. Polymer, 30: 1797 – 1799.

Ward A. G., Courts A., 1977, The Science and Technology of Gelatine (translated by Li Wenyuan et al.). London: Academic Press.

Wber K., Tomandl G., 1997. Preparation of structured ceramics for membranes. Key Engineering Materials, 132 – 136: 1754 – 1757.

Whistler R. L. 1973. Industrial Gums. New York: Academic Press, 45 – 68.

Xie Z. P., Huang Y., Wu J. G., et al., 1995. Microwave debinding of a ceramic injection moulded-body. J. Mater. Sci. Lett., 14: 794.

Xie Z. P., Yang J. L., Chen Y. L., et al., 1998. Gelation forming process of ceramic compacts by agarose. 9th CIMTEC-World Ceramics Congress and Forum on New Materials, Florence, 1998.

Xie Z. P., Yang J. L., Huang D., et al., 1999. Gelation forming of ceramic compacts using agarose. British Ceramic Transactions, 98(2): 58 – 61.

Xie Z. P., Huang Y., 2001. A new gel-casting of ceramics by reaction of sodium alginate and calcium iodate at increased temperature. J. Mater. Sci. Lett., 20: 1255 – 1257.

Yang J. L., Xie Z. P., Tang Q., et al., 1998. Study on rheological behavior and geleasting of α-Al_2O_3 suspension. J. Chin. Ceram. Soc., 26: 41.

Yang X., Rahaman M. N., 1997. Thin films by consolidation and sintering of nanocrystalline powders. J. Eu. Ceram. Soc., 17(4): 525 – 535.

Yi Z. Z., Xie Z. P., Huang Y., et al., 2001. Study on gel-casting and properties of recrystallized silicon carbide. Ceramics International, 28(4): 369 – 376.

Zhang J. G., Edirisinghe M. J., Evans J. R. G., 1989. A catalogue of ceramic injection molding defects and their causes. Ind. Ceram., 9: 72.

Zhang T., Blackburn S., Bridgwater J., 1994. Properties of ceramic suspensions for injection molding based on agar binders. British Ceram. Trans., 93(61): 229 – 233.

Zhou L. J., Huang Y., Xie Z. P., et al., 1998. Study on gel-casting of silicon carbide. In: Proceedings of the First China International Conference on High-Performance Ceramics, Beijing, 1998.

Zhou L. J., Huang Y., Xie Z. P., 2000. Gel-casting of concentrated aqueous silicon carbide suspension. J. Eu. Ceram. Soc., 20(1): 253 – 257.

Zhou Z. K., Gu T. R., Ma J. M., 1996. Fundation of Colloidal Chemistry. Beijing: Beijing University Press, 318 – 327.

Chapter 4 Generation, Development, Inheritance, and Control of the Defects during the Transformation from Suspension to Green Body

Abstract This chapter focuses on the generation, development, inheritance and control of the defects in colloidal forming process, by which the complex near-net shape ceramic products with a homogeneous microstructure and high reliability were consequentially achieved.

The rheological performance of the ceramics suspension was very important to the colloid forming processes. In our studies, the rheological behaviors of aqueous alumina suspensions were investigated, and the various ways of modifying the rheological behaviors of the suspensions were discussed. A critical solid loading between the shear thinning and shear thickening behavior was found in an aqueous alumina suspension. Moreover, the gelation of the suspension with high solid loading of alumina got delayed and the strength of green body decreased, which implies that the fast polymerization of monomers in high solid loading alumina suspension was inhibited, and the flexibility of the gel-casting was improved. We found that there was an optimal solid loading for the ZTA suspension to obtain the highest bending strength and the highest Weibull modulus for the sintered bodies.

Micro mechanisms of agglomerations' generation in ceramic suspensions used in colloidal forming processing were investigated. On the basis of potential energy between particles in liquid, a microscopic model of agglomerations in ceramic suspensions was presented. Two types of agglomerations, tight agglomeration and loose agglomeration, were observed by using an environmental scanning microscope as a quasi-direct method. It is observed that their generation is caused by the deviation of solid phase content in slurry from the stable solid loading.

The influences of the idle time on the microstructures and mechanical properties of green bodies were also described using direct coagulation casting. The vertically laminates cracks were observed in coagulated green bodies after casting using increasing ionic strength. The cracks were attributed to the shrinkage of the particle network (syneresis) during coagulation.

The study indicated that the solid loading in the suspension depended on high valence counter-ions and ion conductivity constants. Further research revealed that the inner stress in the green bodies of ceramics originated from the non-uniform shrinkage during the solidification of the precursor suspension

Y.Huang et al., *Novel Colloidal Forming of Ceramics*
© Tsinghua University Press, Beijing and Springer-Verlag Berlin Heidelberg 2010

and the drying of the green bodies. The gradients of temperature, initiator concentration and moisture are the important original factors causing the inner stress.

It is found that a proper amount of a water-soluble polymer, such as hydroxyethyl acrylate (HEA), polyethylene glycol (PEG) and poly (vinylpyrrolidone) (PVP) added into the concentrated suspension can adjust polymer network structure and thus reduce the inner stresses and cracking in the ceramic green body. The debindering time of large ceramic parts can be significantly shortened by reducing the harmful inner stresses in the green body.

Keywords defect, rheological behavior, counter-ion, inner stress, water-soluble polymer, agglomeration, syneresis, direct coagulation casting (DCC)

The repeatability and the product cost are two major considerations in the applications of advanced ceramics and ceramic composites. Both are directly or indirectly related to the micro-structural defects occurring during the manufacturing process. The random nature of these defects is often responsible for the poor reliability and performance of ceramic products. It is well known that defects like spallation, delamination, micro-cracks or large pores are often present in the green bodies formed by traditional forming processes such as dry-pressing, isostatic pressing and injection molding. These defects can hardly be removed in the subsequent processes and become fracture origins, thereby reducing the properties of the product or leading to higher rejection rate. Since 1990s, more attention has been paid to colloidal forming processes such as gel-casting, direct coagulation casting and colloidal vibration forming due to their potential to improve the homogeneity of ceramic body and reduce harmful micro-structural defects.

Among the colloidal forming processes, gel-casting, by which complex ceramic parts can be prepared (Omatete O. O., et al., 1991b), has been widely studied and applied to various ceramic materials (Nunn S. D., et al., 1996; Waesche B., et al., 1997; Ma L. G., et al., 2002). Homogeneous green bodies with good strength required for machining can be made through this process. The mechanical properties of the materials after sintering are greatly improved (Bossel C., et al., 1995; Liu X. L., et al., 2002; Huang Y., et al., 2000).

In this process, the mold temperature is a parameter that can be controlled to induce polymerization in the suspension. However, a temperature gradient exists in the suspension due to the heat transfer from the mold surface to the concentrated suspension. Consequently, the solidification of the suspension is not uniform, resulting in inner stresses in the green bodies. The inner stresses are often responsible for the cracking of the green bodies in subsequent drying or debindering process. The larger the parts, the more harmful the inner stresses. Therefore, it is of great importance to control the inner stresses in the green bodies in order to prepare large parts by gel-casting.

The inner stresses in the green body originate from differential contraction so that the magnitude of the inner stresses increases with the shrinkage rate and the elastic modulus. There are two sources of shrinkage in gel-casting process: one being the forming shrinkage due to slight contraction during the polymerization of monomers, and the other being the drying shrinkage due to the removal of the moisture. A high solid loading in the suspension is preferred in gel-casting in order to reduce the shrinkage of the green body. However, there is a limit for the solid loading in the suspension so as to achieve the desired viscosity and uniformity. In order to control the shrinkage rate, the wet green body is usually dried slowly under controlled temperature and humidity and then debindered at a very low heating rate. This is not favourable in practice because it leads to a prolonged operation time and high production cost.

The advantages of the fabrication process include dimensional accuracy and complex shaping capabilities, as well as reducing the cost of manufacturing. Fundamental research has been carried out by Young et al. (Young A. C., et al., 1991) and Omatete et al. (Omatete O. O., et al., 1991), showing the general feasibility of the process and its advantages in comparison with other liquid forming processes. Omatete et al. (Omatete O. O., et al., 1997) discussed in detail the key aspects of gel-casting: a premix solution; rheology of gel-casting slurries; drying process of gelcast parts; binder burnout; green strength of dried gel-casting parts, etc. Waesche et al. (Waesche B., et al., 1997) investigated the influence of slip viscosity on the mechanical properties of high purity alumina by gel-casting. They suggested that the interaction of rheological behavior of the suspension and the generation of flaws in the green body are of crucial importance in view of an optimized microstructure for the gel-casting process.

4.1 The Rheological Behaviors of Aqueous Ceramic Suspensions

The ceramic gel-casting is an important colloidal forming processes. It has been rapidly developed in the past decade due to its capability of near-net-shape forming of fine ceramic pieces.

So far, most of the researches on gel-casting has been focused on the preparation of suspensions and the process control on gel-casting, such as the effect of dispersants and pH on the viscosity and rheological behaviors of suspensions (Leong Y. K., et al., 1991), the concentration of monomer, usage of catalysts and initiators, and degassing time (Young Y. K., et al., 1991; Omatete O. O., et al., 1991b, c, a; 1997) etc. The main object of these researches on rheological behavior was to prepare suspensions with high solid loading, as it is believed that the advantages of gel-casting over traditional wet-forming processes can only be achieved with high solid loading suspensions (Omatete O. O., et al., 1997). An exhaustive relationship between the rheological behavior of the suspension and the mechanical properties of the sintered bodies is little known. In the following contents, the rheological behaviors of aqueous ceramic suspensions and their effects on the mechanical properties of the ceramics are expatiated.

4.1.1 The Rheological Behaviors of Aqueous Alumina Suspensions

In this section, the rheological behaviors of aqueous alumina suspensions were explained, and several effective ways to modify the rheological properties of the suspension were discussed. There was a critical solid loading for aqueous alumina suspension. If the solid loading is below the critical value, the suspension shows shear-thinning behaviors all along. On the other hand, when the solid loading reaches or exceeds the critical value, the rheological behaviors change from shear-thinning to shear-thickening. This is explained in detail below.

A high purity alumina powder (diameter 3.39 μm) from Xinyuan Alumina Ltd., China was used in pertinent study. A series of aqueous alumina suspensions at different solid loading, ranging from 35 to 55 vol% were prepared by ball milling in polypropylene jars filled with ZrO_2 milling balls for 24 h at a speed of 150 r/min. Solid loading ϕ of alumina in the suspensions was calculated according to the following formula:

$$\phi = \frac{m / \rho_{\text{th}}}{m / \rho_{\text{th}} + V} \tag{4.1}$$

where m is the mass of alumina powder (g), ρ_{th} is the theoretical density of alumina (g/cm^3), and V is the volume (mL) of the premix solution in the suspension. About 0.25 wt% of tri-ammonium citrate was added as dispersant. The rheological measurements of the suspensions were carried out after the preparation 1 h. The operation was performed at 25℃, using rotational rheometer Physica MCR 300. Each data was obtained logarithmically every 10 s ranging from 0.01 to 250 s^{-1} shear rate (Bergstrom L., 1998; Yang Yunpeng, 2003).

Figure 4.1 shows the rheological curves of aqueous alumina suspensions ranging from 40 to 50 vol% solid loading. Although the viscosity at low shear rate is high, it decreases as the shear rate increases. Moreover, the viscosity decreases faster at low shear rate. When the shear rate increases to a certain value, the viscosity curve becomes smooth. The whole curves show shear thinning behaviors.

Figure 4.2 illustrates the rheological curves of aqueous alumina suspensions ranging from 51 to 55 vol% solid loading. The suspension with 53 vol% solid loading shows shear thinning behavior in the range of 0.08 – 100 s^{-1} shear rate. Thereafter, in the range of shear rate higher than 100 s^{-1}, the viscosity increases distinctly. Furthermore, it increases to the extent of 0.15 Pa·s, which indicates that the whole system presents shear thickening behavior.

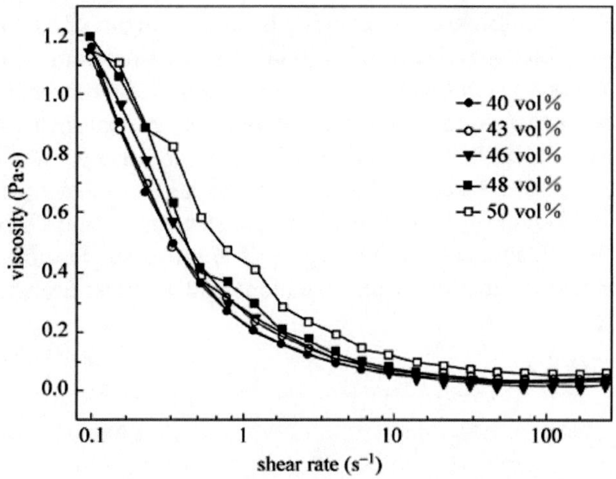

Figure 4.1 The effects of solid loading on rheological behavior of aqueous alumina suspensions

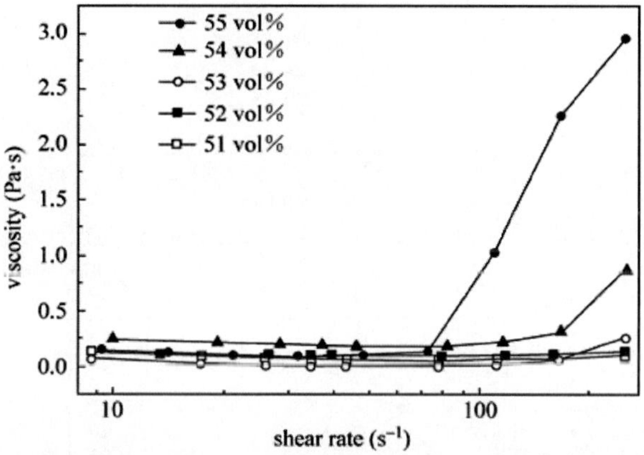

Figure 4.2 The effects of solid loading of alumina on rheological behaviors of aqueous alumina suspensions from 51 to 55 vol%

The aqueous alumina suspension with 54 vol% solid loading indicates that at low shear rate the whole system shows shear thinning behaviors. When the shear rate is in excess of 51.8 s^{-1}, the viscosity increases slightly at the beginning. When in excess of 200 s^{-1}, the curve becomes steeper, and the viscosity increases to the extent of 0.5 Pa·s which suggests more shear thickening behaviors of the whole system.

For the rheological curve of the aqueous alumina suspension with 55 vol% solid loading, the whole system exhibits shear thinning behavior at low shear rate of 0.08 – 48.1 s^{-1}. When in excess of 72.7 s^{-1}, the viscosity proliferates and is more than 3 Pa·s which indicates a prominent shear thickening behavior.

The suspensions exhibit shear thickening behavior when their solid loading are more than 53 vol%, and 55 vol% show typical shear-thickening characteristics. The viscosity increases significantly with the increase of the diminutive shear rate.

Figure 4.3 shows that viscosity corresponding to the critical shear rates increases with the increase of solid loading, and the increment speeds up when solid loading over 50 vol%. Figure 4.4 shows that the critical shear rate decreases with the increase of solid loading of the aqueous alumina suspension and decreases swiftly over 50 vol%, which suggests that the aqueous alumina suspension at high solid loading undergoes a transition from pseudoplastic to plastic characteristics at lower shear rate.

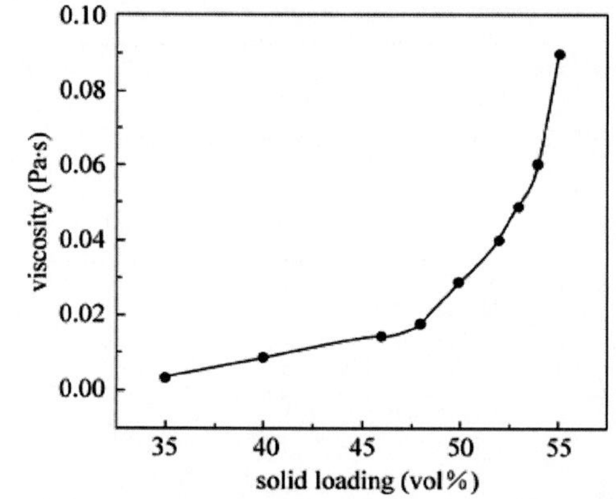

Figure 4.3 The variation of the viscosity at critical shear rate

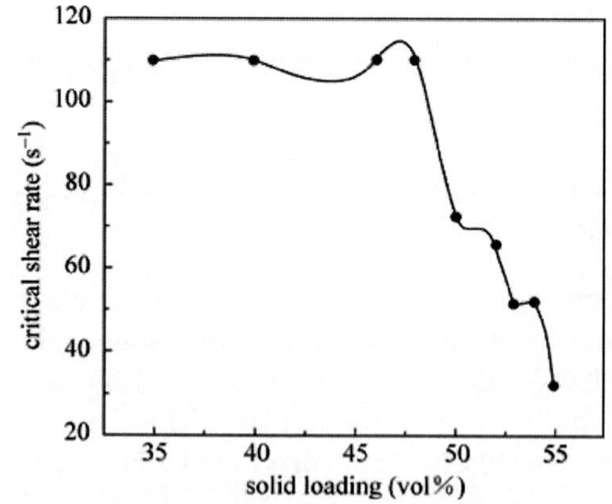

Figure 4.4 The variation of the critical shear rate at different solid loading

Generally, the higher solid loading the suspension has, the better macroscopic performance will be achieved. According to the above explanation, there exists a critical point in rheological performance of aqueous alumina suspension, below which the whole system presents shear-thinning behavior all along, and above which the whole system exhibits shear-thinning at the beginning, followed by shear-thickening afterwards. The grain packing mode inside the aqueous suspension determines the critical solid loading value. One feasible method to adjust the very critical point is to accommodate the interaction between grains and modify the grain packing mode.

4.1.2 The Effect of Rheological Properties of Suspension on Mechanical Strength of Ceramics

There is no doubt that the rheological properties of the suspension affect the mechanical strength of the final sintered ceramic body. Recent research reveals that the highest flexural strength and the largest Weibull modulus were not achieved at the highest solid loading of the suspensions. The mechanical performance deteriorated when the solid loading reached a certain value. The reason for this phenomenon is that the microstructure of the sintered bodies is to a large extent determined by the structure and the rheological behavior of the suspension because the colloidal forming is an *in-situ* occurring process.

The commercially available α-Al_2O_3 (0.44 μm, Ceralox APA-0.5 with MgO, USA) and t-ZrO_2 ($Ce_{0.03}Y_{0.016}Zr_{0.954}O_2$, 0.4 μm, Beijing Founder Ceramics Technology Co. Ltd., China) were used in the investigation. The polyelectrolytes (ammonium polyacrylate, PAA, molecular weight is 3000 – 5000, Tianjin Chemical Industry Research Institute, China) was used as a dispersant. The standardized analytical grade $(CH_3)_4NOH$ (TMAH) was used for adjusting pH of the suspension.

The premix solution was prepared with AM : MBAM = 23.3 : 1 (AM: 14.0 wt%, MBAM: 0.6 wt%). After 1 wt% dispersant was added, the solution was mixed with alumina powder and zirconia powder, and resultant suspensions with 40, 45, 50, 53, and 55 vol% solid loadings, respectively, were adjusted to pH = 10 by adding a TMAH aqueous solution. The suspensions were then ball-milled for 48 h to promote dispersion and admixing process. The suspensions were degassed in a rotary evaporator under vacuum until no further release of air bubbles was observed. Both the initiator, a 5 wt% aqueous solution of ammonium persulfate, and catalyst, TEMED, were then added and stirred slowly to avoid sucking in air. The suspensions were degassed for another 5 min. During the degassing process, the temperature was kept at 0 – 1℃ to prevent excess evaporation of water and polymerization of the monomers. Each suspension was then cast into an $80 \times 60 \times 8$ mm^3 stainless steel mold, whose temperature was kept at about 70℃. After the monomers polymerized, the green body was demoulded and dried at room temperature under controlled humidity to avoid cracking and non-uniform shrinkage due to rapid drying. The dried green bodies were sintered at 1580℃ for 2 h. Each sintered body was cut and ground into specimens of about $3 \times 4 \times 36$ mm^3

for the flexural strength measurement and specimens of about 4×6×30 mm^3 for the single edge notch beam (SENB) toughness measurement. Before their measurement, each of the specimens for the bending strength was ground by using SiC abrasive of 28, 14, 7 and 3.5 μm, respectively, for 30 min. A notch was cut in each of the specimens for the toughness at depth of about 0.4 – 0.5 times of the specimen height.

The zeta potentials of Al_2O_3 and ZrO_2 were measured individually, as shown in Fig. 4.5. The IEPs were 5.8 and 4.9 for Al_2O_3 and ZrO_2, respectively. At low pH, far from the IEP of either powder, only the alumina particles have a high positive zeta potential and are colloidally stable. In the vicinity of the pH_{IEP}, the particles have a low zeta potential that may be either positive ($pH<pH_{IEP}$) or negative ($pH>pH_{IEP}$); suspensions prepared within this region are colloidally unstable and consist of large agglomerates that are likely to lead to porous gelcast ceramics. Lastly, at high pH, far from the pH_{IEP} of both powders, the particles have a high negative zeta potential and are colloidally stable. The profiles in Fig. 4.5 show that the saturation zeta potential and the breadth of the pH range over which the potential is nearly maximized, provide better colloidal stability at high pH for both powders.

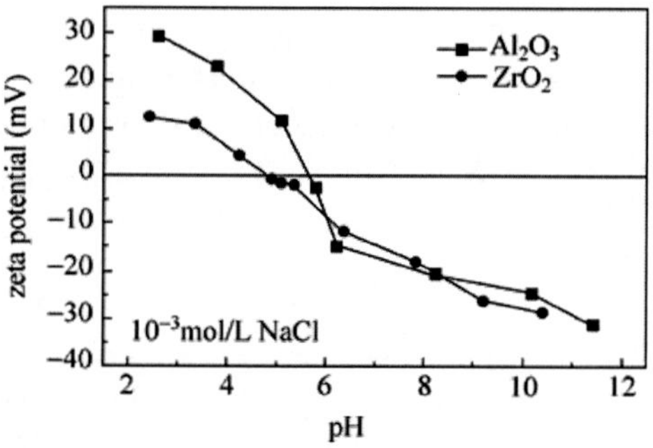

Figure 4.5 The zeta potentials of Al_2O_3 and ZrO_2 as a function of pH

The low viscosity and high solid loading are beneficial for both mixing and casting in slurry processing. It is, therefore, important to maintain slurry fluidity while optimizing solid loading. The study had been performed on suspensions containing 40 – 55 vol% of composite powders. The typical plots of apparent viscosity (η) versus shear rate (γ) after different times of ball milling for a 55 vol% suspension are given in Fig. 4.6. When the milling time was less than 48 h, the viscosity of slurries decreased gradually as the time of milling increased. The results revealed that the absorption of the dispersant on particles did not reach equilibrium and the suspension was unstable until the ball milling time became equal to or more than 48 h. After 48 h of milling, the viscosity of the suspension stabilized. Therefore, the ball milling time should be equal to or more than 48 h to obtain a stable suspension at equilibrium.

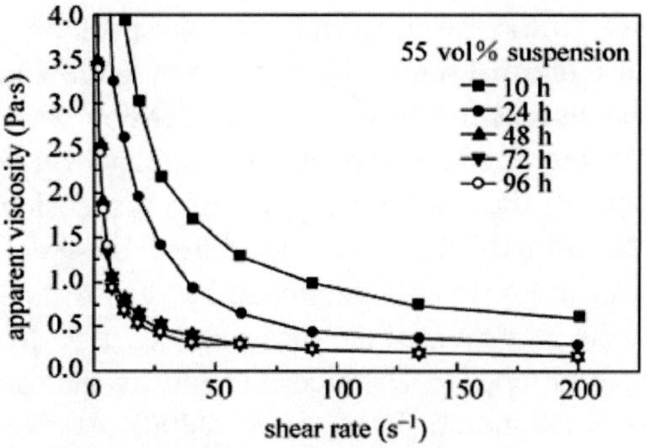

Figure 4.6 The effect of ball-milling time on rheological behavior of 55 vol% suspension

Figure 4.7(a) gives the plots of apparent viscosity versus applied shear rate of stable suspensions after milling for 48 h. It can be seen that all suspensions (40 – 55 vol% solid loading) exhibited a shear thinning behavior and relatively low viscosity (of less than 1 Pa·s at the shear rate of 10 s^{-1}), which was suitable for casting. Figure 4.7(b) shows the plots of shear stress versus the applied shear rate of stable suspensions after milling for 48 h. It can be seen that 53 and 55 vol% suspensions possess a thixotropy hysteresis and their yield stresses were 1 and 3.5 Pa, respectively. This phenomenon was associated with the shear-thinning behavior of the suspensions and indicated a flocculated state of particles in the liquid. The decrease in thixotropy, viscosity and degree of shear-thinning with the decrease of the solid loading of a suspension implied that the degree of powder agglomerate decreased. The thixotropy disappeared, when the solid loading of the suspension was less than 50 vol%.

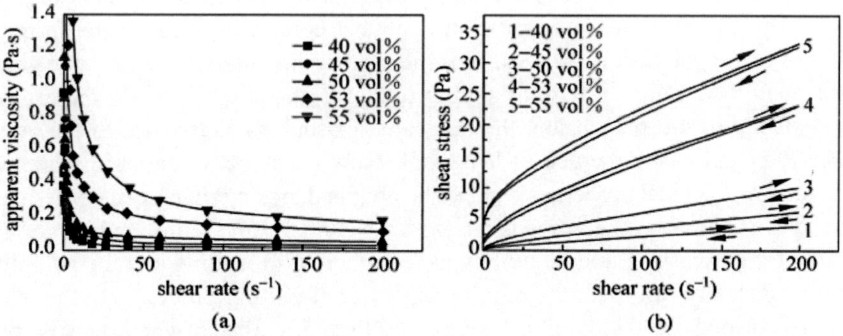

Figure 4.7 The rheological behavior of the suspension with different solid loading (ball-milling for 48 h)

(a) apparent viscosity versus shear rate; (b) shear stress versus shear rate

Concentrated colloidally stable suspensions exhibited shear-thinning behavior in steady shear because of a perturbation of the suspension structure by shear

(Bergstrom L., 1998). At the low shear rates, the suspension structure was close to equilibrium because thermal motion dominated over the viscous forces. At the higher shear rates, the viscous forces affect the suspension structure and thus shear thinning occurred. At very high shear rates, the viscous forces dominated and the viscosity plateau measured the resistance to flow of a suspension with a completely hydrodynamically controlled structure. The degree of shear thinning and the viscosity at high shear rates increased with the increase of the solid loading.

For most systems, the properties of the suspension change drastically at a certain critical particle concentration, F_g, which corresponds to the formation of a space-filling particle network (Sigmund W. M., et al., 2000). At $F<F_g$, the suspensions have no yield stress. Above F_g, the suspensions can sustain a stress before yielding and the elasticity may be significant.

Based on the above description, each of the first three suspensions with solid loadings of 40, 45 and 50 vol%, respectively, had a particle concentration less than F_g, so no yield stress could be observed for them. However, for 53 and 55 vol% suspensions with particle concentration higher than F_g, its high solid loading resulted in the average separation distance between particles in suspension becoming shorter and formed a space-filling particle network, making flow more difficult. When a stress larger than the yield stress was applied, the gelled structure was broken into smaller units (flocs), which then moved past each other. If the floc attrition was affected by the strength of the hydrodynamic and attractive forces, pseudoplastic behavior prevailed and viscosity decreased with shear rates. The strong shear forces at high shear caused flow units to be smaller and thereby the flow was facilitated. The destruction of flocs released constrained solvent, which resulted in a decrease of the effective solid loading of the flocs. This phenomenon resulted in thixotropic behavior in the system.

The gel-casting is a kind of forming process during which suspensions solidify *in-situ*. The structure of the suspension is almost completely kept in the forming bodies after the gel-casting process. This means that the microstructure of the green bodies would be identical with or similar to that of the suspension. In the case of flocs existing in the suspension, the flocs are "frozen" as aggregates in the green bodies after gel-casting process. It is well known that dense ceramic materials with good mechanical properties can not be obtained by sintering green bodies with aggregates. Therefore, the suspensions with 53 and 55 vol% solid loading are not suitable for gel-casting, although they have higher solid loading and their fluidity and low viscosity can meet the requirements of gel-casting process. On the other hand, the suspensions with solid loading less than 50 vol% in which no or a few flocs existed are more suitable for using in the gel-casting process. Green bodies of high uniformity can be prepared by gel-casting from these suspensions and ZTA materials with high mechanical properties can be obtained after sintering.

Figure 4.8 shows the relationship between mechanical properties of sintered samples and the solid loading of the suspension from which the samples were formed by gel-casting. The flexural strength of sintered bodies, which were sintered at 1580℃ for 2 h, increased gradually from (621.35±71.91) MPa to (791.58±33.42) MPa as the solid loading of the suspensions changed from 40 to 50 vol%. Their Weibull modulus had the same tendency and reached the highest value 24.38, as shown in Fig. 4.9. However, as the solid loading exceeded 50 vol%, both bending strength and Weibull modulus of sintered bodies began to decrease. On the other hand, the fracture toughness of sintered bodies increases monotonously as the solid loading of the suspension increases from 40 to 55 vol%.

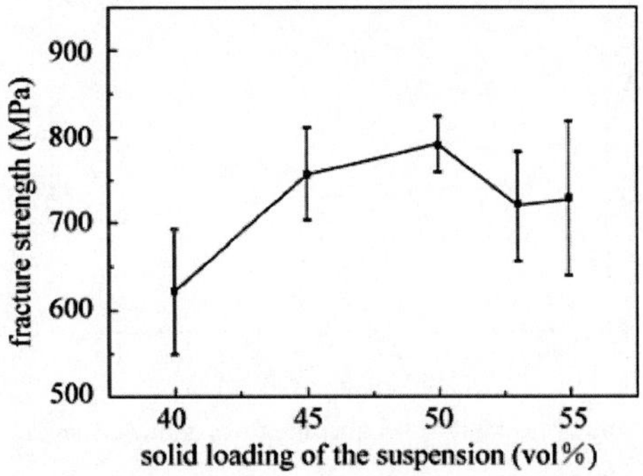

Figure 4.8 The bending strength and fracture toughness of sintering bodies prepared from different suspensions

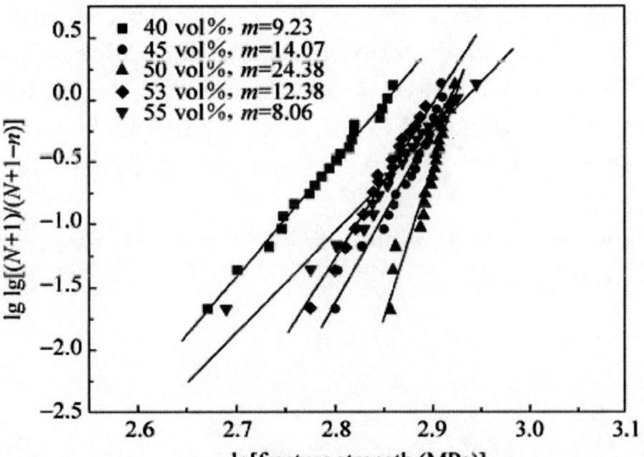

Figure 4.9 The Weibull modulus distribution of sintering bodies prepared from different suspensions

It is well known that the mechanical properties of ZTA ceramics are greatly influenced by the microstructure including relative density (porosity), grain size, ZrO_2 content and its distribution, microstructural flaws (Wang J., et al., 1989). From Fig. 4.10, it can be seen that the sintered sample prepared from the suspension with the lowest solid loading, 40 vol%, had the lowest relative density and highest sintering shrinkage. As the solid loading of suspensions increased, the relative densities of sintered samples increased while the sintering shrinkage decreased. The occurrence of the phenomenon is mainly due to the packing behavior in the suspensions and the gel-casting procedure, as previously discussed. Figure 4.11 also showed that the porosity of the sintered bodies decreased with increasing the solid loading of the suspension.

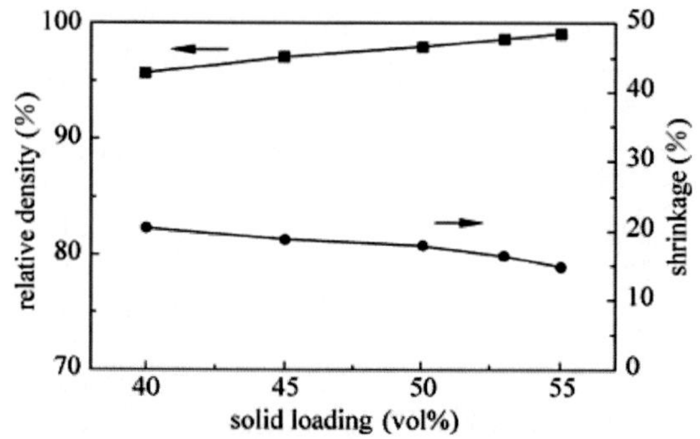

Figure 4.10 The relative density and shrinkage of sintered specimen as a function of solid loading

In general, sintered bodies with higher relative density exhibit better mechanical properties. However, there is also an influence of microstructural uniformity on mechanical properties. In the investigated system, the suspensions with solid loading below 50 vol% had uniform microstructure. When the solid loading exceeded 50 vol%, some kind of flocs developed in the suspension, as discussed above. The flocs were maintained in the green bodies after gel-casting process. These flocs in green bodies often act as sources of defects that occur during sintering.

The critical flaw size c which initiates failure can be estimated from the relation (Kingery W. D., et al., 1976),

$$c = \frac{K_{\mathrm{IC}}^2}{\pi \sigma_{\mathrm{f}}^2} \tag{4.2}$$

where K_{IC} is the fracture toughness of the material and σ_{f} is the flexural strength.

Table 4.1 shows the calculated critical flaw size for each composition. The critical flaw size decreased gradually from 30 to 22 μm as the solid loadings of the suspensions changed from 40 to 50 vol%. However, as the solid loading exceeded 50 vol%, the critical size began to increase. This emphasizes the fact that a suspension with uniform microstructure will result in sintered compacts with relatively small flaws, thus improving mechanical strength and reliability of the ZTA.

(a) (b) (c)

Figure 4.11 The SEM photographs of fracture surfaces of the sintered samples prepared from suspension with solid loading
(a) 40 vol%; (b) 50 vol%; (c) 55 vol%

Table 4.1 The mechanical properties and calculated flaw sizes of ZTA (Al_2O_3/ 15 vol% ZrO_2)

solid loading of the suspensions (vol%)	flexural strength (MPa)	fracture toughness ($MPa \cdot m^{1/2}$)	calculated flaw sizes (μm)
40	621.35±71.91	5.96±0.19	30.0
45	757.05±55.03	6.40±0.28	22.7
50	791.58±33.42	6.58±0.37	22.0
53	719.44±62.12	6.66±0.33	27.3
55	728.56±89.46	6.87±0.27	28.3

Based on the above discussions the following can be concluded: flexural strength and Weibull modulus increase with the increase of solid loading up to a maximum of (791.58±33.42) MPa and, 24.38 at 50 vol% solid loading respectively. All suspensions exhibit a shear-thinning behavior and relatively low viscosity, but suspensions with more than 50 vol% solid loading possess thixotropy hysteresis and yield stresses. Although suspensions with good casting behavior could be prepared also for higher solid concentrations, a decrease of strength, as well as of Weibull modulus were observed beyond a 50 vol% solid loading. This was a result of flocculation occurring in the suspension: the flocs may act as sources of flaws in the sintered compact.

4.1.3 The Effect of Solid Loading on Colloidal Forming

Besides the polymerization of monomers, some new gelation mechanisms such as coordinative chemistry (Morissette S. L., et al., 1999) and protein chemistry (Lyckfeldt O., et al., 1999) were introduced into the system of ceramics gel-casting in recent years. Generally, these researchers mainly focus on the preparation and gelation of dense ceramic suspensions concerned, and most of them seek an optimized circumstance of dispersants, pH and something else to increase solid loading of the suspensions, which is important to obtain ceramic parts with good properties and high performance. However, little attention was paid to how the solid loading affects the technical process, which is also important to ensure good product quality. The effects of solid loading in alumina suspensions on the rheological behavior and gelation of the suspensions, and debindering and sintering of green bodies prepared by colloidal forming of alumina were investigated, and the results suggested that a moderate solid loading is advantageous to both process control and the strength of the final products.

In this section, alumina powders, CT3000SG from Alcoa Qingdao Ltd., were used as received. The powders have a size $D_{50}=0.3$ μm and a specific surface area 5.55 m^2/g. The tri-ammonium citrate (TAC) was used as dispersant. 15 wt% acrylamide (AM) premix solution was prepared with some MBAM (*N, N'*-methylenebisacrylamide, 2 wt% to AM) added. Empirically, TAC added into the suspension was 0.2 wt% of alumina powders. The mixture, consisting of alumina, TAC and premix solution with a certain ratio, was ball-milled for 48 h to prepare the suspension.

To determine the gelation induced time by measuring the temperature (Young A. C., et al., 1991), the suspension (30 mL) was poured into 25℃ thermostatic bath, then an initiator (4 wt% $NH_4S_2O_8$ solution) and catalyst (tetramethylethylenediamine) were added into the desired suspension in the ratio initiator : catalyst : premix solution = 4 μL : 2 μL : 1 mL. The temperature of the suspension was collected and recorded by a computer linked with the temperature detector (Yang J., et al., 2002).

The effect of solid loading on the gelation of alumina suspensions was studied. Because the polymerization of acrylamide monomers is an exothermic reaction, the temperature rising of the suspension denotes that the polymerization has begun. The period from adding initiator and catalyst into the alumina suspension to the commencement of temperature rising of the suspension is called the induced time. From Fig. 4.12, it can be seen that the induced time prolongs as solid loading of alumina suspension increases. After the initiator and catalyst were added into alumina suspension, the initiator decomposed and yielded free radicals gradually. The polymerization of acrylamide monomers was initiated by free radicals. When the free radical met with an interface or another free radical, they combined with each other and lost the ability to initiate the polymerization. As the solid loading of alumina suspension increased, the total surface of alumina particles per unit volume suspension increased, and the probability of the free radicals interacting with the particle surface increased. At the beginning of polymerization, the free radicals interacting with particle surface are unable to initiate the polymerization of acrylamide monomers. As a result, the gelation of the alumina suspension was delayed, and the induced time was prolonged for the high solid loading suspension.

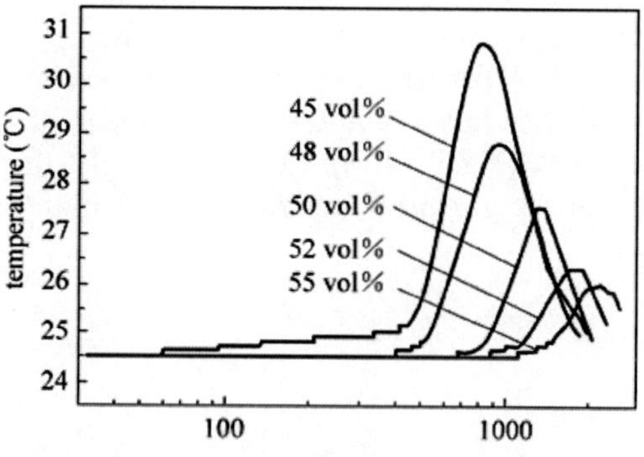

Figure 4.12 The temperature rise of Al_2O_3 suspensions with different solid loadings during gelation

The effect of solid loading on the debindering of alumina green bodies includes: The polymerization of acrylamide monomers in the suspensions terminated when the growing end of polyacrylamide chain collided with the surface of alumina particles. Therefore, the length and the crosslinkage of polyacrylamide chain decreased when the solid loading of the suspension increased. This resulted in the strength of the green bodies prepared by gel-casting decreasing with the solid loading rising (Fig. 4.13), because the strength of the green bodies depended on the polyacrylamide chains formed, and the C—C bond in polyacrylamide chains with longer length or higher crosslinkage have a higher strength.

Because the strength of C—C bond between the branches and principal chain is lower than that in principal chain, the branches of polyacrylamide chains separated from the principal chains and burned out firstly during the debindering of the green bodies. This produced the first exothermal peak in the DTA pattern (Fig. 4.14(a)). When the temperature was elevated continuously, the C—C bonds in principal

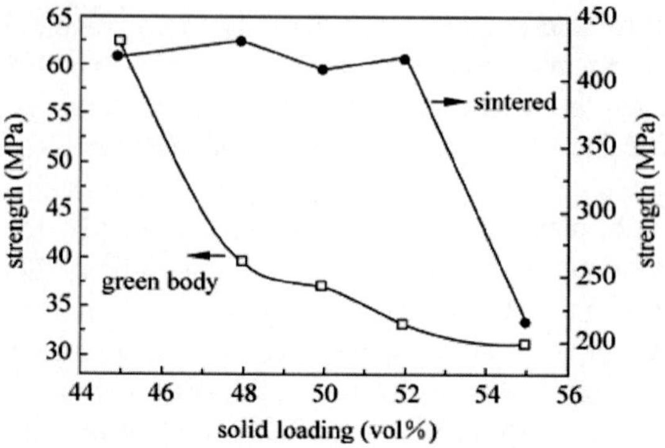

Figure 4.13 The effects of suspension solid loading on the strength of green bodies and sintered bodies

chains began to break and burn, and this resulted in the second exothermal peak in the DTA pattern (Fig. 4.14(a)). Because the length of the principal chain affects the C—C bonds in principal chains more than that in branches, the position of the exothermal peaks in the DTA pattern of the green body prepared with higher solid loading suspension shifts to a lower temperature, and the shift is more obvious for the second exothermal peak (Fig. 4.14(b)). Correspondingly, the weight-loss ended at lower temperature for the green bodies from higher solid loading suspensions (Fig. 4.14(b)).

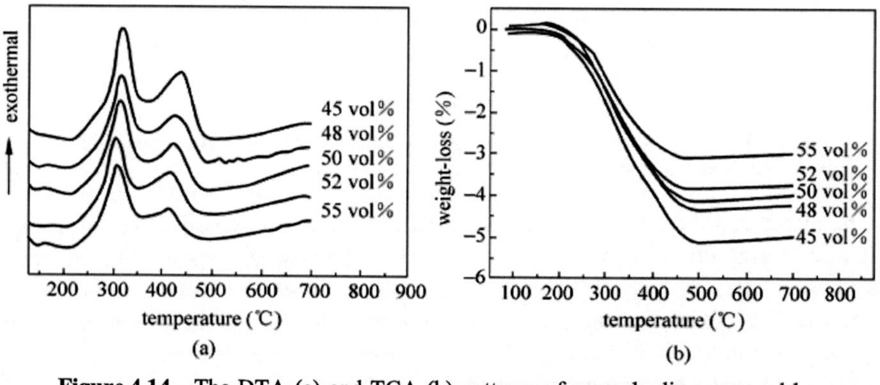

Figure 4.14 The DTA (a) and TGA (b) patterns of green bodies prepared by colloidal forming of Al_2O_3 suspensions with different solid loadings

The effect of solid loading on the sintering of green bodies was: Figure 4.15 shows that all alumina samples had the similar density after being sintered at 1580℃. However, a steep fall of the strength for alumina samples prepared from 55 vol% suspensions can be seen from Fig. 4.13. The steep fall of the strength may result from the microstructure of the sintered bodies, corresponding to coarse crystal grains, which represent the characteristics of overfiring (Fig. 4.17(d)). From the rheological curves of alumina suspensions (Fig. 4.16), it could be found that the yield stress of the suspensions increased with increasing solid loading, and 55 vol% suspension showed obvious shear-thickening behavior. Because shear-thinning is the result of breakage of particle aggregates in suspensions, a higher yield stress of alumina suspension corresponded to a higher strength of particle aggregates in it. When sheared, particle aggregates in a 55 vol% suspension couldn't be broken entirely before shear-thickening occurred, which indicated the formation of hydrodynamic aggregates. Therefore, the microstructures of 55 vol% suspension and the green body prepared from it were far from homogeneous when compare to that of suspensions with less solid loading. Like the pores, the aggregates in the green bodies also hindered the densification during sintering; and more disadvantageously, the aggregates resulted in the overgrowth of local crystal grains and then a coarse microstructure of the sintered body. As a result, the strength of alumina sample from 55 vol% suspension decreased, although it had the highest density. In addition, the results showed that the microstructure rather than the relative density played a more important role in affecting the strength of sintered bodies. Choosing a better dispersant may overcome the aggregates affecting the sintering of the green bodies, and this aspect should be paid attention to in further studies.

The results indicated that the polymerization of acrylamide monomers was delayed and hindered in high solid loading alumina suspensions for gel-casting, because alumina particles provided a surface to adsorb free radicals. Therefore, the induced time was prolonged for the high solid loading suspensions during gelation; and when heated, the exothermal peaks shifts to lower temperature for the green bodies from high solid loading suspensions. Although the strength from high solid loading suspensions decreased, it is high enough to sustain the configuration of the green bodies. So, the flexibility of gel-casting could be

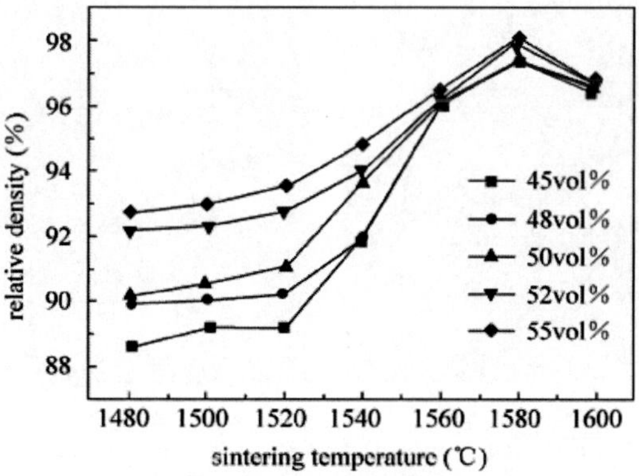

Figure 4.15 The effects of sintering temperature on the relative density of Al_2O_3 samples from suspensions with different solid loadings

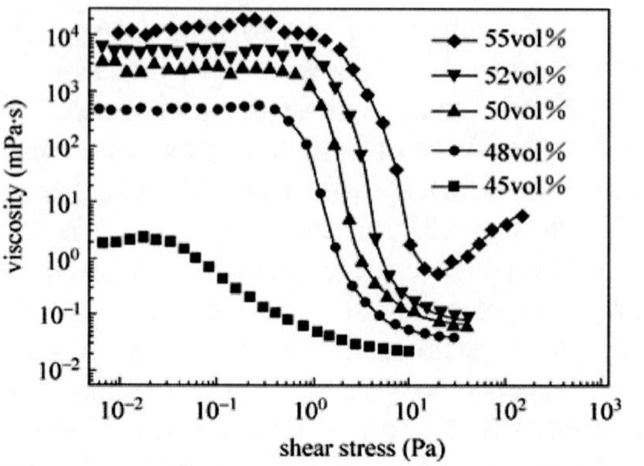

Figure 4.16 The rheological curves of Al_2O_3 suspensions with different solid loadings

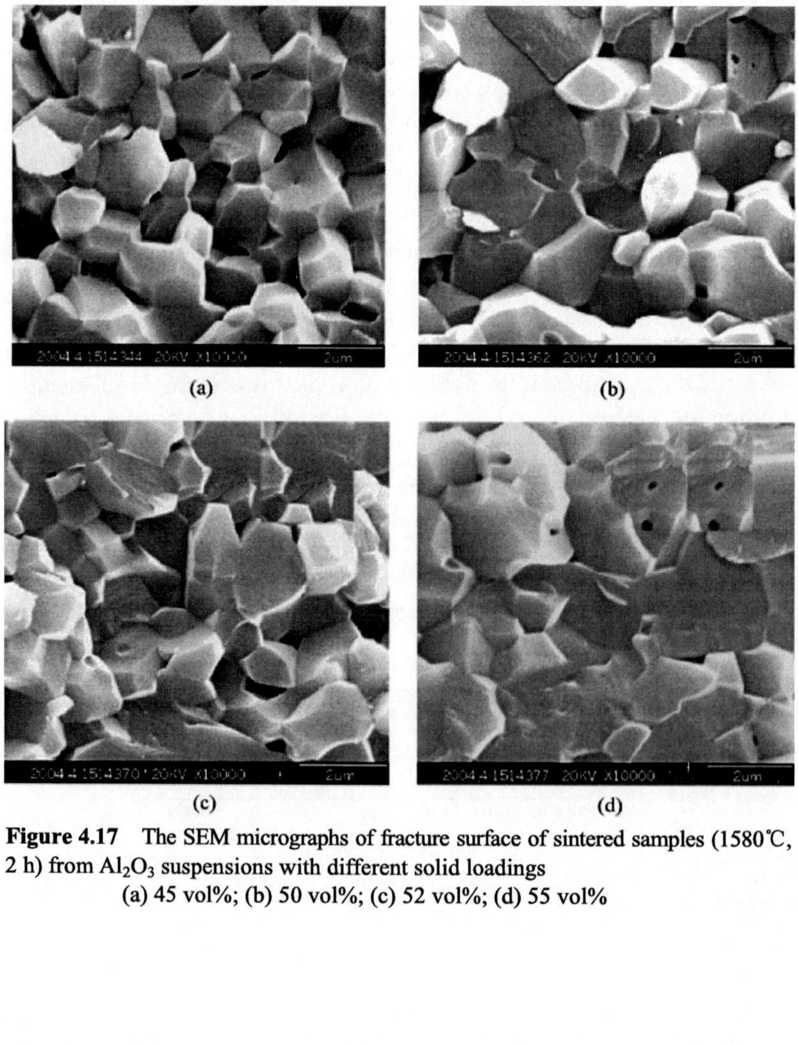

Figure 4.17 The SEM micrographs of fracture surface of sintered samples (1580℃, 2 h) from Al_2O_3 suspensions with different solid loadings
(a) 45 vol%; (b) 50 vol%; (c) 52 vol%; (d) 55 vol%

improved with high solid loading suspensions. However, excessive solid loading suspensions had aggregates resulting in a coarse microstructure and decreased strength of sintered samples, which should be avoided.

4.2 The Generation and Development of Defects

4.2.1 The Generation Mechanisms of Agglomerations in Ceramic Suspensions

The novel colloidal forming technology with *in-situ* solidifying and near-net-size forming, can significantly improve the microstructure and uniformity of the ceramic green body, and reduce the production cost of advanced ceramics (Huang Yong, et al., 2007).

The characteristics of new colloidal forming technology are based on a highly dispersed ceramic suspension that can be converted into uniform structural green bodies and products through *in-situ* curing yet still retain original structures of the suspension (Giuliano T., 2003). However, if there are agglomerations in the suspension, they can cause uneven microstructure in the ceramic products, leading to a decrease in the reliability of their performance (Lang F. F., 1989). In addition, agglomerations also reduce the fluidity of the suspension and restrict the solid loading so that the realization of the net size molding is affected. Therefore, the eradication and control of agglomerations are quite necessary to optimize colloidal forming process and its further applications.

The hard aggregates often exist in ceramic raw materials, and are formed due to the chemical bonding of powder particles, while soft aggregates are formed by van der Waals force between particles (Sigmund W. M., et al., 2000). By selecting high-quality powder or effectively screening the powder prior to colloidal forming processes, the formation of aggregates could be avoided. The soft aggregates are usually decomposed in the preparation process of suspension because in most liquid mediums the attraction between the powder particles in water is far weaker than that in air. For example, the Hamaker constant of alumina is 3.67×10^{-20} J in water, but 15.2×10^{-20} J in air, while the van der Waals attraction potential between the particles is directly proportional to the Hamaker constant (Bergstrom L., 1997). The soft aggregates in the suspension can be further eliminated by adding a dispersant and adjusting pH. Even then, many studies (Rahaman M. N., 2003; Liu Den-Mo, 1998) found that the grain clusters or unusual large-size grains can often be found in actual sintered ceramics; this phenomenon is more observable in products made with nano- or micro-powders (Lim L. C., et al., 2000). The reason for this is that different degrees of aggregates in suspension still exist.

At present, the research on agglomerations in suspension focused on the two aspects: the effects of reunion on the rheological characteristics of suspension (Zhang Liming, et al., 2004; Bender J., et al., 1996) and the dispersibility of powder which can be effectively improved by using dispersing agents and by other methods (Tari G., et al., 1998). However, a few studies on the causes of agglomerations and on the effect of microscopic characteristics of ceramic have been conducted. Thus, the categories and generation mechanisms of agglomerations were investigated by analyzing the basic role of potential energy between ceramic powder particles, observing agglomerations in suspension, and studying the effect of agglomerations on ultimate properties of materials. The objectives of these studies are to control and eradicate agglomerations, to improve colloidal processes, as well as to obtain a more complete theoretical basis on the subject.

1) The Pertinent Experimental Procedures

The ceramic powder used in this research was fine alumina (Al_2O_3 powder, type CT3000, Alcoa Chemical) with a mean particle size of 0.2 mm. Deionized water was used as solvent and ammonium citrate (type TAC, AR, the Beijing Chemical Reagents Company) was used as dispersing agent. Alumina slurries with alumina solid loading of 20%, 45%, 50% and 55% were prepared with 0.2 wt% TAC dispersant (relative to alumina powder), and the adjusted pH = 10. The unmolded wet green body was heat treated to dry and the organic contents were removed. Subsequently, the dried bodies were sintered into ceramic products at 1500℃ for 2 h.

The distribution and morphology of agglomerations were directly observed by using an environmental scanning electron microscope (ESEM, Model Quanta 200, FEI Company, the Netherlands). The microstructures of wet green bodies after colloidal forming and solidifying *in-situ* displayed a true microstructure of the suspension, when observed under an ESEM in a low vacuum, high humidity mode with a pressure of 600 Pa and a humidity of 98% on the dissociation plane of specimens. The microstructures of sintered bodies were observed by using an SEM (Model JSM-6460, JEOL Ltd., Japan).

2) The Generation Mechanisms of Agglomerations in Ceramic Suspensions

According to DLVO theory the motivating force for coagulation is the ever-present van der Waals attraction force which is a function of the dielectric constant of the medium and the mass and distance between the particles. For two small spherical particles of radius a, the potential energy of attraction is V_{vdw}; the repulsive force is provided by the interaction of two electrical double layers. The magnitude of the repulsive force depends on the size and shape of the particles, the distance D between their surfaces, thickness of the double layer κ^{-1} and dielectric constant of the liquid medium. The total potential energy V_{total} is the algebraic sum of the attractive van der Waals potential energy V_{vdw}, and the repulsive energy V_{elect}:

$$
\begin{aligned}
V_{\text{total}} &= V_{\text{vdw}} + V_{\text{elect}} \\
&= -\frac{Aa}{12D} + 2\pi\varepsilon_0\varepsilon_r a\Psi_0^2 \exp(-\kappa h)
\end{aligned} \tag{4.3}
$$

where A represents the Hamaker constant, ε_0 is vacuum dielectric constant, ε_r is the relative dielectric constant of solvent, and Ψ_0 is the particle surface potential energy. The potential role between particles changes with the change in the distance between particle surfaces, as shown in Fig. 4.18.

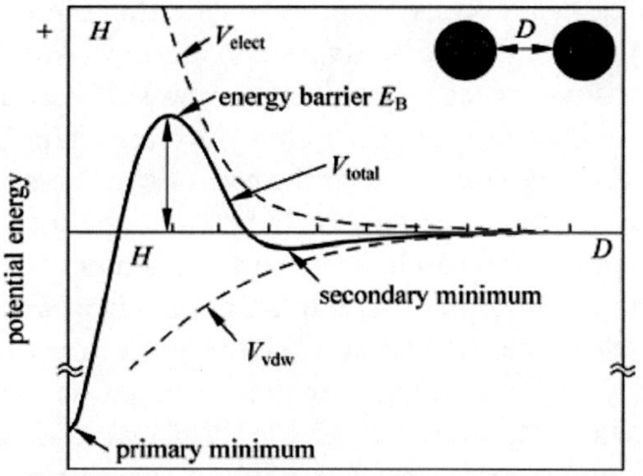

Figure 4.18 The sketch of potential energy between two particles in medium

It is found from Fig. 4.18 that there are minima at two locations on the particles' potential energy curve *i.e.*, the 'primary minimum' corresponding to very short distances in between the particles and the 'secondary minimum' corresponding to the relatively bigger distances between the particles. When the surface distances between powder particles correspond to the trap of potential energy (i.e., the primary minimum), the system's energy is in its lowest energy level. In still or slow flow suspensions, Brownian motion is the main motion form of the powder particles, and it's $K_b(T)$ energy is of 4×10^{-21} J at T =300 K. The energy difference between the potential E_B of powder particles and the 'secondary minimum' is usually about 10^{-19} J, thus, it is very difficult for particles of suspension to cross over the barrier E_B to get into the 'first trap'. Thus, in general terms, the 'secondary minimum' represents that the system is in its minimum energy state. Therefore the interparticle separation corresponding to the secondary minimum can be called 'stable separation D_0'. For a given powder particle size, the solid loading in suspension is determined directly by the distance between the particles. The solid loading corresponding to 'stable interval D_0' of the powder particles is defined as the stable solid loading ϕ_0 of this suspension. When the solid loading is lower than ϕ_0 in suspension, surface interval of the powder particles is greater than D_0 and potential energy between the particles is not at the lowest level, deviating from the secondary potential well. Under such circumstances, the particles spontaneously close up to each other and restore stability at pitch D_0. As a result, aggregate structures are formed in the local area and reduce the system energy to a minimum. Corresponding to the depth of secondary potential well, the attraction force in aggregates is usually relatively weaker. However, in this process a 'loose reunion' forms because of the attraction due to the large gap between the particles. Large gaps are also left between aggregates. The average distance between particles in suspension is equal to D_0 when solid loading reaches ϕ_0. Despite the surface distance between particles deviating from space-balanced positions because of Brownian motion, the system spontaneously restores the particle distance to D_0 again on its own to maintain the minimum energy. In this manner, the suspension takes on a stable and uniform structure. If the solid loading of the suspension exceeds ϕ_0 while the average distance between particles is shorter than D_0, the energy of the system increases rapidly, leading to the potential energy of particles crossing over the barrier E_B. Therefore, the probability for the potential energy level falling into the 'first trap' increases significantly due to Brownian motion. As a result, some particles strongly close up to form the reunion structure and fall into the 'first trap'. On the other hand, the rest particles then restore the 'stable interval D_0' and the system returns back to its lowest energy state. The particle reunion structure formed is defined as a tight agglomeration because of its short particle distances and the strong mutual attraction potential among particles. These three conditions are displayed in Fig. 4.19 (Chen Ruifeng, 2007).

3) The Effect of Solid Loading on the Behavior of Alumina Suspension

The solid loadings in the alumina slurries selected are 20 vol%, 50 vol% and 55 vol%. The observation results of the agglomerations in alumina suspensions with different solid loadings are shown in Fig. 4.20.

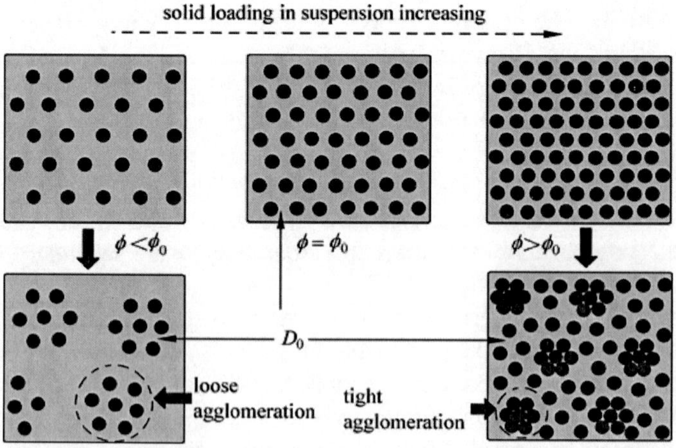

Figure 4.19 The microscopic model of agglomerations in suspension

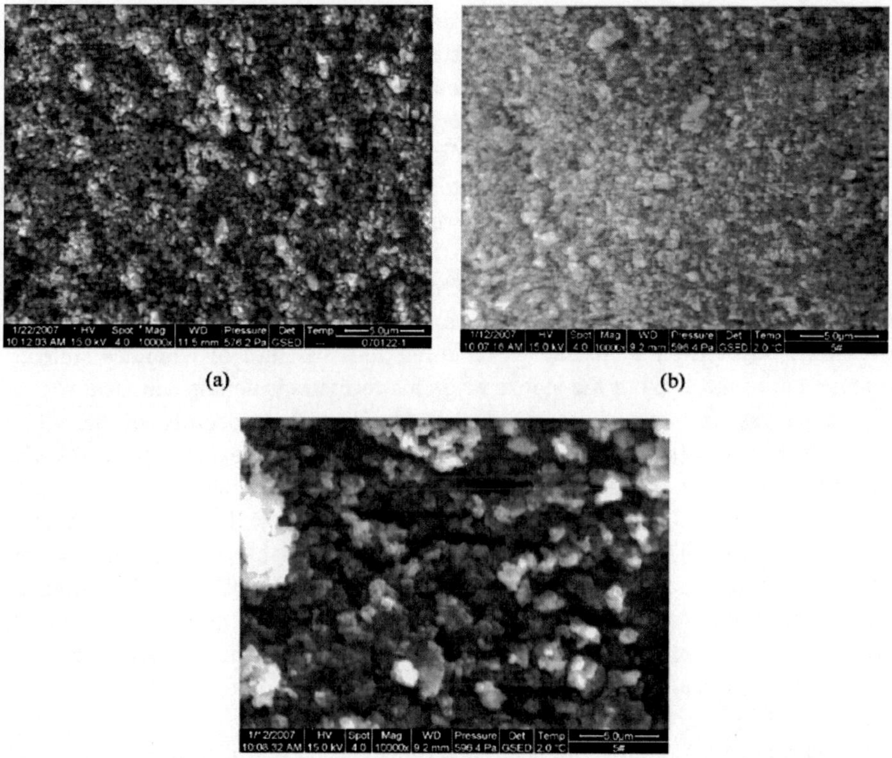

(a) (b)

(c)

Figure 4.20 The agglomerations in alumina suspensions with different solid loadings (a) 20 vol%; (b) 50 vol%; (c) 55 vol%

From Fig. 4.20(a) it can be seen that the ceramic powders at the sub-micron level in the suspension with very low solid loading (20 vol%) did not distribute evenly despite the addition of an appropriate proportion of dispersant. Instead, a partial reunion of loose state could exist. When the solid loading was increased up to 50 vol%, a suspension with a highly uniform structure was observed, as shown in Fig. 4.20(b). It can be explained as the powder particles in suspension achieved a stable distribution. When the solid loading was increased to 55 vol% (Fig. 4.20(c)), large agglomerations with diameters of 1 – 2 μm can be clearly observed. This phenomenon results from the formation of aggregates in the suspension due to the powder particles are fully closed up each other. Based on these observations the optional alumina solid loading is about 50 vol% to form a stable uniform dispersion in suspension.

The agglomerations generation mechanisms in an alumina ceramic suspension can be summarized as follows: when the solid loading are less than ϕ_0, a loose agglomeration is produced. When the solid loading is higher than ϕ_0, a tight agglomeration is formed. When solid loading is equal to ϕ_0, the particles in suspension take on a uniform and stable dispersion, but no reunions appear. If the solid loading is too low in a ceramic suspension used for colloidal forming, it causes not only a heavy shrinkage but also an uneven shrinkage of the ware during the later heat treatment.

4) The Effect of Agglomerations on Microstructures of Ceramic Materials

The microscopic structures of the sintered ceramics are shown in Fig. 4.21. It can be seen from Fig. 4.21(a) that there were the locally uneven cluster structures of 5 – 10 μm sizes in the sintered alumina ceramics which were made of alumina slurry with a solid loading of 20 vol%. Because solid loading in suspension was less than ϕ_0 for a stable suspension, the looser reunions were formed. When solid loading of suspension increased to 45 vol%, the structure of ceramics sintered became more uniform, but the stocks of grains formed clusters in the local region with larger gaps between the grain clusters (see locations specified by the white arrows in Fig. 4.21(b)). When the solid loading was increased to 50 vol%, the grains and pores of the ceramics were evenly distributed with the grain sizes mostly in the range of 1.0 – 2.0 μm as shown in Fig. 4.21(c). It is apparent that the solid loading in this suspension was consistent with stable solid loading ϕ_0 on which the powder particles were distributed evenly. When the solid loading of suspension continued to increase up to 55 vol%, large-size grains of more than 5 μm appeared again (black arrows in Fig. 4.21(d)) in the sintered alumina. This showed the solid loading in the suspension exceeded ϕ_0 and resulted in formation of reunions (see Fig. 4.19 and Fig. 4.20).

5) The Effect of Agglomerations on Mechanical Properties of Alumina Ceramics

The strength and toughness of sintered alumina ceramics shown in Figs. 4.21(b), 4.21(c) and 4.21(d) are measured and listed in Table 4.2. The data in Table 4.2 corresponding to Fig. 4.21 show that when the solid loading in suspension is 50 vol%, the prepared ceramic displays the highest strength with the best particle distribution uniformity in the sample. This can be explained that the solid loading is approaching stable, solid loading ϕ_0 and the powder particles in suspension are optimally dispersed. Owing to this, the ceramic obtains the most uniform microstructure and its strength is significantly improved. When the solid loading in suspension is reduced to 45 vol%, both the strength and density of the prepared ceramics decrease. This is due to the formation of loose reunions, which is because the solid loading in suspension is lower than ϕ_0. Although the internally particle distributions of aggregates maintain stable states, the distances between loose reunions are relatively great.

The existence of loose reunions affects the mass transferring and sintering densification due to the great distance between loose reunions. The ceramic characteristics of lower density and less uniformity are the direct results of large pores and uneven microstructure.

When using the suspension with a solid loading of 55 vol%, the density of the sintered ceramic increases by 0.7%, while the breaking strength decreases.

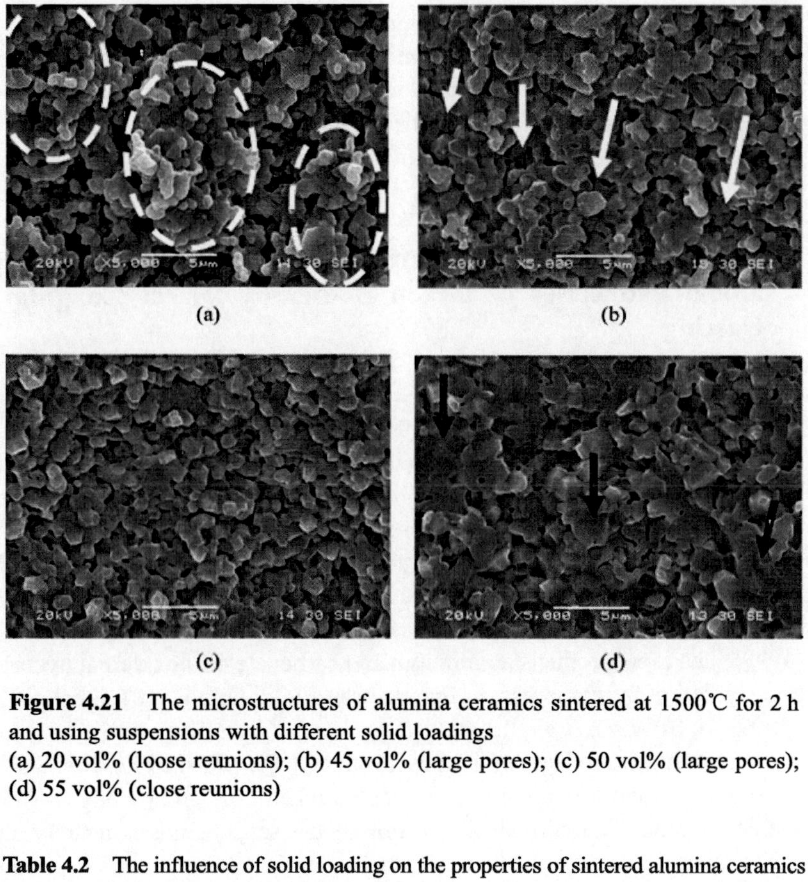

(a) (b)

(c) (d)

Figure 4.21 The microstructures of alumina ceramics sintered at 1500℃ for 2 h and using suspensions with different solid loadings
(a) 20 vol% (loose reunions); (b) 45 vol% (large pores); (c) 50 vol% (large pores); (d) 55 vol% (close reunions)

Table 4.2 The influence of solid loading on the properties of sintered alumina ceramics

solid loading in slurry (vol%)	relative density of sintered body (%)	bending strength (MPa)	K_{IC} ($MPa{\cdot}m^{1/2}$)
45	96.6±0.5	429±28	4.1±0.6
50	97.5±0.3	483±16	4.2±0.2
55	98.2±0.3	456±32	4.1±0.5

This is mainly because tight reunions are formed as the solid loading in suspension is greater than ϕ_0. The particles within the reunion regions are very close to each other, and so sintering takes place among them first and thereafter the particles grow into large grains when the sintering temperature increases.

The toughness K_{IC} is not significantly different among the ceramics prepared by the slurries with three different solid loadings, but the bending strengths are relative lower for those prepared by the suspensions with solid loadings of 45 vol% and 55 vol%.

This shows that the aggregates generated in suspension can reduce the microstructure uniformity and performance reliability of final ceramic material. Therefore, the suspensions used in colloidal forming fabrication must reach to the stable solid loading ϕ_0 in order to avoid the formations of agglomeration which can ultimately affect the performance of ceramic materials.

4.2.2 The Influences of Idle Time on Microstructures and Mechanical Properties of Green Bodies by Direct Coagulation Casting

Ceramic green bodies can be produced by enzyme-catalyzed reactions from suspensions with high solid loading (>50 vol%). This forming process is known as direct coagulation casting DCC (Gaucker L. J., et al., 1999a, b). It is a near-net-shape forming method, which is suitable for ceramic parts having complex geometry. In this paper, we will focus on discussing the coagulation method i.e. increasing ionic strength. This method can produce much higher wet strength of the green bodies by aging for 1 – 3 days than the pH-shift method and may be suitable for other ceramic systems, such as zirconia or silicon carbide (Si W. J., et al., 1999). The ionic products ammonium and carbonate of the internal hydrolysis lead to coagulation by compressing the electrostatic double layer (Gaucker L. J., et al., 1999a, b; Si W. J., et al., 1999).

The microstructures of green bodies produced by pH-shift to the IEP are always very dense and homogeneous without cracks in the green body as well as in the sintered state. The consolidation time of the suspensions is usually about 0.5 – 2 h, depending on temperature and urease concentration. In this coagulation method, no syneresis or sedimentation occurs. This means that the particle network is not contracted during coagulation, although the separation distance between neighboring particles is decreased (Helbig M. H., et al., 2000).

The coagulation time by increasing ionic strength is longer than that of pH-shift. In general, several hours or up to 3 days are necessary to get a maximum strength of the wet green bodies. Its compressive wet strength is higher by a factor of up to 10 compared to that of the pH-shift method but it strongly depends on the coagulation time. This is explained by gel network aging during which the number of particle contacts is increased, thereby improving the strength of the gel network (Balzer B., et al., 1999). In our study, the microstructure of wet or dry ceramic green bodies produced by increasing the ionic-strength always showed cracks on the surface and in the interior of the green bodies.

In this section, the author investigated the microstructures of coagulated alumina suspensions by increasing the ionic strength in aqueous suspensions by the enzymatic urea/urease reaction. We observed three stages of the coagulation process by increasing ionic strength: the initial stage where no shrinking was found, followed by a stage of syneresis in combination with sedimentation, and finally the aging stage, where the compressive strength of the wet green body increased. We discovered that the cracks were formed during the first two stages of the coagulation. They can be reduced or even avoided by introducing an idle time before casting the slurries or by adding salts during preparation of the slurries. By increasing the idle time or the ionic strength prior to casting, the microstructure of the green bodies gradually changes from an inhomogeneous structure with many macroscopic cracks to a homogeneous microstructure without cracks.

1) Starting Materials

High purity alumina powder (HPA-0.5 MgO, CONDEA Vista Company, USA) containing 0.5% MgO in the form of $MgAl_2O_4$, was used. The powder has a d_{50} of 0.5 μm and a surface area of 9.0 m^2/g. The other chemicals were deionized water (>18 MΩ/cm), urea, p.a. (Sigma), 2 mol/L HCl (Titrisol, Merck), ammonium carbonate $CH_8N_2O_3$(Fluka, FLUKA Chemie AG, Switzerland), diammonium citrate, purum (Alfa, Johnson & Mathey Brandenberger AG, Switzerland), and urease (activity: 71000 units/g, Lyo. Special Quality, Roche, Diagnostics Mannheim, Germany). The solid loading of slurries, if not explicitly mentioned, was 57 vol%.

2) The Preparation of the Slurries

The slurries for the pH-shift method were prepared as follows: first urea and HCl were added to deionized water, and then the alumina powder was slowly introduced during stirring. After mixing homogeneously, the slurries were ball-milled for 24 h by using the alumina ball milling. The pH of the ball-milled shurries was measured and adjusted to 3.5 – 4.5 with 2 mol/L HCl. An urease water solution was dispersed by ultrasonic treatment, and the slurry was cooled to around 5℃ and degassed. The urease solution was then added to the cooled slurry. After mixing for about 5 min, the slurries were cast into the mold. After casting for 2 h, the coagulated slurries were demoulded.

After careful drying, the body's microstructure was examined. The following parameters were used in the pH-shift experiment: 0.3 mol/L urea (referred to liquid), urease of 1 unit/g (to alumina).

The pH-shift slurry is prepared at pH=4 and destabilized by the urea hydrolysis reaction to shift pH=9. The pH of the slurry buffered by the produced ammonium carbonate is pH=9.

Due to the adsorption of diammonium citrate on the particles, the surface charge on the alumina particles changed: the IEP shifted from 9.2 to 3.4 (Helbig M. H., et al., 2000). Well-dispersed ball-milled slurries can be prepared within a pH range from neutral to basic. The suspension is stable at pH=9 and at low ionic strength. Then it is destabilized at pH=9 by increasing the ionic strength. The preparation process of the ionic-strength slurries before casting is similar to the pH-shift slurry except for replacing HCl by diammonium citrate at the beginning. After adding urease, the degassed and cooled slurry is idled at different time. Before casting, the idle slurries were shaken for 5 min by hand. One day after casting the slurries were demoulded and carefully dried. Then the microstructures were examined. The following parameters were used during the preparation of the slurries: 2.2 mol/L urea (referred to liquid), 0.45 wt% diammonium citrate (referred to alumina), and 10 unit/g urease (referred to alumina). Diammonium citrate is about 150% of the amount needed for a monolayer on a powder with 9.0 m^2/g, the ionic strength after completion of the hydrolysis reaction is adjusted by the amount of urea added.

3) The Characterization of Coagulating Slurries and Green Bodies

A light microscope and stereo lens (from Leica Co., Germany) was used to observe the microstructure of the green bodies. An electronic image capturing system was used to store the pictures.

In order to determine the volume shrinkage of the liquid phase during the hydrolysis of urea, the prepared slurry was degassed and cooled down to about 5℃ and was cast into two 100 mL graduated cylinders with height: 25 cm and diameter: 2.5 cm. One was used for measuring the temperature of the slurry. The other one was used to monitor the volume change vs. the time during the reaction.

The rheological behavior of the suspensions was measured with a controlled-stress rheometer by using a vane tool. The drying of the sampled was prevented by the application of a silicon oil layer on top of the slurry surface.

4) The Cracks Formation during ΔI Coagulation

The coagulation of ceramic suspensions with high solid loading (>50 vol%) was investigated. The coagulation experiments were carried out using a special type of direct coagulation casting (DCC) procedure. This method uses the enzyme-catalyzed hydrolysis of urea to increase the ionic strength (ΔI) in the suspension at pH = 9. The microstructures of the achieved green bodies were examined by photographing the fracture surfaces of dried green bodies. The volume shrinkage (syneresis) and the concentrated suspension were monitored during coagulation.

The fracture microstructure of all of the dried green bodies coagulation by ΔI showed vertically oriented cracks in the carefully dried green body. Samples coagulated by ΔpH are dense and crack-free. The crack to crack distance of the laminated cracks in the ΔI coagulated samples was about 0.3 to 2 mm.

The influence of several experiment parameters on the crack formation was investigated (Table 4.3). First the coagulation kinetics was changed. The amount of added urea was varied from 0.55 to 2.2 mol/L urea, and the used amount of urease was varied from 2.5 to 40 unit/g urease (2.2 mol/L urea) in slurries of 53 – 62 vol% alumina (see Table 4.4). All samples also showed curved lines on their surface as demonatrated in Fig. 4.22(a). After three days the surface of the slurry could still be plastically deformed through touching (Fig. 4.22(b)). As all samples showed these cracks, we conclude that the cracks are not related to any of the different conditions and ratios of urea/urease and solid loading of the slurries.

Inhomogeneously distributed urease as a cause for the cracks could be ruled out also. Using high power ultrasonic treatment in the suspension after adding urease as well as an experiment where urea and urease were added to the slurry and milled in a low-temperature mixing container for 15 – 30 min did not influence the crack microstructures (see No.2 and No.3 in Table 4.3).

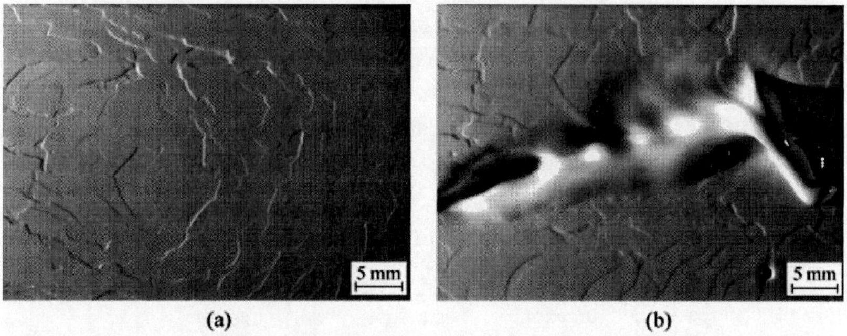

Figure 4.22 The surface pattern (a) and surface after touching (b) of non coagulated alumina slurry after 3 days by increasing ionic strength

One might speculate, that the crack formation is caused by the ammonium citrate which is used for the electrically stabilization. Therefore, the alumina particles were stabilized by polyacrylic acid (Aldrich 2000 g/mol) which stabilizes by electrostatic and steric reasons. However, no difference was observed and an influence by the dispersant can be ruled out (see No.4 in Table 4.3).

Table 4.3 The experimental details for ΔI coagulation

No.	experimental conditions	fracture microstructure	notes
1	different ratios of urea/urease	cracks	see Table 4.4
2	mixing the slurry with urea/urease in low-temperature ball-milled jar	cracks	distribution of urease
3	using high power ultrasonicto distribute urease in the ball-milled slurry	cracks	distribution of urease
4	replace diammomium cirated by PAA	cracks	influence of dispersant
5	casting slurry onto the surface of liquid with high specific density (2.96 g/mL)	cracks	avoiding wall effects

We also wanted to check the mold surface on the crack formation. We cast the slurry onto the surface of a hydrophobic liquid of high specific density (1, 1, 2, 2-tetrabromoethane) (see No.5 in Table 4.3). The fracture surface of these two wet green bodies did not differ from the usually observed ones after coagulation. Therefore, cracks were not generated by the contact of the weak slurry sticking to the wall.

5) The Influence of an "Idle Time" on the Microstructure of Green Bodies

To follow the crack development and to observe in which stage cracks are formed, we introduced an idle time before casting. The idle time is defined as the period of time starting with the addition of urease and before casting of the slurry. After the idle time before casting, the slurry was shaken well by hand in order to destroy any larger coagulated structures in the slurry. The microstructures of the fracture surface of dried green bodies with different idle time are shown in Fig. 4.23. It is very surprising that the idle time strongly influences the formation of cracks in the green bodies coagulated by increasing ionic strength. The results clearly show that no cracks occur in samples that were cast after 150 min idle time.

Table 4.4 The effect of different ratios of urea/urease on microstructure of green bodies

sample No.	urea (mol/L)	urease (unit/g)	fracture microstructure	solid loading (vol%)	notes
JL008	2.2	10	cracks	57	
JL064	2.2	5	cracks	57	
JL066	2.2	2.5	cracks	57	
JL065	1.1	5	cracks	57	
JL067	1.1	2.5	—	57	no coagulation
JL068	0.55	2.5	—	57	no coagulation
JL025	2.2	15	cracks	55	
JL026	2.2	15	cracks	53	
JL084	2.2	20	cracks	57	
JL088	2.2	40	cracks	57	
JL184	2.2	10	cracks	62	

The sample shown in Fig. 4.23(a) was directly cast after the dispersion of the urease in the slurry which took about 5 min and which is the minimum idle time tested. A vertically laminated-cracks structure in the green body was observed. With increasing idle time, the distance between the vertically laminated cracks became considerably larger (see Fig. 4.23(a) to (g)). On the other hand, the number of cracks in the green body decreased gradually (see Fig. 4.23(b) to (g)), until finally the cracks completely disappeared (see Fig. 4.23(h)). The structure became dense and homogeneous on a millimeter to centimeter length scale when the idle time reached about 150 min (see Fig. 4.24(h)). The shortest necessary idle time to avoid cracks is defined as the critical idle time.

Actually, when the idle time increases, the ionic strength and the viscosity in the slurry also increased. Thus, those factors that determine the critical idle time include the starting temperature, ionic strength in the suspension, the amount of added urea and urease as well as the solid loading of the suspension.

6) The Influence of Addition of Different Salt on the Microstructures of the Green Bodies

Increasing the idle time means not only increasing the ionic strength within the suspensions, but also shearing the slurry before casting. In order to simulate the effect of increasing the idle time on the microstructure of green bodies we

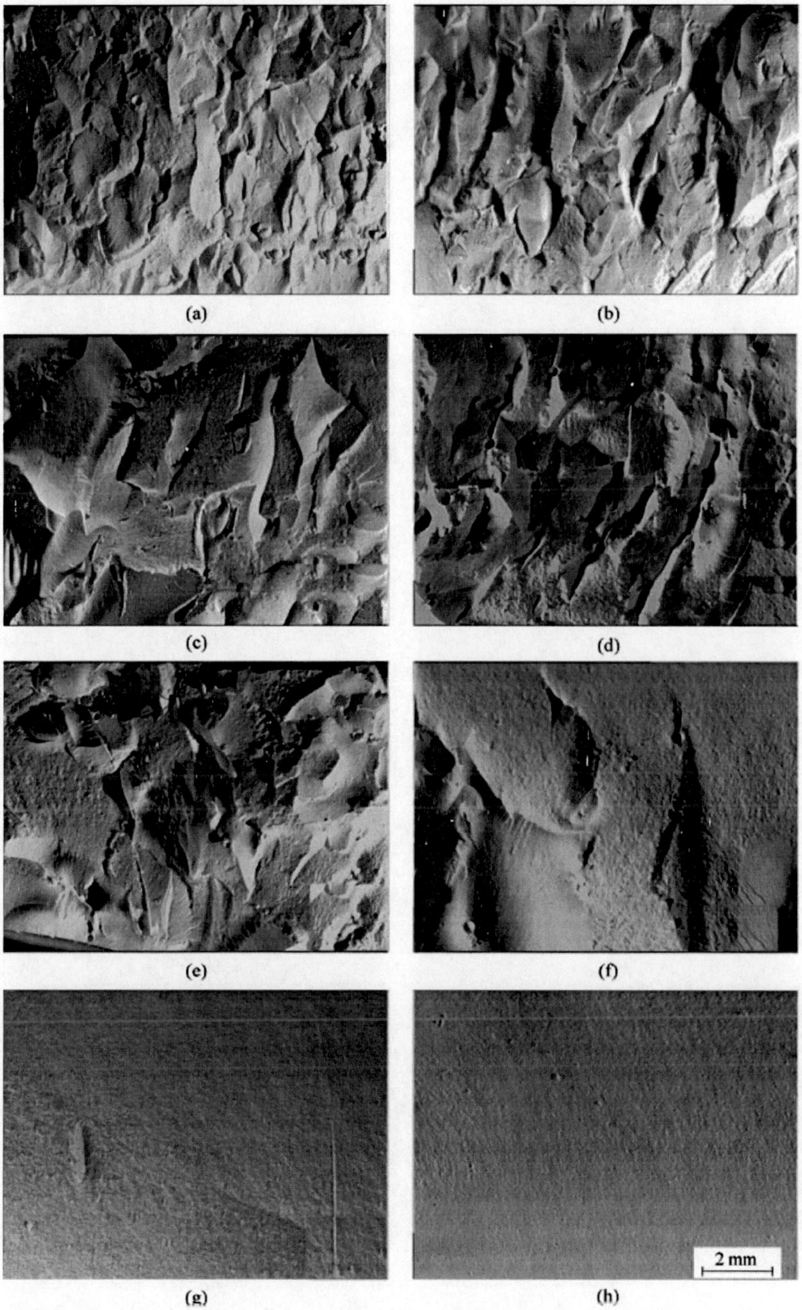

Figure 4.23 The influence of different idle times on the microstructures of green bodies produced by increasing ionic strength: (a) 5 min; (b) 20 min; (c) 40 min; (d) 60 min; (e) 80 min; (f) 90 min; (g) 120 min; (h) 150 min

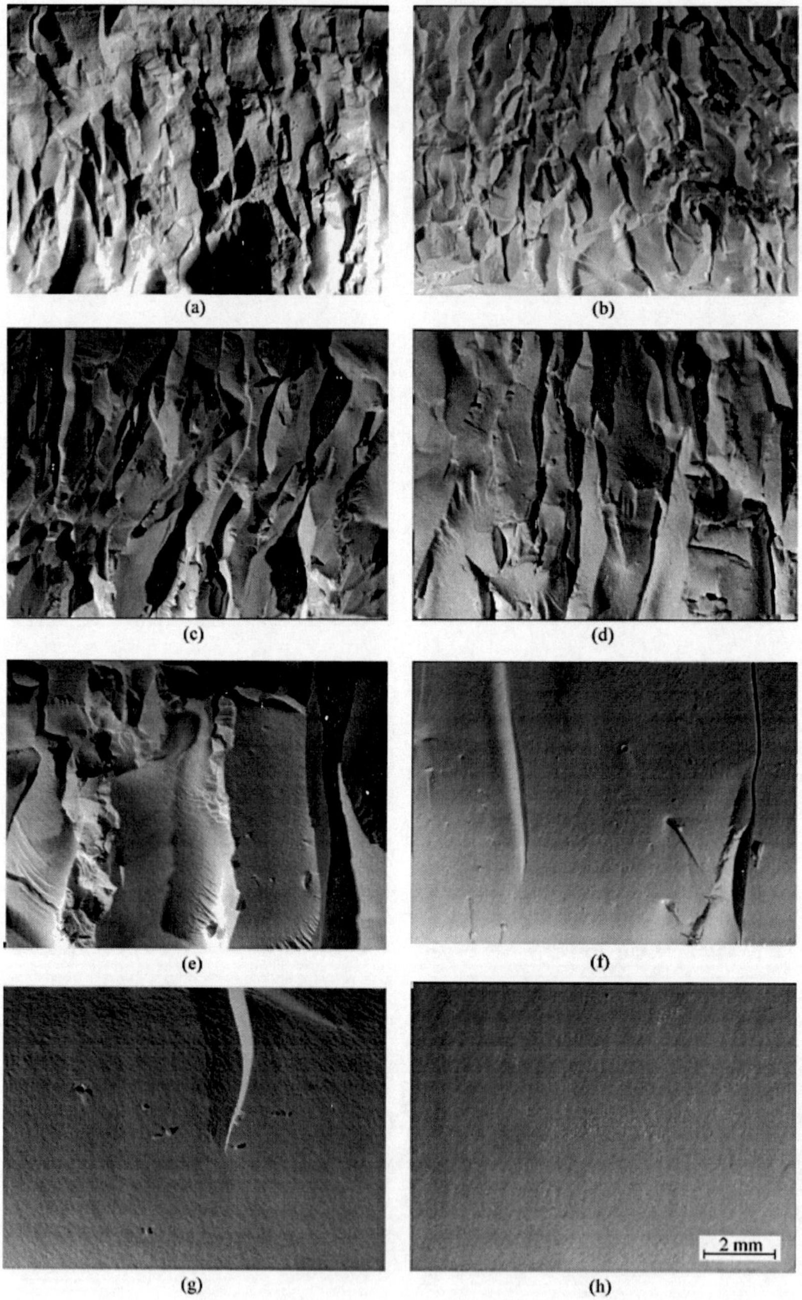

Figure 4.24 The influence of different amounts of hydrated ammonium carbonate on microstructures of the green body by increasing ionic strength: (a) 0.001 mol/L; (b) 0.01 mol/L; (c) 0.05 mol/L; (d) 0.1 mol/L; (e) 0.2 mol/L; (f) 0.3 mol/L; (g) 0.35 mol/L; (h) 0.4 mol/L

introduced different amounts of salt into the suspensions prior to ball milling. For simulating the real reaction system, an ammonium carbonate ($CH_8N_2O_3$) was chosen. It was added to the initial slurry containing the urea to result in salt concentrations of 0.001 to 0.4 mol/L at pH=9. The microstructures of the green bodies with varying amounts of salt are shown in Fig. 4.24. The slurries were cast right after the addition of urease without any idle time. The results clearly indicate that the crack development is connected to the salt concentration in the suspensions. The resulting microstructures of the green bodies are the same at salt concentrations up to 0.05 mol/L of added salt (Fig. 4.24(c)). The micrograph of the sample with a salt addition of 0.1 mol/L is comparable to that sample with 40 min of idle time (Fig. 4.24(d)). A crack-free structure was achieved by the addition of 0.4 mol/L salt (Fig. 4.24(h)). This is comparable to the microstructure of the sample that was cast after 150 min of idle time. This experiment documents that increasing the idle time means increasing the ionic strength inside the suspension. It is reasonable to derive that the crack formation is connected to the salt concentration within the slurry prior to casting.

To identify the reasons of crack formation, we investigated the volume shrinkage of the slurry during coagulation.

7) The Syneresis during Coagulation

In order to characterize the network formed during coagulation after casting we measured the volume change of the suspension. The shrinkage (syneresis) of the particle network as well as a decrease of the water phase due to the urea hydrolysis were monitored. The volume shrinkage of the coagulating alumina particle network as a function of the reaction time at different solid loadings is shown in Fig 4.25. There are three stages that are observed in our experiment which lead to the final coagulated particle network structure. It has to be mentioned that the different time stages depend on the urease/urea content as well as the slurry temperature.

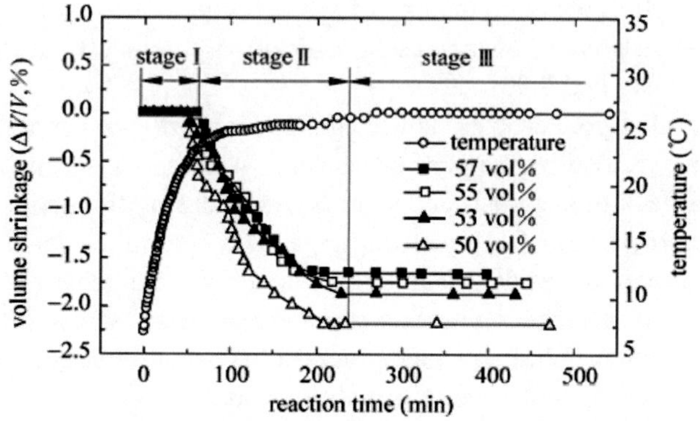

Figure 4.25 The volume shrinkage of slurries vs. the coagulation time at different solid loadings

The volume shrinkage vs. the reaction time clearly shows three stages.

Stage Ⅰ: The temperature of the slurry increases to 26.5℃ (room temperature) and the urea decomposition starts. The volume of the slurry remains constant during the first 60 min. In this stage, particles gradually start to agglomerate as the ionic strength increases. The gel network formed by the particle agglomerates is very weak. The viscosity of the slurry remains low and no syneresis is observable.

Stage Ⅱ: In this stage a remarkable shrinkage of the coagulating particle network up to 240 min is observed (Fig. 4.25). The volume shrinkage is a consequence of syneresis and sedimentation. Syneresis results from the spontaneous contraction of the gel without evaporation of the constituent pore fluid (Brinker C. J., et al., 1990; Scherer G. W., 1989). In our experiments, we only observed one-dimensional compaction along gravity direction and no lateral shrinkage. It resulted from the combination of both sedimentation and syneresis of the destabilized suspension.

Stage Ⅲ: The syneresis of the suspension has stopped. The viscosity is high and the particle network starts to age. During this period the strength of the gel network is still increasing as shown in previous works (Balzer B., et al., 1999; Dewhurst D. N., et al., 1999).

The different idle time means actually the different ionic strength in the suspension. Before casting at every idle time, slurry must be shaken homogeneously in order to destroy its formed microstructure and then rebuild its new structure of slurry. By increasing the idle time or ion strength prior to casting, the microstructure of the green bodies gradually changes from an inhomogeneous structure with many macroscopic cracks to a homogeneous microstructure without cracks. The microstructure of green body becomes homogeneous and crack-free when the idle time is about 150 min which corresponds to 0.4 mol/L of added ammonium carbonate in this study.

The influence of the idle time on the microstructure is related to the ionic strength in the slurry, and the starting temperature of slurry, the amount of urea and urease, as well as the solid loading of suspension.

The coagulation process of increasing ionic strength consists of the following three stages as described below.

Stage Ⅰ: liquid stage. Particles start to agglomerate with ion strength increasing. The gel network formed by the particles is very weak. The viscosity of the slurry is low. There is lack of syneresis and sedimentation during this stage.

Stage Ⅱ: sedimentation and syneresis stage. The one-dimensional compaction along gravity direction was dominant and no lateral shrinkage was observed. The main parameter to influence syneresis strain rate is gel viscosity. The high gel viscosity will eventually shut down the process of syneresis, but not until relatively high strains have been attained after destabilization.

Stage Ⅲ: aging stage. This stage involves improving the strength of the gel network by increasing the number of contacts in the contracted structure.

4.3 The Effect of Ionic Conductance on Preparation of Highly Concentrated Suspension

4.3.1 The Academic Idea and Research Program

With the development of the *in-situ* consolidation forming processing, such as direct coagulation casting (Graule T. J., et al., 1994), gel-casting (Omatete O. O., et al., 1991b), temperature induced flocculation (Bergstrom L., 1994), etc., highly concentrated ceramic suspension with low viscosity is the prerequisite to achieve high density and good homogeneity of green bodies for their application.

Polyelectrolytes are widely used as dispersants to prepare highly concentrated ceramic suspensions in industrial application. A 62 vol% aqueous alumina suspension could be prepared with polymethacrylic acid and polyacrylic acid (Cesarano III J., et al., 1988). However, the use of polyelectrolytes (or dispersants) makes a great difference for the same materials due to different preparation processes. The selection of dispersants is very important for a variety of commercial powders. On the other hand, a suitable graduation of fine and coarse particles is usually an effective method to achieve a highly concentrated suspension (Chong J. S., et al., 1971), but it will lead to an increase of processing cost for the commercial powders.

According to rheological theory, macro-rheological behaviors of a suspension depend on the micro-interaction forces between particles. In an aqueous medium, interaction forces between particles usually include the van der Waals attractive force and the electrostatic repulsive force, which depends on ionic valence number and concentration that affect the thickness of electrical double layer and critical coagulation concentration. The effects of ion valence and the ion concentration on the solid loading in alumina suspension were studied by the authors. For the first time, an ion chromatogram technique was introduced to determine a variety of ions in the ceramic suspension and the ion conductivity constant related to solid loading in the suspension was used to characterize the quality of powders from around the world.

The starting materials used in our study were α-Al_2O_3 powder (produced by a ceramic factory, Wu county, Jiangsu, China) of high-purity (>99.99%), of submicrometer-size, and relatively regular shape, which was prepared by pyrolysis of aluminium ammonium sulphate. The chemical reactions are as follows:

$$Al_2(SO_4)_3 + (NH_4)_2SO_4 + 24H_2O \longrightarrow 2NH_4Al(SO_4)_2 \cdot 12H_2O \tag{4.4}$$

$$2NH_4Al(SO_4)_2 \cdot 12H_2O \xrightarrow{1100\text{–}1300^\circ C} \alpha\text{-}Al_2O_3 + 2NH_3\uparrow + 4SO_3\uparrow + 13H_2O \tag{4.5}$$

The particle size distribution measured by an X-ray centrifugal sedimentation technique (Instrument type: BI-XDC, made by Brookhaven Instrument Co., New

York, USA) was shown in Fig. 4.26.

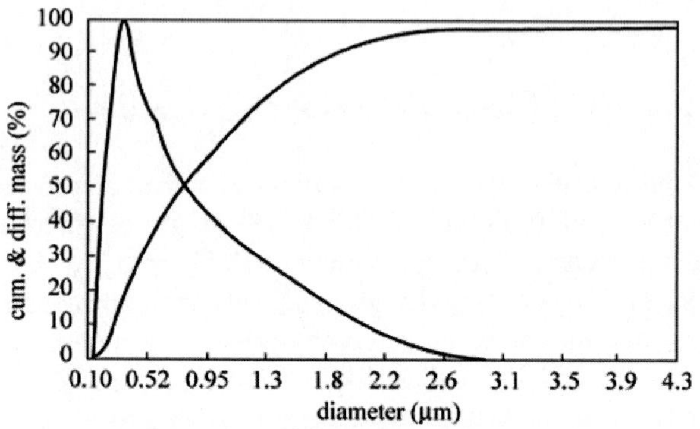

Figure 4.26 The particle size cumulative and differential distribution of α-Al_2O_3

The viscosity of the suspensions was measured by a rotary viscometer with a small sample chamber. All of the samples measured were milled to disperse particles in an alumina jar with alumina balls at pH=4 for 10 h. The shear rate of each suspension used was fixed to 60 s^{-1} for comparison.

The zeta potential was calculated from the measured electrophoretic mobility by using electrophoretic light scattering (ELS) with a 5 mW laser source. The design of a novel electrode assembly to avoid electroosmosis phenomenon due to no need to focus at any stationary plane is shown in Fig. 4.27. The suspension of 0.05 vol% concentration was chosen.

To avoid the effect of particles on ion conductivity measurements, suspensions with different solid loadings were first centrifuged to obtain samples of the supernatant mother liquor. Then, its conductivities were measured. These measured values, which were the ionic conductivities of mother liquor (namely free ion conductivity of suspension), were assumed to be ion conductivities of the suspensions since there was no way to measure the actual suspension.

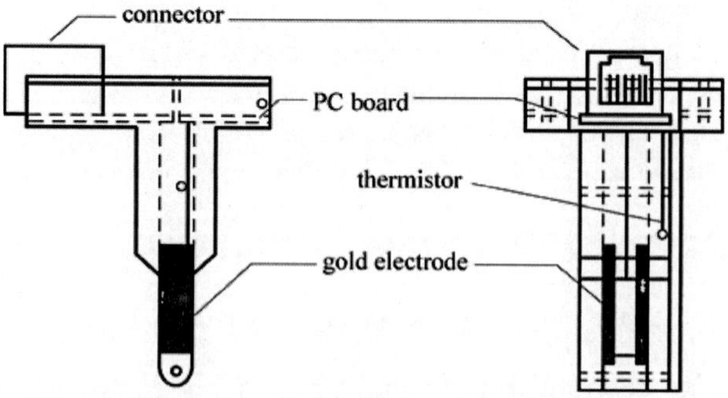

Figure 4.27 The design of zeta plus electrode assembly

An ion chromatogram technique (Type: DX-100, made by Dionex Co., U.S.A.) was used to determine ions in the suspension. An AS4A-SC analytical column and AG4A guard column were used to determine anions with an AMM-2 anion micromembrane suppressor. Typically a 1.8 mmol Na_2CO_3/1.7 mmol $NaHCO_3$ eluent flow rate of 1 mL/min was used. The injection volume used was 25 μL. A CS-12 analytical column and CG-12 guard column were designed for cations with a CMMS-Ⅱ cation micromembrane suppressor. The CS-12 analytical column should have generated a system back pressure of approximately 800 psi when operated with a 20 mmol HCl eluent at a typical flow rate of 0.5 mL/min. 50 mmol KOH was used as a regenerant. The injection volume used was 25 μL. The anions and cations in the 5 vol% alumina suspension were given in Fig. 4.28.

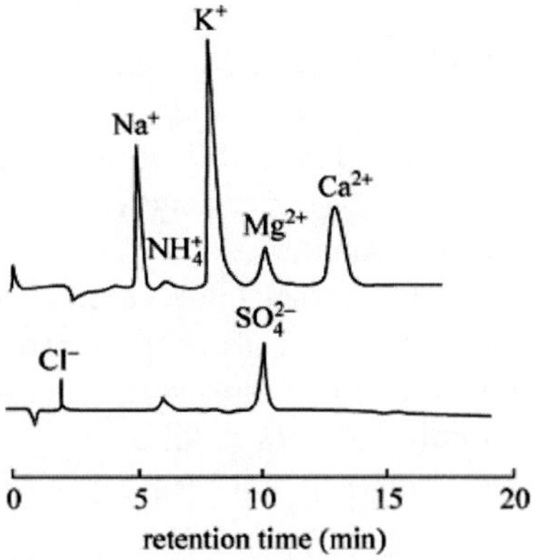

Figure 4.28 The ion chromatogram of anions and cations in the alumina suspension

4.3.2 The Relationship Between Ion Conductivity Constants and Solid Loading

The variation of zeta potential with pH of raw materials and hydrated alumina that had been aged at 160 h in the acid medium was shown in Fig. 4.29. The figure shows that IEPs of alumina increase from 5.6 to 9.2 with hydration time increasing. When hydration time reached 160 h, the IEP of alumina became a constant. This is because an aluminum mono-hydrate layer on alumina particle was formed in an aqueous medium (Carniglia S. C., 1981). This hydration process is due to the thermodynamical instability of alumina in water.

By attrition during the wet-milling process, the aluminum monolayer is continuously removed from the surface and aluminum ions are released from the particles into the water. The unhydrolyzed Al^{3+} is stable below pH = 3 in the acid medium. With increasing pH, the Al^{3+} ion hydrolyzes and condenses, forming a variety of charged mono- and polynuclear complexes (Baes C. F., et al., 1976).

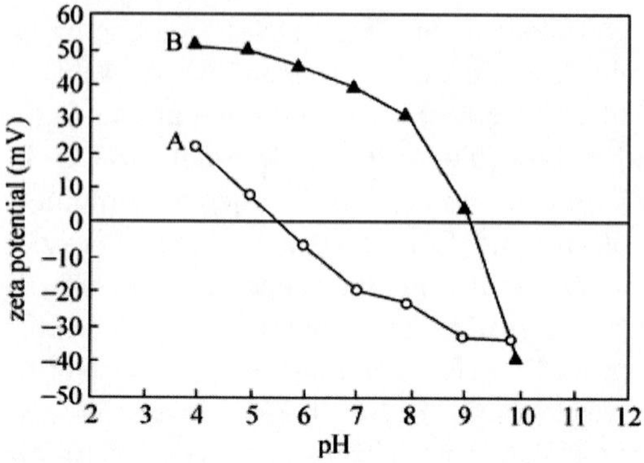

Figure 4.29 The zeta potential vs. the pH of alumina
A: raw materials; B: hydrated alumina (aging 160 h)

Many of these large polynuclear hydroxides and oxo-hydroxides are unstable. This is why the IEP of alumina increases with hydration time. When hydration process reaches to equilibrium, the IEP of the alumina gradually become stable. Therefore, all of the alumina used to prepare highly concentrated suspension was aged at 160 h in the aqueous acid medium.

The ion concentration in the suspension: ceramic suspension consists of suspension medium and ceramic powder. Ion concentration (n_t) in the suspension consists of ion concentrations (n_m) of suspension mediums and the soluble ion concentration (n_p) in the ceramic powders as follows:

$$\begin{aligned} n_t &= n_m + n_p \\ &= (n_{m1} + n_{m2} + \cdots + n_{mi} + \cdots) + (n_{p1} + n_{p2} + \cdots + n_{pi} + \cdots) \end{aligned} \tag{4.6}$$

where n_{mi} is the ion concentration of the ith suspension medium; n_{pi} is the soluble ion concentration of the ith ceramic powder. Two or more than two kinds of ceramic starting materials are usually doped as sintering aids or composition of ceramics in the research and production of ceramic materials. A variety of ceramic powders have different soluble ions. Sometimes suspension media consists of manifold mediums, and so the effects of ion and ion concentration on highly concentrated suspension become very complex. Better understanding of ions and ionic concentrations in every powder and medium is essential in the preparation of a highly concentrated suspension.

The high-valence counter-ions: counter-ions are the reverse charged ions to particle charge. The highly valence counter-ions mean counter-ions whose charge number is $\geqslant 2$. The critical coagulation concentration depends inversely on the sixth power of the charge numbers (ze) according to:

$$n_c = \frac{107\varepsilon^3 (kT)^5 \gamma^4}{A^2 (ze)^6} \tag{4.7}$$

where ε is a dielectric constant, k Botlzmann constant, T the absolute temperature, γ a interfacial energy, A a Hamaker constant, z an electrovalent number, e a electrical charge. If there were the soluble high valence counter-ions in the ceramic powder, they would increase with the increase of solid contents of suspension, the critical coagulation concentration would drop dramatically. In other words, the van der Waals attractive potentials would increase and electrostatic repulsion between particles would reduce, to some extent particles would occur to agglomerate, the viscosity of suspension would raise rapidly. Thus, a highly concentrated suspension could not be prepared. If high valence counter-ions were removed from ceramic suspension, the critical coagulation concentration would not drop rapidly with increase of solid loading. Hence, there was a the possibility to prepare a concentrated suspension. Figure 4.30 shows an anion chromatogram curve of 5 vol% suspension after removing SO_4^{2-} from the alumina powder.

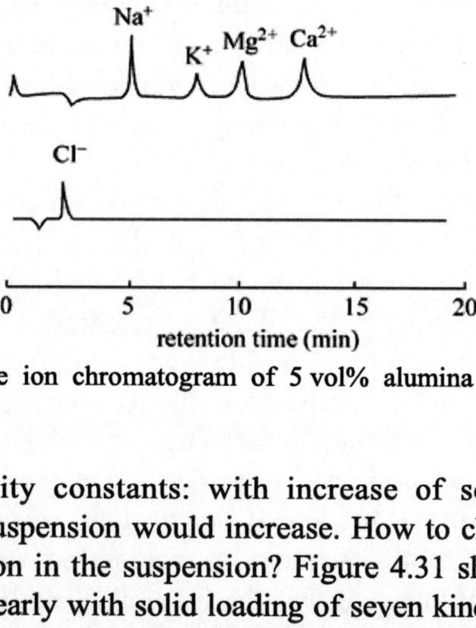

Figure 4.30 The ion chromatogram of 5 vol% alumina suspension after removing SO_4^{2-}

The ion conductivity constants: with increase of solid loading, the ion concentration in the suspension would increase. How to characterize the change of the ion concentration in the suspension? Figure 4.31 shows that the ion conductivities increase linearly with solid loading of seven kinds of alumina powders containing different ion concentrations(A, B, C, D, E, F, G) after removing SO_4^{2-}. For a concentrated suspension, these lines have a greater deviation. This is because the free ions in the suspension are constrained by the overlapping of the electrical double layers around the particles. The inclination slopes of these lines were determined according to the following formula:

$$K = \frac{\Delta C}{\Delta \phi_V} \tag{4.8}$$

where K is a constant, with a unit of mS/cm. It is defined as the ion conductivity constant. For a given ceramic powder, because ion concentration of suspension medium does not change with solid loading, ion conductivity constant never varies even though the suspension medium used is different. For a fixed starting

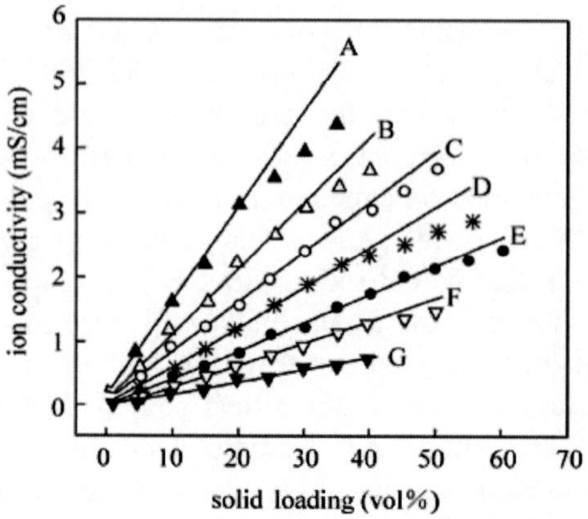

Figure 4.31 The ion conductivity vs. the solid loading

material with the same particle size, distribution, shape, specific surface area, etc., if there were different ion concentrations, ion conductivity constants would be different. Thus, it can be used to characterize the soluble ion concentration in the ceramic powders. It is a very important parameter related to the successful preparation of a concentrated suspension.

The relationship between the ion conductivity constants and the solid loading: after removing high-valence counter-ions, with the increase of the solid loading, ion concentration in the suspension would not immediately cause the reduction of critical coagulation concentration. If ion concentration in the suspension was very low, the thickness of electrical double layer would be very thick resulting in the increase of the viscosity of suspension due to overlapping between electrical double layers. If ion concentration was very large, the thickness of electrical double layer would be very thin so as to result in particle agglomeration due to the increase of attractive potential and reduction of electrostatic repulsion. The above two situations cannot lead to the successful preparation of highly concentrated suspension with low viscosity. Only with a suitable ion concentration to a thin thickness of electrical double layer that would maintain enough repulsion to disperse particles, a highly concentrated suspension with low viscosity could be achieved.

At pH=4, seven alumina powders with different ion conductivity constants as stated above (A, B, C, D, E, F, G) were used to prepare aqueous suspensions with varied solid loading after ball milling for 10 h. The viscosity of each suspensions was measured. The apparent viscosities of suspensions at shear rate 60 s^{-1} (as y-axis) and solid loading (as x-axis) were shown in Fig. 4.32. These ion conductivity constants as a function of solid loading of alumina were discovered after removing SO_4^{2-} from alumina. If viscosity were 1 Pa·s at shear rate 60 s^{-1}, these results were apparently shown in Fig. 4.33. At K=2.7 – 8.1 mS/cm, a highly concentrated suspensions of more than 50 vol% with apparent viscosity of 1 Pa·s was successfully prepared and a maximum solid loading of alumina was achieved at K=4.5 mS/cm.

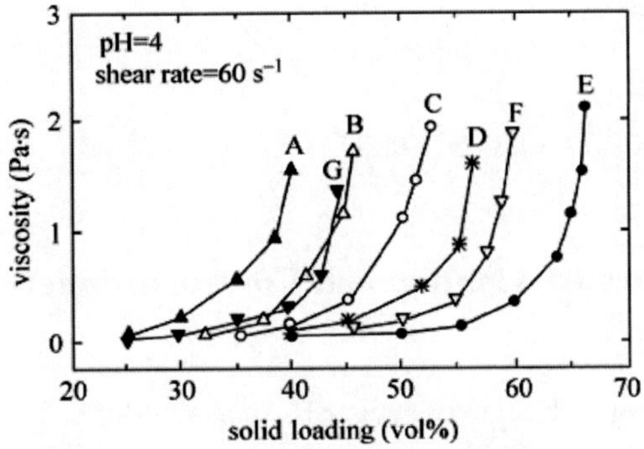

Figure 4.32 The viscosity vs. the solid loading of alumina suspensions with different ion conductivity constants

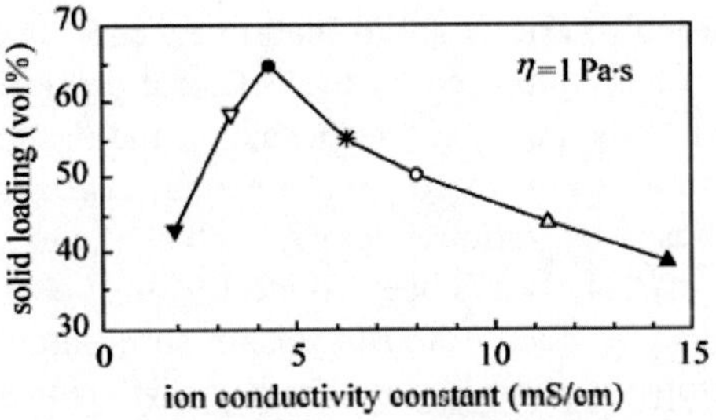

Figure 4.33 The solid loading vs. the ion conductivity constant (K) of alumina suspension

Based on the above results, the following conclusions can be achieved.

The IEPs of alumina are a function of hydration time of particle surface. In an acidic medium, IEPs of alumina changed gradually from 5.6 to 9.2 and then stabilized when hydration time reached to 160 h.

The ion conductivities increase linearly with solid loading of suspensions, its slopes is defined as ion conductivity constant which is an important parameter to characterize the quality of ceramic powder directly related to the preparation of concentrated suspension.

At pH=4, particle surface of alumina is charged positively, a highly concentrated suspension (>50 vol%) cannot be prepared due to the presence of high valence counter-ion of SO_4^{2-}.

When the ion conductivity constant K=4.5 mS/cm, a concentrated suspension of 65 vol% with a viscosity of 1 Pa·s at a shear rate of 60 s^{-1} was successfully prepared.

A novel way to prepare a highly concentrated suspension of ceramics is to control the ion conductivity constants suitably to reduce the thickness of electrical double layer and to maintain repulsion between particles, after removing high valence counter-ions.

4.4 Control of Inner Stress in Green Body

4.4.1 Origin, Transformation and Control of Inner Stress in Green Body

The colloidal forming technique (Gaucker L. J., et al., 1999a, b; Young A. C., et al., 1991; Sasa Novak, et al., 2002), which was characterized by *in-situ* solidification of concentrated ceramic suspension, is a novel process to form ceramic green bodies. The typical route of colloidal process is to prepare a suspension with high solid loading and low viscosity, and then solidify the suspension which is cast in a pore-free mold through various solidification mechanisms such as DCC and gel-casting. The green body prepared by the colloidal process has a homogeneous microstructure similar to the precursor suspensions, and the structure homogeneity and reliability of ceramics are improved. However, asynchronous solidification of the suspensions caused by various factors, such as the gradient of temperature or initiator distribution, has always been found for most colloidal processes. The asynchronous solidification makes the suspension shrink non-uniformly, and then the inner stress develops in ceramic green body, which is liable to be an origin for cracks during subsequent handling. In this section, the occurrence and transformation of inner stress in ceramic green bodies prepared by colloidal forming were expatiated, and some methods to control and eliminate the inner stress in ceramic gel-casting green bodies were proposed.

The alumina powder and organic monomer used in the gel-casting are similar to the previous study. The temperature of the suspension during gelation was collected and recorded by a computer linked with the temperature detector (Yang J., et al., 2002). The pictures of fracture surface of green bodies were taken with digital camera.

Inner stress caused by the concentration gradient of initiator during solidification: The gradient of initiator in the suspension results in an uneven distribution of free radicals, and then the asynchronous solidification occurs. A previous study (Zhao L., et al., 2002) showed that Cu^+ ions could accelerate the decomposition of ammonium persulfate (AP) due to the following reaction:

$$S_2O_8^{2-} + Cu^+ \longrightarrow SO_4^{2-} + SO_4^{x-} + Cu^{2+} \tag{4.9}$$

Thus, a copper beaker was used to give a gradient of free radicals in Al_2O_3

suspension (Fig. 4.34). After the suspension containing initiator was cast into the copper beaker, Cu^+ diffused from the beaker wall to the center of the suspension. As a result, a gradient of Cu^+ ions and then a gradient of the free radicals appeared. To avoid the influence of axial diffusion of Cu^+ ions, a nylon bottom was used. During the polymerization, the temperature changes at three points (point A, B and C) were detected and collected, and the corresponding results were shown in Fig. 4.35. From the temperature rise in Fig. 4.35, it is found that the suspension near the beaker wall solidifies within a short time of about 200 s. This is a result of the effects of high Cu^+ concentration on the polymerization of AM monomers. However, the high Cu^+ concentration near the beaker wall has very little effect on the solidification of the suspension in the center. Therefore, the temperature rise of the center point in the copper beaker has no obvious difference from that

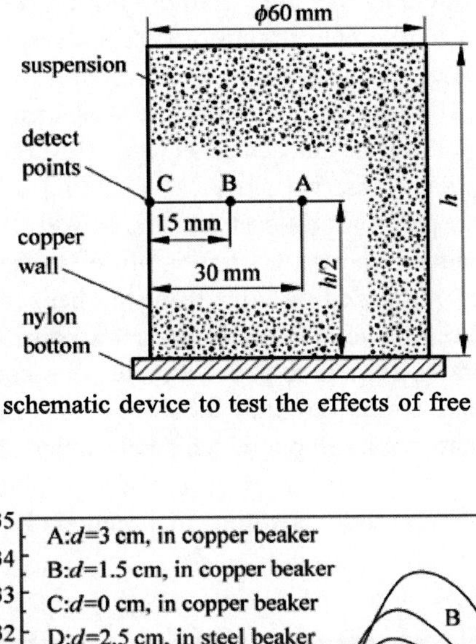

Figure 4.34 The schematic device to test the effects of free radicals gradient on the solidification

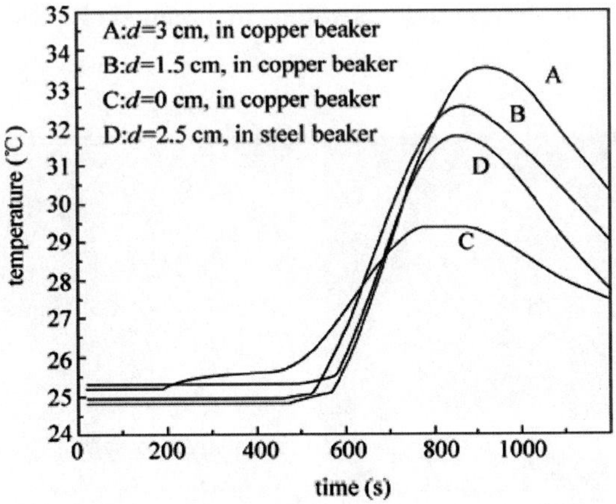

Figure 4.35 The effects of free radicals gradient on the solidification of Al_2O_3 suspension

in stainless steel beaker. Because the diffusion of ions is slow in a dense Al_2O_3 suspension, the enrichment zones of initiator or free radicals, once they appear, will not be easy to eliminate. The suspensions in the enrichment zones solidify in the first instance and become contraction centers, so that the shrinkage of subsequent solidifying zones is constrained. As a result, a complex distribution of inner stress occurs. To analyze the effect of enrichment of initiator in the suspension on the green body, AP and TEMED were added into a 300 mL suspension, and then the suspension was cast without sufficient stirring to prepare a green body. During de-binding, the macro-cracks that appeared in the green body were shown in Fig. 4.36(a). From the orientation of the cracks, it could be determined that there exist at least two origins in the green body. The reason for the random cracks is the non-homogeneous distribution of initiator or free radicals, which yields multiple centers of contraction and causes stress concentration.

Inner stress caused by the temperature gradient during solidification: The effect of temperature gradient on the solidification of Al_2O_3 suspension was studied by using a double-layer beaker (Fig. 4.37). The outer ring of the beaker was filled with 60℃ water, and 100 mL 25℃ suspension was cast into the core of the beaker. Then the temperatures at three points (A, B, C in Fig. 4.37) were detected and collected. The temperature rise at point A, B, C during solidification was shown in Fig. 4.38. After the suspension was cast into the beaker, the 60℃ water in outer ring of the beaker produced an annular temperature gradient in the suspension. Thus, the solidification proceeded radially from the beaker wall to the center of the suspension. Therefore, the solidification of the suspension near the center was constrained by the outer solidified ring. As a result, the center part of the green body endured radial and circumferential tensile stress, and the circumferential tensile stress was dominant for a sample with a small radius of curvature. Therefore, radial cracks in the green body were observed (Fig. 4.36(b)) after the polymer chains, which integrate with the ceramic powders in green body, were broken during de-binding.

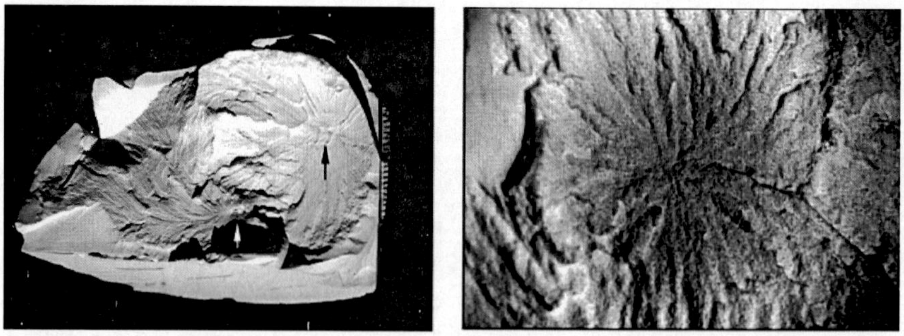

(a) (b)

Figure 4.36 The cracks in the green bodies caused by the gradient of free radicals (a) and temperature (b)

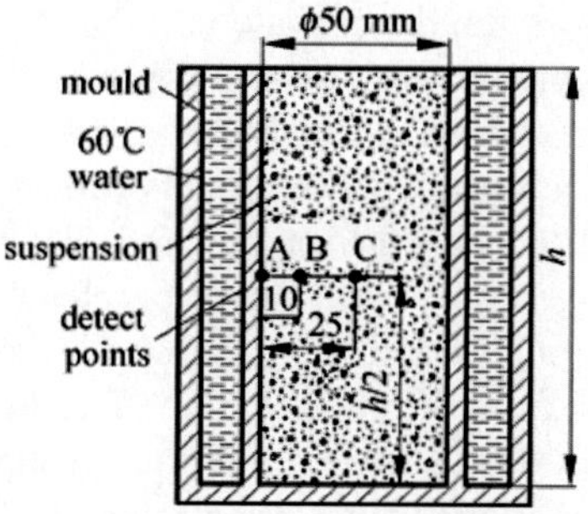

Figure 4.37 The schematic device to test the effects of temperature gradient on the solidification

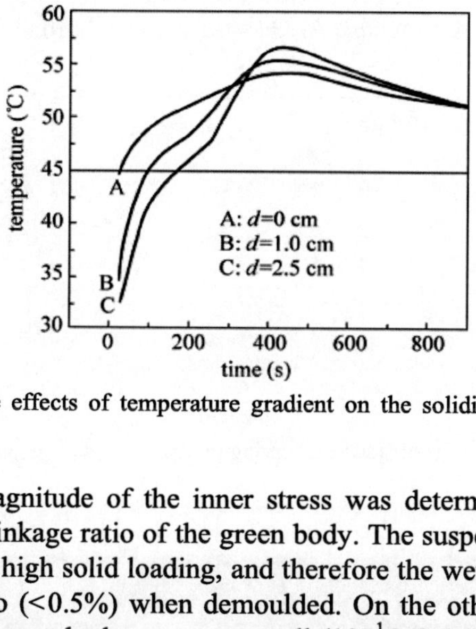

Figure 4.38 The effects of temperature gradient on the solidification of Al_2O_3 suspension

Generally, the magnitude of the inner stress was determined by the elastic modulus and the shrinkage ratio of the green body. The suspensions for colloidal forming usually had high solid loading, and therefore the wet green bodies had a small shrinkage ratio (<0.5%) when demoulded. On the other hand, the elastic modulus of the wet green body was very small (10^{-1} MPa) (Guo Ruisong, et al., 1994), and hence, the inner stress in wet green bodies was also very small. However, the wet green body may shrink more than 3 – 4% and the elastic modulus may rise to the magnitude of 1 GPa during drying. This results in a considerable magnitude of inner stress. Because of the binding of the polymer chains, the inner stress did not cause a macro-cracks during drying, but could result in the damage during de-binding and sintering.

Inner stress caused by moisture gradient in green body during drying: Although the moisture-distribution in wet green body was homogeneous, the moisture gradient existed in the green body during drying, because the moisture drained at different rate at the surface and the center of the wet green body. The moisture gradient caused the green body to shrink unevenly, and then the inner stress occurred. Figure 4.39 shows the drying rate and the shrinkage ratio of an Al_2O_3 disk (ϕ 10.5 cm×3.3 cm) prepared by gel-casting of a 50 vol% Al_2O_3 suspension. As shown in Fig. 4.39, the shrinkage finished about 80% during the constant-rate drying period (Abbas E. N., et al., 1999) which varied from 0 to 2.5 h, but the moisture drained only 7 wt% during the same period. The rearrangement of particles on surface and the shrinkage of the surface are completed at the end of this period. After the constant-rate drying period was completed, the interface between solid and liquid water withdrew from the surface to the interior of the green body. This means that the remaining 18 wt% moisture in the green body would be removed by diffusion through a capillary. The "dried" superficial layer constrained the shrinkage of the interior part of the green body, and therefore the inner stress appeared in the drying green body. During de-binding, the radial and annular cracks appeared in the Al_2O_3 disk (see Fig. 4.40). The influence of temperature gradient could be ignored because Al_2O_3 suspension solidified at room temperature.

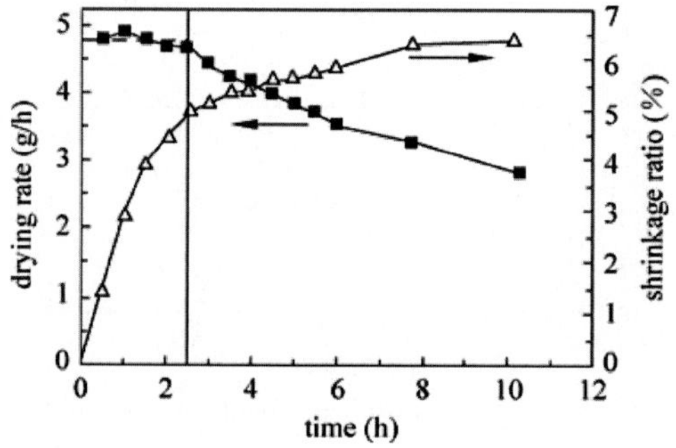

Figure 4.39 The shrinkage ratio and drying rate of Al_2O_3 disk prepared by gel-casting during drying

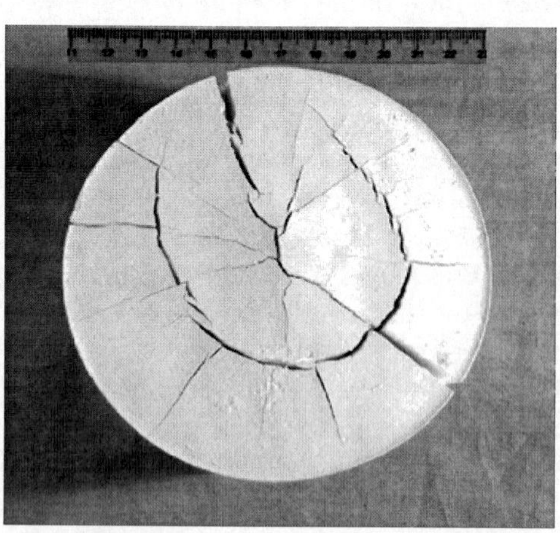

Figure 4.40 The cracks in Al_2O_3 disk prepared by gel-casting after de-binding

The non-uniform shrinkage during drying was likely to be an origin of the inner stress in green body. For the disk sample, radial and circumferential tensile stresses were introduced into the green body at the interface between solid and liquid water during drying. The radial cracks in the green body were caused by circumferential stress, and the annular cracks were caused by radial stress.

The inner stress in the green body prepared by gel-casting mainly resulted from the asynchronous solidification of the precursor suspension, and remained in the green body. The gradients of temperature and initiator concentration in the ceramic suspensions, and the moisture in the green bodies are the major origins of the inner stress. The inner stress could result in macro-cracks during de-bindering. To eliminate or reduce the inner stress in the green bodies, temperature induced solidification should be avoided, and drying the green body at controlled moisture and temperature is required.

4.4.2 The Release and Control of Inner Stresses in Ceramic Green Body

Among the colloidal forming processes, gel-casting, by which complex ceramic parts can be prepared (Omatete O. O., et al., 1991a, b), has been widely studied and applied to various ceramic materials (Nunn S. D., et al., 1996; Waesche B., et al., 1997; Ma L. G., et al., 2002). Homogeneous green bodies with good strength required for machining can be made through this process. The mechanical properties of the materials after sintering are greatly improved (Bossel C., et al., 1995; Liu Xiaolin, et al., 2002; Huang Y., et al., 2000). The main steps of gel-casting are as follows (Young A. C., et al., 1991): First, the ceramic powder (e.g., alumina or silicon nitride) is thoroughly mixed with small quantities of gel initiators, catalysts, monomers, cross-linkers and sintering additives to form a homogeneous suspension with high solid loading and low viscosity. Next, gelation is initiated by casting the suspension into a non-porous mould at elevated temperature. A green body is then formed by *in-situ* polymerization with a 3D gel network holding all the ceramic particles together. The green bodies are then demoulded, dried, debindered and finally sintered at a high temperature.

In recent years, a pressure-assisted gel-casting process using injection molding of colloidal suspension is developed (Huang Y., et al., 2004). The mould temperature and the temperature gradient can be reduced owing to the pressure-induced solidification mechanism. The reliability of the products is improved due to more homogeneous solidification but involves extra cost. We found that the cracking of the green bodies could be reduced by the addition of a plasticizer or a moderator in the suspension. The inner stresses are correlated with the elastic modulus of the green body that can be controlled by the modification of the gel network by using moderators. This mechanism has been explored by using lab simulations. The strategies to minimize the inner stresses will be reported and the implications on the cracking of green body and the efficiency of debindering process will be discussed.

Premix solutions were prepared by dissolving MBAM, AM and HEA in deionized water. Since the moderator (HEA) was in fact a kind of monomer, the total concentration of HEA and AM was kept constant at 14 wt% while the ratio of HEA to AM varied. The concentration of the cross-linker, MBAM was kept at 0.6 wt%. The concentrations of initiator and catalyst are referred to reference (Huang Y., et al., 2004). Six premix solutions with decreasing HEA concentration, M_1, a, b, c, d, e, were prepared, as summarized in Table 4.5. Al_2O_3 powder was dispersed into the premix solutions at a solid loading of 50 vol%. In order to obtain a low viscosity, 1 wt% of ammonium citrate of the ceramic powder was added into the suspensions as dispersant. The suspensions were subjected to ball milling for 24 h and then poured into a mould to form ceramic green bodies. The moulds are listed in Table 4.6. The green bodies were dried for 48 h at the room temperature and then put into an oven at 80℃ until constant weight was obtained. The dried green bodies were debindered for 40 – 50 h by heating slowly from the room temperature to 600℃ and then sintered in a furnace at 1560℃ for 2 h.

Table 4.5 Weight percent of HEA in the monomers in different premix solutions

premix solution	HEA/(HEA+AM)
M_1	1
a	1/3
b	1/5
c	1/9
d	1/11
e	0

Table 4.6 The moulds used to form green bodies

No.	name	dimensions
1	simulation mold	shown in Fig. 4.41
2	a glass beaker of 150 mL	diameter 5.6 cm
3	a plastics beaker of 1000 mL	diameter 11.3 cm
4	an annular stainless steel mould	external: diameter 12.0 cm inner: diameter 5.3 cm
5	a spherical nylon mould	diameter 7.0 cm

To simulate the asynchronous solidification of ceramic suspensions, an organic glass mould was made. The mould was divided into three slots (L, M and R)

using two dummy plates, as shown in Fig. 4.41. The suspensions with different solidification rates were cast into different slots, respectively, and then merged by withdrawing the two dummy plates immediately after casting. In order to facilitate demoulding, a trapezoidal section of the mould was used.

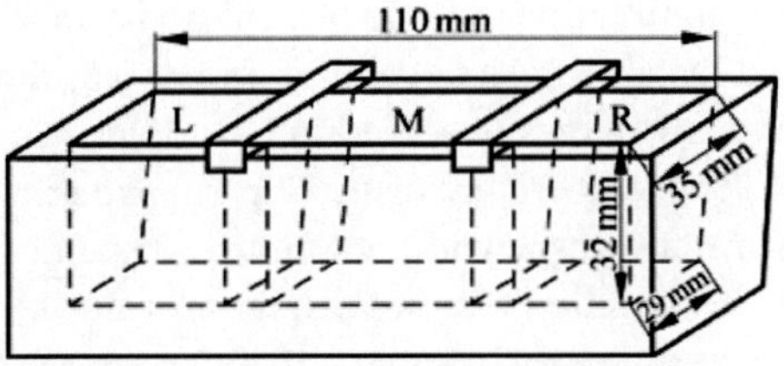

Figure 4.41 The mould to simulate non-uniform solidification of ceramic suspension

Initiation of inner stresses: The solidification rate of a ceramic suspension increases with the temperature and the concentrations of the solidification agents including initiator and catalyst (Huang Y., et al., 2004). Using the equipment shown in Fig. 4.41, a series of experiments were designed, as summarized in Table 4.7, to examine the effect of non-uniform solidification rate on the cracking of ceramic green bodies. In experiment #1, the dose of the catalyst in the middle slot was less than that in the left one and the right one. In experiment #3, the dose of the catalyst decreased in turn from the left to right. In experiment #2, the initial temperature in the middle slot is lower than that in the other two. The dose of initiator is the same in all suspensions as shown in Table 4.7.

Table 4.7 The experiments to simulate non-uniform solidification in ceramic suspension

test number		temperature (℃)	catalyst (μL)	initiator (μL)
#1	L	18	150	150
	M	18	50	150
	R	18	150	150
#2	L	18	150	150
	M	4	150	150
	R	18	150	150
#3	L	18	150	150
	M	18	100	150
	R	18	50	150

Note: Room temperature is 18℃; each slot with 50 mL of 50 vol% Al_2O_3 suspension which was made from premix solution "e".

Under the conditions in experiments #1 and #2, the suspension in the right and the left slots solidified faster than that in the middle slot. In experiment #3, the suspension in the left slot solidified first, then the middle one and finally the right one. No cracks were observed in all three samples after drying and debindering. However, there were cracks in samples #1 and #2 after sintering, as shown in Fig. 4.42. The cracking of samples #1 and #2 is related to the development of inner stresses in the green bodies. Since the suspensions in the left and right slots solidified earlier, the contraction of the green body in the middle slot was restricted. Therefore, a tensile stress developed in vertical direction in the middle layer. The maximum tensile stress occurred in half depth where the cracking was observed in sample #1. The temperature distribution in experiment #2 is complicated by the heat transfer from the suspensions on both sides and from the atmosphere to the suspension in the middle. The solidification rate is not uniform in each layer. This leads to a complex distribution of inner stresses and thus the crack occurred at the one-third regions in the length direction. No cracks were found in sample #3 because the inner stresses are significantly reduced by a gradual transition in the solidification rate from the left to the right. These simulations confirm that the inner stresses can be traced back to the forming stage due to the gradient of initial temperature or the concentrations of solidification agents. In the subsequent drying stage, the inner stresses are inherited and magnified due to further contraction. Control of the inner stresses developed during the forming process is critical to the success of the subsequent drying and debindering processes after gel-casting.

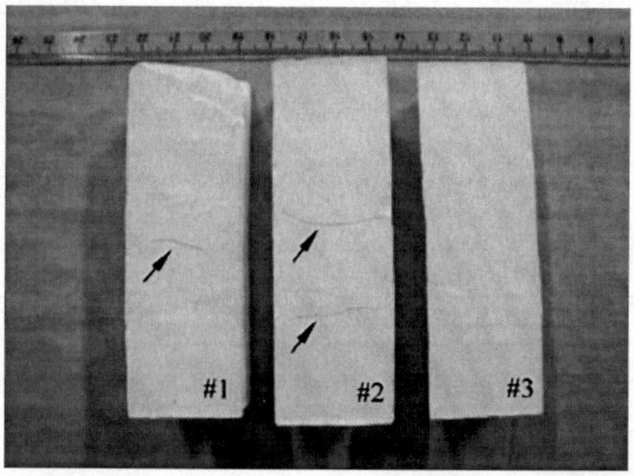

Figure 4.42 The samples to simulate the non-uniform solidification of ceramic suspensions

Control of inner stresses: Figure 4.43 shows the effect of the amount of the moderator on the strength of ceramic green goodies. It is found that the strength of the green bodies decreases almost linearly with increasing concentration of the moderator. The strength of the green bodies (M_1) is less than 2 MPa in which the moderator replaces the original acrylamide (AM) monomer entirely. While the strength of green bodies with a concentration of moderator of 20 wt% to 33 wt% (samples "b" and "a") is about 20 MPa, which is slightly less than that of those

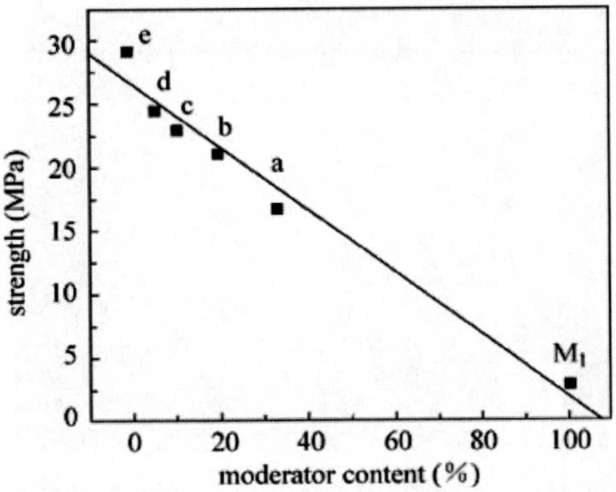

Figure 4.43 The strength of ceramic green bodies vs. the amount of HEA

Table 4.8 The elastic modulus of green bodies vs. the concentration of moderator

test sample	M_1	a	c	e
content of the moderator (%)	100	33.3	11	0
elastic modulus (GPa)	0.84	3.57	5.00	8.63

Note: The concentration of the moderator can be referred to Table 4.5.

produced without any moderator (sample "e"). The elastic modulus of green body vs. different concentration of the moderator was listed in Table 4.8. This indicated that the elastic modulus of the green bodies also decreases with the concentration of moderator. The elastic modulus of green body "e" (prepared from original AM premix solution) is 8.63 GPa while the green body "M_1" has an elastic modulus of 0.84 GPa in which the moderator replaces the original acrylamide monomer entirely. This is because the elastic modulus of a green body is determined by the strength of the gel network structure. The results indicate that the structure of the polymer network can be adjusted by the co-polymerization of the acrymide and moderator (HEA) monomers.

Five Al_2O_3 green bodies a, b, c, d and e were produced by casting 100 mL of the suspensions into a 150 mL beaker respectively. The amount of moderator was varied from 33.3 wt% to 0 in the total weight of monomers as summarized in Table 4.5. The surfaces of these Al_2O_3 green bodies after debindering are shown in Fig. 4.44. The green bodies c, d and e cracked when the amount of the moderator is not sufficient to soften the polymer network. However, the inner stresses were effectively reduced and cracking was suppressed when the concentration of moderator was more than 20 wt%. By choosing a proper amount of the moderator, cracking of green body can be avoided and sufficient strength of the green bodies can be retained.

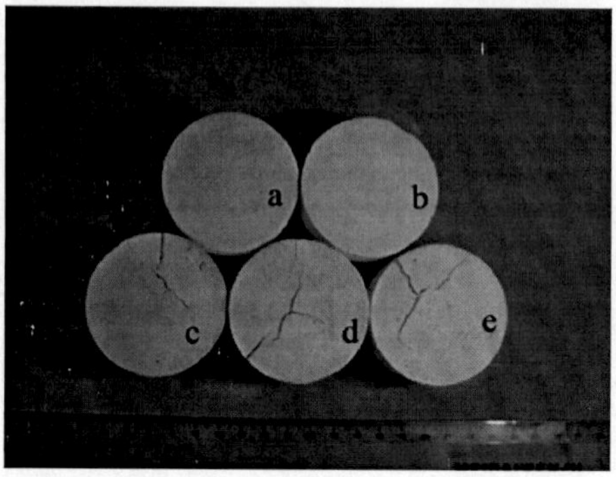

Figure 4.44 The surface patterns of the Al_2O_3 green bodies after debindering

Figure 4.45 indicates the influence of HEA on the viscosity of ceramic suspension with 50 vol% solid loading. It is found that the viscosity increases slightly with the amount of HEA. Nevertheless, the viscosity was less than 1 Pa·s, even if the amount of HEA was one-third of the total weight of HEA and AM (sample "a"). The suspension was still suitable for gel-casting.

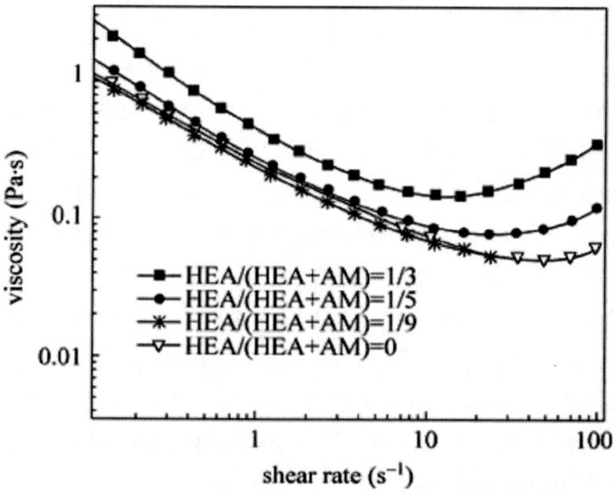

Figure 4.45 The effect of HEA on the viscosity of ceramic suspension

The effect on cracking of large ceramic green bodies: Two annular Al_2O_3 green bodies with diameter 12 cm were obtained by casting suspensions, one containing 33 wt% moderator while the other had no moderator, into hot moulds (the annular type, as shown Table 4.6) respectively. It is shown in Fig. 4.46 that the green body without HEA cracked annularly, while the other containing HEA did not crack at all. Disks were also prepared from these suspensions using a plastic beaker with a diameter 11.3 cm. Similarly, the disk with moderator was free of cracks as shown in Fig. 4.47. The results confirm that relatively large ceramic green bodies can be made successfully by gel-casting with the addition of a proper moderator.

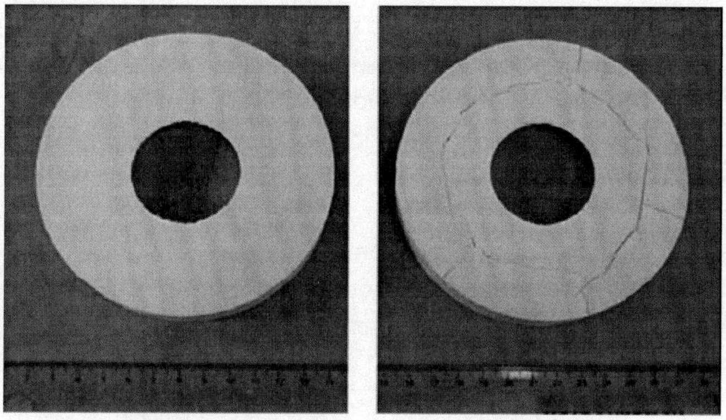

(a) (b)

Figure 4.46 The annular alumina green bodies (a) with 33.3 wt% moderator; (b) no moderator

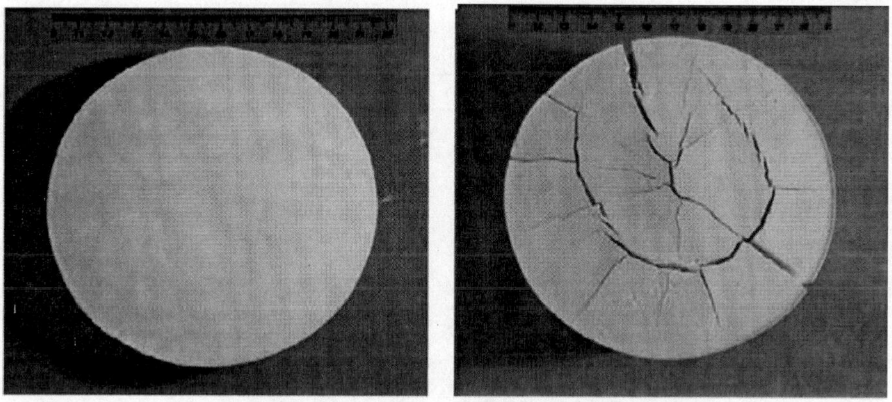

(a) (b)

Figure 4.47 The green bodies of alumina disks (a) with 33.3 wt% moderator; (b) no moderator

The improvement in the efficiency of debindering: As shown in Table 4.9, a series of large-size ceramic balls were prepared by using a spherical nylon mould under various conditions. The green bodies were dried, debindered and sintered. It is found that both samples without HEA (sample #1 and #2) cracked after debindering even if the solidification was induced by microwave heating to minimize the gradient of temperature. It is inferred that the gel network with no moderator is very stiff which results in significant inner stresses in the green body. The inner stresses are significantly reduced with the addition of HEA and this is why sample #3 survived the debindering process. However, the sample #3 cracked during the sintering process. This indicated that a certain level of inner stresses existed in the green body due to the temperature gradient and increased with further shrinkage during sintering. In experiment #4, both strategies were explored: the gradient of temperature was minimized by using microwave and the gel-network was softened by the addition of HEA. Ceramic balls prepared using this process were successfully debindered and sintered without cracking, as shown in Fig. 4.48 (the linear shrinkage rate was 16.4% after sintering and the final diameter was 5.6 cm). Further experiments showed that the debindering time could be reduced to 9 h (from room temperature to 600℃) without cracking, compared to 40 – 50 h by using conventional heating process (Young A. C., et al., 1991). This significantly increases productivity and reduces the production cost.

Table 4.9 The results of large-size ceramic balls

No.	#1	#2	#3	#4
gel system	without HEA	without HEA	adding HEA	adding HEA
casting method	hot mould	microwave	hot mould	microwave
result	cracked after debindering	cracked after debindering	not cracked after debindering, cracked after sintering	good

Figure 4.48 The sintered alumina balls (the linear shrinkage rate was 16.4% after sintering and the final diameter was 5.6 cm)

The initiation and evolution of inner stresses in the green body formed by gel-casting have been discussed above. It can be concluded that the important factors responsible for the inner stresses were the temperature gradient and the distribution of solidification agents. Adding proper amount of HEA into the concentrated suspension improves the flexibility of the polymer network and thus reduces the inner stresses in the green body. By reducing the inner stresses in the green body at the forming stage, the debindering time of large ceramic parts can be significantly shortened.

4.5 The Suppression of Surface-Exfoliation with the Addition of Organic Agents

4.5.1 The Suppression of Surface-Exfoliation by Introducing PAM into Monomer System in Suspension

The gel-casting involves the casting of the slurry that consists of ceramic powder, water and water-soluble organic monomers. After casting, the monomer in the slurry is polymerized to form a gelled network which holds the ceramic powders together in a mold. Compared to other casting methods, gel-casting produces results in much more homogeneous green bodies with little density variation over the parts and higher flexural strength. However, the gel-casting technique has a main problem: surface spallation.

The Oak Ridge National Laboratory (USA) (Young A. C., et al., 1991) first reported flaws and spallation on the surface of dried green bodies prepared by gel-casting in air. Their study concluded that the surface spallation phenomenon is due to the presence of oxygen, which hinders polymerization and gelation of the monomer in the system (Odian G. G., 1991; Landham R. R., et al., 1987). Surface spallation could result in green bodies with a powdery surface that can be of a few millimeters thick. This will reduce the strength and make it difficult to control the dimension of the final products, and thus it must be avoided in production. It was found that gel-casting in a nitrogen atmosphere could suppress the effect of oxygen inhibition and hence prevent the occurrence of surface spallation (Janncy M. A., ct al., 1992; Gilissen R., et al., 2000). In industrial production, however, a technical operation under nitrogen introduces complexity with significant increase in the production cost.

In our research, we had attempted to resolve the surface spallation problem by introducing a water-soluble polymer, polyacrylamide (PAM), into the acrylamide based system. The effects of the polymer on gel-casting of alumina and the surface spallation phenomenon of green bodies were investigated.

Polyacrylamide (Aldrich Chemical Company, USA) used in the study has an average molecular weight of 10000. The determination of PAM amount: a proper amount of PAM may eliminate the surface spallation phenomenon of green bodies cast in air, but decrease the fluidity of the suspension. To determine the optimum amount of PAM, which is enough to eliminate the surface spallation phenomenon of green bodies cast in air, the following experiments were performed. The suspensions with different amounts of PAM were prepared. The amount of PAM and the degree of surface spallation are shown in Table 4.10. The suspension with 2.8 wt% PAM (based on alumina) has a better fluidity and has no surface spallation phenomenon. Thus 2.8 wt% PAM was chosen as additive for the suspensions in following experiments.

Table 4.10 The relationship between sample surface and polymer amount

amount of PAM (wt%)	surface of sample	viscosity (Pa·s)	
		shear rate 1.18 s^{-1}	shear rate 48.1 s^{-1}
0	spallation	0.45	0.10
1.0	spallation	0.65	0.26
2.0	spallation	1.68	2.81
2.5	spallation	4.78	4.24
2.8	without spallation	5.91	5.43
3.0	without spallation	7.30	9.42
4.0	without spallation	33.5	37.8

Note: PAM content is based on alumina.

Process of gel-casting with PAM: The preparation of suspensions is similar to that of reference (Omatte O. O., et al., 1991b) The 2.8 wt% PAM (based on alumina) and 0.25 wt% dispersant (based on alumina) were first completely dissolved in a premix solution which was prepared by dissolving proper amounts of AM and MBAM in deionized water (the mass ratio is 50:1:315), to produce a complex solution. Alumina powder and its sintering aids were then added into the solution. Suspensions with solid loading of 50 vol% were mixed manually first and then milled for 24 h in a nylon resin jar using alumina balls as ball-milling media to break down the agglomerates formed during solvent evaporation and to achieve a good homogeneity. The prepared suspensions were subjected to measurements for the rheological property. 100 μL initiator and 50 μL catalyst were added to the suspension (30 mL) to form a slurry which was used for measurement of gelation characteristics and to fabricate cast samples for microstructure and mechanical characterization.

Binder burnout and subsequent sintering were carried out in stationary air and conducted separately. Binder burnout was done at 700℃ for 2 h, with a heating rate of 2℃/min and a natural cooling. Sintering was carried out at 1550℃ for 2 h with a heating rate of 1℃/min from room temperature to 1250℃ and a rate of 0.5℃/min from 1250℃ to 1550℃ and a cooling rate of 1 to 1000℃/min, followed by natural cooling.

The dispersion of Al_2O_3 powder: Figure 4.49 shows the zeta potential of the Al_2O_3 powder in water with and without PAM as a function of pH. Error bars of ±2 mV indicate the reproducibility of the values. Adding PAM affected the surface charge of the particles in dispersion and their zeta potential. With 2.8 wt% PAM added to the slurry, the zeta potential-pH curve shifted toward a lower pH region, and the IEP of the slurry moved from pH=7.4 to pH=3.6. Increased zeta potential absolute value (higher than –30 mV) were observed at pH>10.5, and the slurry with PAM could be well-dispersed in a much wider pH range compared to that without PAM. The shift of the IEP of the oxide powders toward lower pH is typical of the PAM function and similar behavior has been observed for Al_2O_3 slurries with the addition of NH_4-PMA (Christos A., et al., 2000).

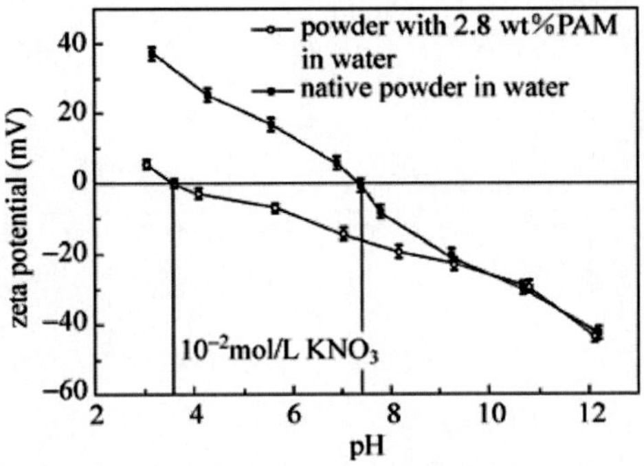

Figure 4.49 The zeta potential of the Al_2O_3 powder in water with and without 2.8 wt% PAM as a function of pH

The rheological properties of Al_2O_3 suspensions with PAM: In order to further examine the colloidal behavior, authors present the curves of the rheological behavior in this section. Figure 4.50(a) shows the variation of steady-state viscosity as a function of shear rate for the suspensions with and without PAM. It can be seen that the suspension without the addition of PAM displays relatively strong shear thinning behavior. The degree of shear thinning decreases significantly after the addition of PAM. Using PAM, highly concentrated suspensions can be prepared but their fluidity is likely to be compromised. Concentrated stable suspensions usually exhibit the shear thinning behavior due to perturbation of the suspension structure by shearing (De Kruif C. G., 1990). Another characteristic shown in Fig. 4.50(a) for the suspension with the addition of PAM is the shear thickening (dilatancy). At high shear rates (>4 s^{-1}, depending on solid loading), the viscosity increases as the shear rate increases. Hoffman (1972) suggested that shear thickening is a consequence of an order-to-disorder transition of the particle microstructure.

Figure 4.50(b) shows the corresponding variation of shear stress as a function of shear rate for the suspensions with and without PAM. The rheological behavior depicted in Fig. 4.50(b) has been analyzed using the Herschel-Bulkley model (Herschel H., et al., 1999):

$$t = t_y + k\gamma^n \tag{4.10}$$

where t_y is the yield stress, t the shear stress, γ the shear rate, n the shear rate exponent and k a constant. The yield stress values presented in Table 4.11 are calculated from this model. The suspension with the addition of PAM exhibits a higher yield stress compared to the suspension without PAM.

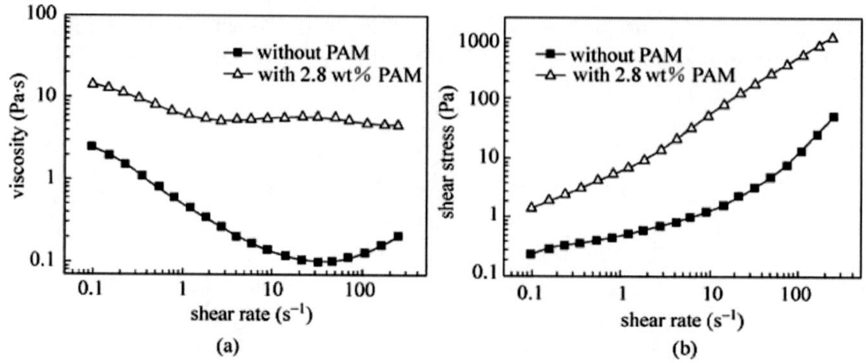

Figure 4.50 The curves of the rheological properties for the suspensions with and without 2.8 wt% PAM

Shih et al. (1999) recently reported that the suspension yield stress (t_y) scales with the solid loading (ϕ) in the following form:

$$t_y \sim \left(1-\frac{3}{2}\alpha\zeta^2\right)\frac{A}{24S_0^{3/2}}\frac{1}{R^{d-3/2}}\phi^m \tag{4.11}$$

where α is a defined constant; ζ the zeta potential; A the Hamaker constant; R the particles radius; S_0 the surface-to-surface distance between two particles; ϕ the solid loading; d the Euclidean dimension; and $m=(d+X)/(d-D)$, D and X the fractal dimension of the clusters and the backbone of the clusters, respectively. It can be concluded that shear stress (t_y) increases with the decrease in the zeta potential at given the solid loading from Eq. (4.11). The result of the present research is fit to this equation.

Table 4.11 The Hershel-Bulkley parameters t_y, and n of alumina suspensions (50 vol%, 0.25 wt% dispersant, 25℃) for 2.8 wt% PAM

sample	monomer content (wt%)	polymer content (wt%)	yield stress t_y (Pa)	constant k	shear rate exponent n	correlation ratio R_{xy}
A	3.2	0	0.3238	0.0863	0.9698	0.9214
B	3.2	2.8	1.3252	4.7269	0.9973	0.9987

Note: Contents of the monomer and polymer are based on alumina.

The gelation characteristics of Al_2O_3 slurries with PAM: as is well-known that the reaction of free radical polymerization is exothermal. When the polymerization starts, the temperature of the system is enhanced. Thus, the gelation characteristics of the suspension may be described by the variation of temperature with time.

Figure 4.51 shows how the temperature of alumina slurry with and without PAM varies with time. It can be seen that the increase in the temperature of slurry without PAM is faster and reaches a higher level at a shorter time, compared to the slurries with PAM. The initial increase in temperature (i.e. the consolidation starting point) occurs at different time between the PAM-containing system (about 800 s) and the one without PAM (about 700 s). Above 700 s, the monomers start to polymerize, resulting in a temperature rise because of exothermal reaction. Besides, the increase in temperature terminates at about 1300 s (i.e. the consolidation completing point) for the PAM-free system, but for the PAM-containing system the consolidation continues at least at 1650 s. A possible reason for the suspension with PAM requiring a longer time for initial rise of the temperature compared to the PAM-free system is that the contact among monomers AM would be interfered and hindered by PAM in the solution. Hence, longer time would be required for AM to become a hydrogel in the presence of PAM. Further work is required to confirm this.

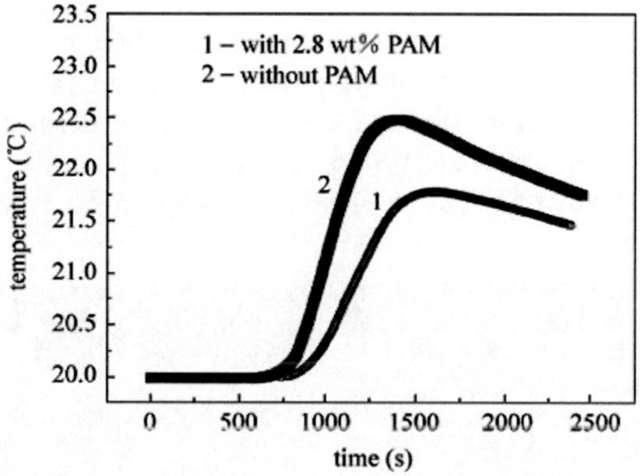

Figure 4.51 The relationship between temperature and time

The surface of gelcast alumina green bodies: Fig. 4.52 shows annular cylinders of alumina obtained by gel-casting. Figure 4.52(a) and (b) are green bodies gelcast from slurry without addition of PAM and slurry with addition of 2.8 wt% PAM, respectively. The surface of the green body gelcast in air from the slurry without addition of the polymer shows surface spallation phenomenon, but a surface without spallation is observed in the green bodies gelcast in air from the slurry containing 2.8 wt% PAM. The mechanism of eliminating surface spallation by PAM is similar to the one in our previous study (Ma J. T., et al., 2003).

The properties of alumina green and sintered bodies: Figure 4.53(b) presents the SEM micrograph of an alumina green body obtained from the system by adding PAM, while Fig. 4.53(a) gives the SEM micrograph of a green body without PAM. It can be seen that a satisfactory microstructure with very few pores has been achieved for both systems and a polymer network can be observed in both green bodies. The variation in density and green strength of dried bodies with and without 2.8 wt% PAM were shown in Table 4.12. The density of the green body

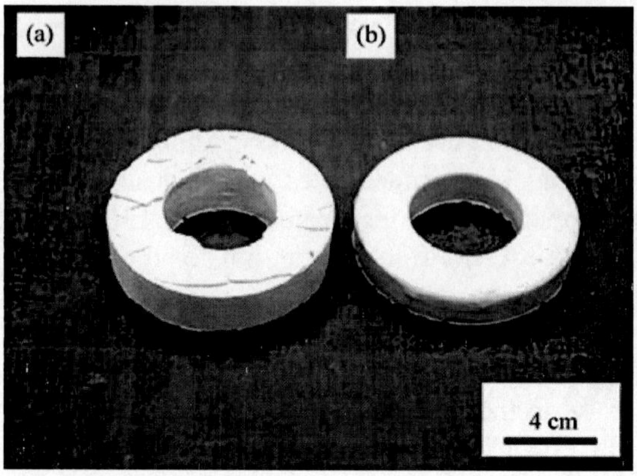

Figure 4.52 The alumina green body surfaces gelcast in air from slurries (a) without PAM; (b) with 2.8 wt% PAM

prepared from the suspension with PAM increased obviously compared to that without PAM. The green strength, on the other hand, showed a significant increase after addition of PAM. After adding 2.8 wt% PAM, the density and strength of the green body were approximately 58.6% of theoretical density (3.85 g/cm^3, the mixed density of alumina and sintering aids) and 43.38 MPa, respectively.

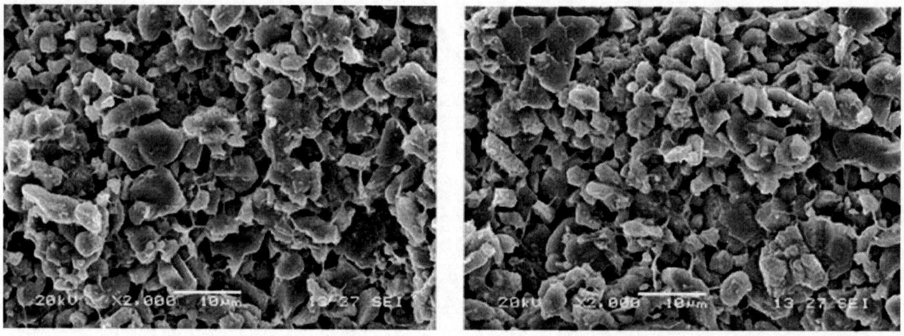

(a) (b)

Figure 4.53 The microstructures of green bodies prepared from suspension (a) without PAM; (b) with 2.8 wt% PAM

Table 4.12 The bulk density and flexural strength of dried green and sintered bodies with and without 2.8 wt% PAM

PAM content	flexural strength of green body (MPa)	density of green body (g/cm^3)	flexural strength of sintered body (MPa)	density of sintered body (g/cm^3)
0	37.11 ± 1.67	2.176	351.85 ± 19.10	3.71
2.8 wt%	43.38 ± 4.58	2.323	347.63 ± 24.72	3.68

The previous studies indicated that the strength of a green body formed by gel-casting was provided by the polymer gel (Asad U. K., et al., 2000; Bauer W., et al., 1999). But for the monomer solution with 2.8 wt% PAM, as the polymerization proceeds, the native PAM with free amino and keto groups are likely to form a more complicated network structure through hydrogen bonding with the keto groups of the PAM polymerized by monomers, resulting in an intimate mix of polymer chains (as shown in Fig. 4.54). The relatively high gel strength obtained for the monomer-polymer system with respect to a pure monomer system is attributed to the enhanced polymer network due to this cross-linking through hydrogen bonding. Hence, the flexural strength of the green body prepared from the suspension with addition of PAM increased significantly compared to those without PAM.

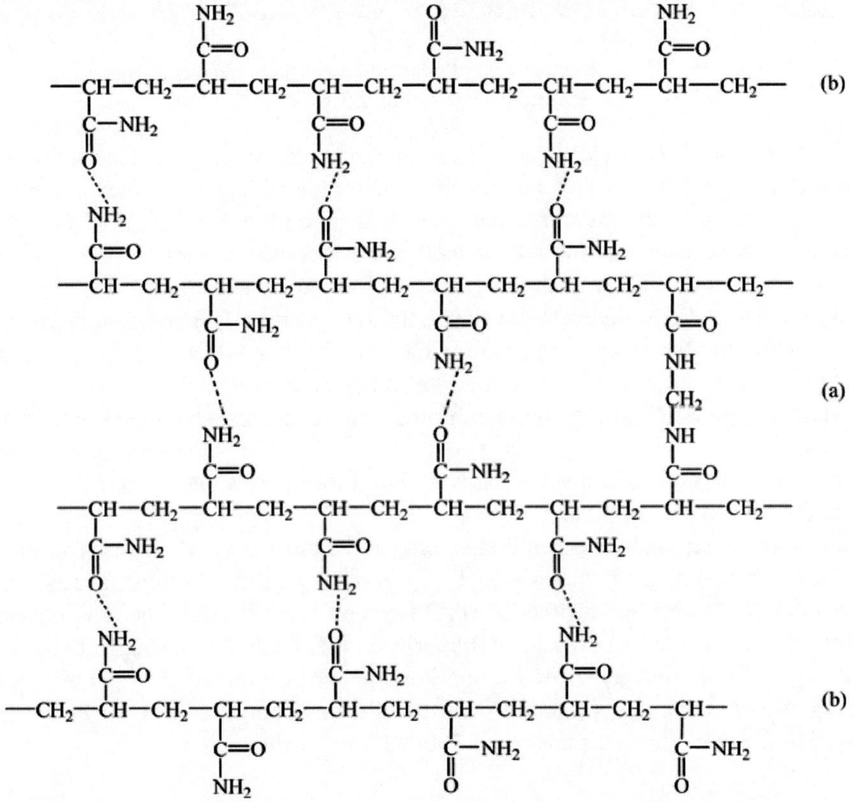

Figure 4.54 (a) The proposed structure of cross-linked PAM; (b) Gel after an addition of native PAM

Figure 4.55 shows the SEM micrograph of alumina ceramics. It should be noted that the ceramics have a high density and uniform microstructure without developing a significant grain growth. The flexural strengths are about 350 MPa for both ceramics obtained from the sole AM system and ones from the mixed PAM and AM system (as shown in Table 4.12).

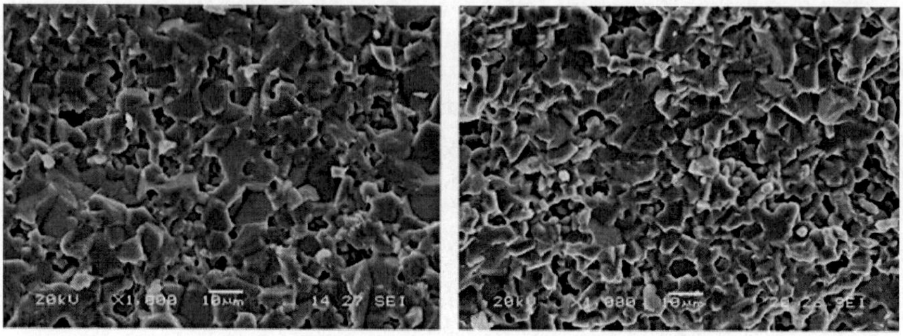

(a) (b)

Figure 4.55 The microstructures of sintered bodies prepared from suspension (a) without PAM; (b) with 2.8 wt% PAM

Based on the above explanation, it can be concluded that the surface spallation phenomenon of green bodies gelcast in air can be eliminated by adding a small amount of PAM to the monomer solution. It is found that by using PAM, highly concentrated suspensions (about 50 vol% solid loading) can be prepared with a small reduction in fluidity of the slurries (0.1 Pa·s, at shear rate 48.1 s^{-1}).

With 2.8 wt% PAM added to the slurry, the zeta potential-pH curve shifts toward a lower pH, and the isoelectric point of the slurry shifts to pH = 3.5 as the added PAM results in a reduction of the positive shear-plane potential.

After adding PAM to the AM-containing slurry, the suspension exhibits shear thinning behavior with yield stress. The increase of the temperature becomes more rapid and reaches a higher level at a lower period for the sole AM system compared to the PAM/AM mixed system.

The flexural strength of green bodies prepared from the PAM/AM mixed slurry is higher than that from the AM only system. The relatively high green body strength 43 MPa obtained for the PAM/AM system is attributed to the cross-linking between the native PAM and polymerized PAM chains through hydrogen bonding and polymer gel. A dense and homogeneous microstructure is observed for the alumina samples prepared from the slurries containing mixed PAM and AM. The flexural strengths are about 350 MPa for both ceramics.

4.5.2 The Suppression of Surface-Exfoliation by Introducing Polyethylene Glycol into Monomer System in Suspension

This study is intended to resolve the problem of oxygen inhibition by chemical methods. In an attempt to develop components of an aqueous system, authors had investigated the gel-casting with adding non-ionized water-soluble polymers and noticed that the surface-exfoliation phenomenon of green bodies gelcast in air was eliminated by the addition of a proper amount of polymers to the monomer solution. A new introduced component is nontoxic, which provides the possibility for further industrialization.

Polyethylene glycol (PEG) has the simplest structure of the non-ionized water-soluble polymers. Its physical conformation with 10000 molecular weight is a white wax-state solid. PEG has good stability, adhesional wetting, low toxicity, and it burns out completely, resulting in wide application in the casting of ceramics.

The purpose of this study is to investigate and discuss the effect of water-soluble polymer PEG with a molecular weight of 10000 on gel-casting of alumina, based on the precondition of that the surface-exfoliation phenomenon of green bodies cast in air is inhibited.

Preparation of suspensions: A proper amount of PEG may eliminate the surface-exfoliation phenomenon of green bodies cast in air, but decreases the fluidity of the suspension. To determine the optimum amount of PEG, which is enough to eliminate the surface-exfoliation phenomenon of green bodies cast in air, the following experiments were performed. The suspensions with various amounts of PEG were gelcast. The amount of PEG and the degree of surface-exfoliation are shown in Table 4.13. The suspension with 1.3 wt% PEG (based on alumina) has a better fluidity and exhibits no surface-exfoliation phenomenon. Thus, 1.3 wt% PEG was chosen as the additive for the suspensions in following experiments.

The PEG (1.3 wt%, based on alumina) and dispersant (0.25 wt%, based on alumina) were first completely dissolved using mechanical stirring for 5 min in a premix solution, which was prepared by dissolving AM and MBAM in demonized water. The premix solution serves as a dispersing medium for the ceramic powder. Next, the alumina powder was added. The suspensions, with solid loading of 50 vol% were mixed manually and then milled for 24 h in a nylon resin jar by using alumina balls as milling media to break down the agglomerates and to achieve a good homogeneity. The prepared suspensions were used to evaluate

Table 4.13 The relationship between the amount of PEG and the degree of surface-exfoliation of green body

amount of PEG (wt%)	surface of sample	viscosity (Pa·s)	
		shear rate 11.8 s^{-1}	shear rate 21.1 s^{-1}
0	exfoliation	0.447	0.105
0.5	exfoliation	0.684	0.874
1.0	exfoliation	0.709	0.981
1.3	no exfoliation	0.730	1.09
1.5	no exfoliation	0.882	1.55
2.0	no exfoliation	1.6	3.57

their rheological properties and viscoelastic properties. The prepared suspension (25 mL) with 100 μL initiator and 50 μL catalyst was used to evaluate the gelling characteristics and to fabricate samples. The surface of the samples were inspected and their flexural strength was measured.

Figure 4.56 presents the measured results of zeta potential with or without PEG in a 0.01 wt% alumina suspension. The zeta potential for alumina powder only ranged from 38 mV at pH = 3 to about –40 mV at pH = 12 with an IEP at pH = 7.3, while that in the case of the suspension containing 1.3 wt% PEG ranged from 24 mV at pH = 3 to –38 mV at pH = 12 with an IEP at pH = 7.1.

It was found that the PEG did not dissociate and had hardly any effect on the alumina surface charges.

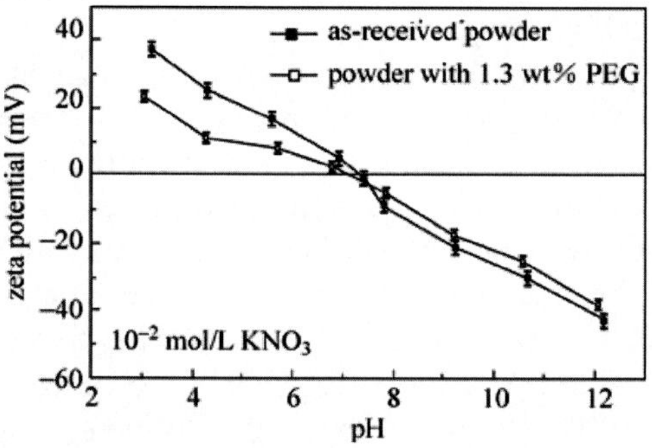

Figure 4.56 The zeta potential of the Al_2O_3 powder in water without and with 1.3 wt% PEG

Figure 4.57 presents that the viscosity of a suspension is strongly dependent on the solid loading, with the viscosity approaching infinity at a maximum solid loading, ϕ_m, where ϕ_m relates to the particle concentration at which the average separation distance between the particles approaches zero and the particles pack together, making flow impossible.

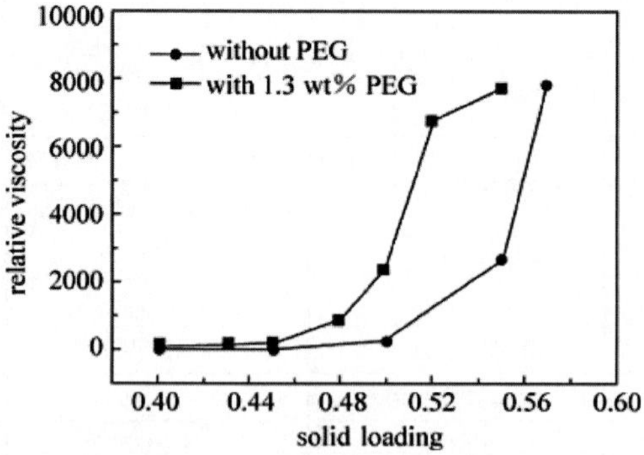

Figure 4.57 The relative viscosity of Al_2O_3 suspension without and with 1.3 wt% PEG vs. the solid loading at a shear rate of 110 s^{-1}

The relative viscosity is the ratio between the viscosity of the suspension and the viscosity of monomer solution (1.45 mPa·s). The experimental points are fitted to a modified Krieger-Dougherty equation (Lyckfeldt O., et al., 1998):

$$\eta_r = (1-\phi/\phi_m)^{-n} \tag{4.12}$$

Where ϕ_m and n are used as fitting parameters. The best fit of the experimental data to the above equation shows that ϕ_m is lower for the suspension with PEG ($\phi_m = 0.52$) compared with the Al_2O_3 suspension ($\phi_m = 0.58$). The difference of value of ϕ_m illustrates that the addition of PEG has an influence on packing behavior of particles, but this influence is not significant.

The steady shear data of the suspensions without and with PEG are shown in Fig. 4.58. Both of these suspensions exhibit the same type of rheological behavior. Shear-thinning occurs at low shear rates and shear-thickening occurs at very high shear rates. However, these suspensions showed significant differences in viscosities at the same solid loading of solids as shown in Fig. 4.58. The suspension without PEG has lower viscosity and the critical point from shear thinning to shear-thickening occurs at higher shear rate ($\gamma = 32\ s^{-1}$), while the suspension with PEG exhibits a higher viscosity and the critical point from shear thinning to shear thickening occurs at lower shear rate ($\gamma = 4\ s^{-1}$).

The stable concentrated suspensions display shear thinning because of a perturbation of the suspension structure by shear (Lyckfeldt O., et al., 1998). At low shear rates, the suspension structure is close to equilibrium because thermal motion dominates over the viscous forces. At higher shear rates, viscous forces affect the suspension structure, and shear thinning occurs. At very high shear rates, the viscosity increases as the shear rate increases.

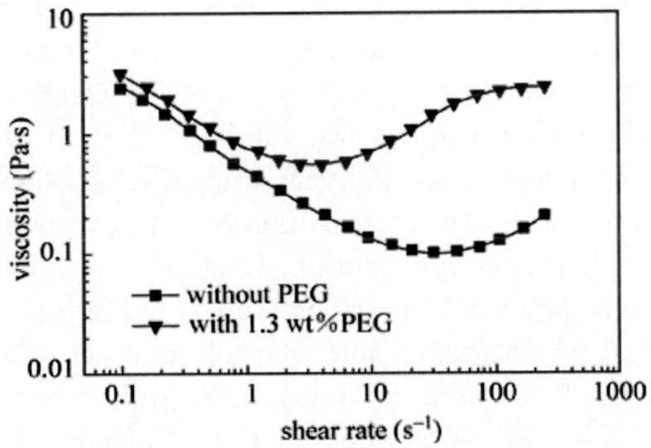

Figure 4.58 Viscosity of the Al_2O_3 suspension without and with 1.3 wt% PEG

With the addition of polymers such as PEG into monomer solution, the system becomes a mixture of organic monomer and polymer. It is important that the PEG is absorbed on the ceramic particle surface as homogeneously as possible. For this reason, ball milling of ceramics with PEG has been commonly used.

Steady shear properties of Al_2O_3 suspensions with PEG at various ball milling periods are shown in Fig. 4.59. It can be seen that the optimum ball milling period is 24 h for the alumina suspension to which PEG has been added. All of the suspensions are characterized by a shear thinning behavior at low shear rates and then a thickening behavior at high shear rates except the suspension with 0 h ball milling period. These results imply that the PEG molecule would be absorbed on the alumina particles, and influence the interaction between particles during ball milling.

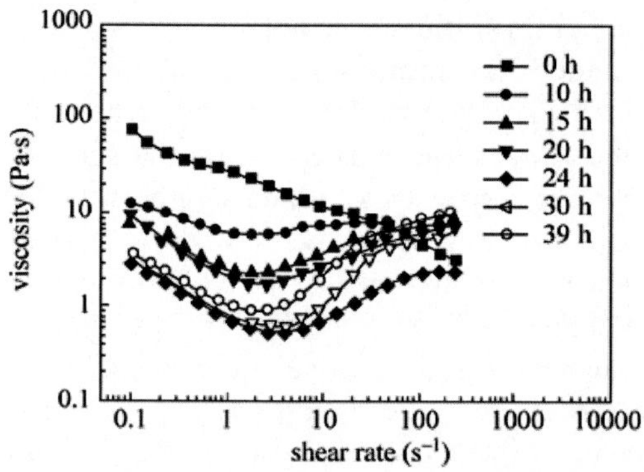

Figure 4.59 Effect of ball milling time on the viscosity of Al_2O_3 suspension without and with 1.3 wt% PEG

Figure 4.60 shows the log-log plot of the variation of G' and G'' with the solid loading of alumina suspension with 1.3 wt% PEG, where G' is the storage modulus and G'' is the loss modulus. G', which is proportional to the elastic modulus of the gel, characterizes the elastic response of the gel. G'', which is proportional to the viscous modulus of the gel, characterizes the viscous response of the gel. The modulus of the suspension shows a gradual change from more viscous ($G'<G''$) to more elastic ($G'>G''$) with increasing solid loading. At relatively low solid loading, there is a weak interaction between the particles or/and PEG, which are only slightly compressed. Under these conditions, the suspension behaves as a viscous fluid with little elastic contribution. When the solid loading is increased, G″ becomes relatively large until a critical solid loading ($\phi_m = 0.47$) is reached where G', G'', above which greater inter particle interaction occurs. Above the critical solid loading, G' continues to increase until it becomes close to G^* (complex modulus, $G^* = G' + \mathrm{i}\, G''$, where $\mathrm{i} = (-1)^{1/2}$). Under these conditions, the system becomes predominantly elastic, with significant interpenetration and compression of the chains occurring. This is also reflected in a rapid increase in dynamic viscosity η' and the whole suspension behaves like a gel.

For gel-casting the polymerization was induced by an initiator and the reaction can be accelerated by a catalyst and temperature. By using an initiator and at the

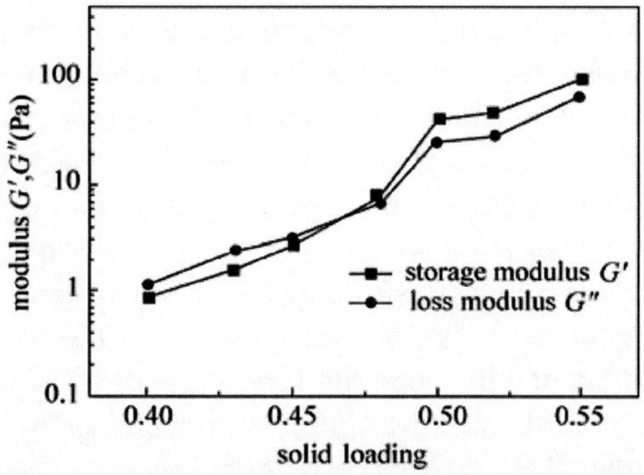

Figure 4.60 The variation of G' and G'' with the solid loading of alumina suspension with 1.3 wt% PEG at constant frequency1.27 Hz

same time fixing the amount of catalyst and the temperature were fixed, the structural change of the suspension vs. time could be observed (Lyckfeldt O., et al., 1998).

Figure 4.61 shows how the storage modulus of alumina/PEG slurry varies with time. It can be seen that the increase of the storage modulus is more rapid and reaches a higher level at the same temperature without PEG, compared to those of slurry with PEG. The storage modulus of the slurry with PEG is lower than that of the slurry without PEG, which indicates the strength of green body with PEG is less than that of the green body without PEG.

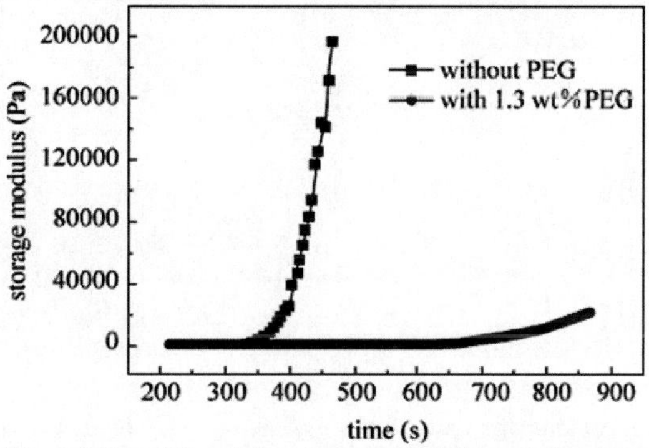

Figure 4.61 The relationship of the storage modulus of alumina suspension vs. solidification time

The reason for the fact that the storage modulus of the suspension with PEG varies earlier and higher than that of the suspension without PEG is probably due to the fact that the contact among monomers AM might be interfered and hindered by PEG in the solution. Hence, longer time would be required for AM to become a hydrogel in the presence of PEG. Above 350 s, the monomer starts to polymerize, resulting in an increased viscosity and an elastic viscoelastic response (see Fig. 4.60).

The physical properties of gelled sample prepared from the alumina suspension with PEG were studied. An acrylamide/PEG-based gel-casting system is a viable alternative to one based on *in-situ* polymerization of monomeric systems. This study highlights the importance of suspension composition (i.e., polymer, monomer, cross-linking agent, and solid loading) and rheological and gelatin behavior. A minimum PEG concentration in solution is required to inhibit the surface-exfoliation phenomenon. By a series of experiments it is concluded that 1.3 wt% PEG is the optimum concentration. Figure 4.62 shows the surface of an annular cylinder of alumina obtained by using the present method. Figure 4.62(a) and (b) shows green bodies gelcast from the slurry with additions of 1.3 wt% PEG and the other slurry without PEG respectively. It can be seen that the surface of the green bodies without PEG has a surface-exfoliation phenomenon. However, the surface-exfoliation phenomenon of green bodies gelcast in air was eliminated by adding 1.3 wt% PEG to the monomer solution. The mechanism that inhibits surface-exfoliation phenomenon of green bodies gelcast in air by the addition of polymer is referred to our further study.

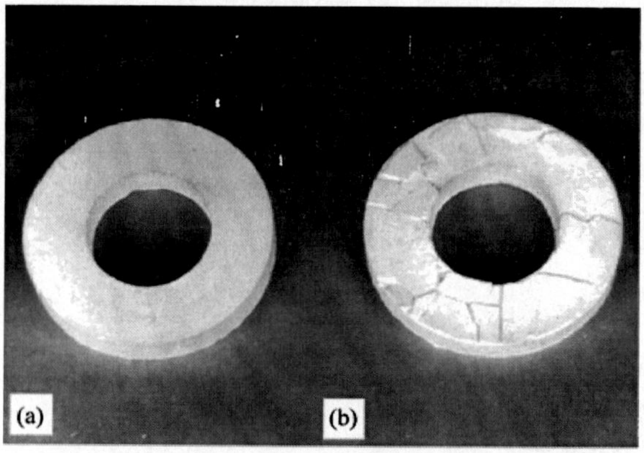

Figure 4.62 The surface of alumina green body prepared from the suspensions (a) with 1.3 wt% PEG; (b) without PEG

An advantage of gel-casting over slip casting is the high strength of the dried green body. In this study, after adding 1.3 wt% PEG the density and strength measured were approximately 53.1% of theoretical density and 9.7 MPa as seen in Fig. 4.63. The addition of 1.3 wt% PEG decreased the green density and the green strength of the dried bodies compared to the strength of green body without PEG (37 MPa). However, the reduced strength is enough to enable inexpensively green body machining into even more complex shapes. The decrease in strength can be explained by the change in microstructure of the green body.

Figure 4.64(a) and (b) show the microstructures obtained by the gel formed by polymerization of monomer without and with PEG mixed with alumina. There is

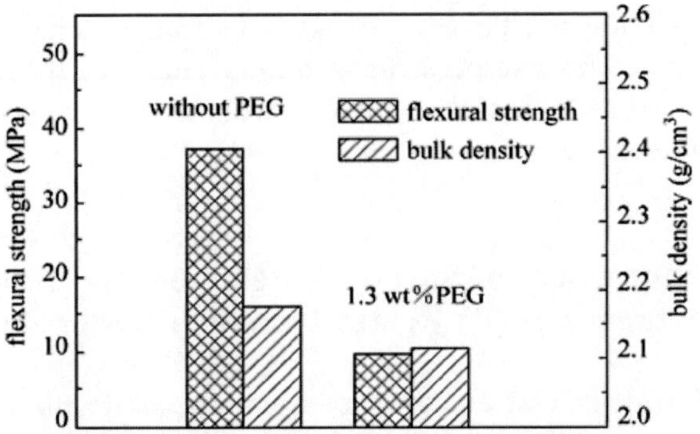

Figure 4.63 The bulk density and flexural strength of dried green bodies without and with 1.3 wt% PEG

(a) (b)

Figure 4.64 The microstructures of green bodies prepared from suspension (a) without PEG; (b) with 1.3 wt% PEG

a good microstructure with very few pores in both green bodies without PEG and also in those with PEG. But it is to be noted that there is an obvious structure of network in green bodies without PEG compared to those with PEG.

Then, a new gel-casting system based on aqueous Al_2O_3-PEG-AM suspensions cross-linked with a coupling agent MBAM was successfully developed. An advantage of this system over monomer-based system is that it eliminates the surface-exfoliation phenomenon of green bodies gelcast in air and makes gel-casting an easy process. Addition of 1.3 wt% PEG has little influence on dispersion of alumina slurry and on its rheological properties. A marked decrease in both, gelation rates and flexural strength of ceramic parts, is observed for green bodies prepared from suspensions with PEG. However, the flexural strength of ceramic parts (9.7 MPa) is high enough for them to be machined inexpensively into more complex shapes.

4.5.3 The Suppression of Surface-Exfoliation by Introducing Polyvinylpyrrolidone (PVP) into Monomer System in Suspension

Surface spallation of gelcast alumina green bodies due to the exposure of the surface to air during gelation was successfully eliminated by adding a proper amount of PVP into the aqueous acrylamide premix solution. The influences of PVP on the colloidal characteristics of alumina in aqueous solutions, the rheological properties of alumina slurries, the gelation process, and the properties and microstructures of the gelcast green bodies and the sintered samples were systematically investigated.

The influence of PVP on colloidal characteristic of alumina powder in aqueous solutions: Zeta potential curves of alumina powder vs. the pH in deionized and premix solutions without and with PVP are shown in Fig. 4.65. It can be seen from the figure that the isoelectric points (IEPs) and shapes of curves of alumina powder in premix solutions with dispersant are obviously different from those of powder in deionized water. The differences can be attributed to the absorption of dispersant and monomers on the powder. However, in the same solution, either deionized water or premix solution, curve shapes of zeta potentials for the powder almost remain the same with or without PVP, with little differences in IEPs. The IEP changed from 8.2 to 7.5 when 15 wt% of PVP was added into the deionized

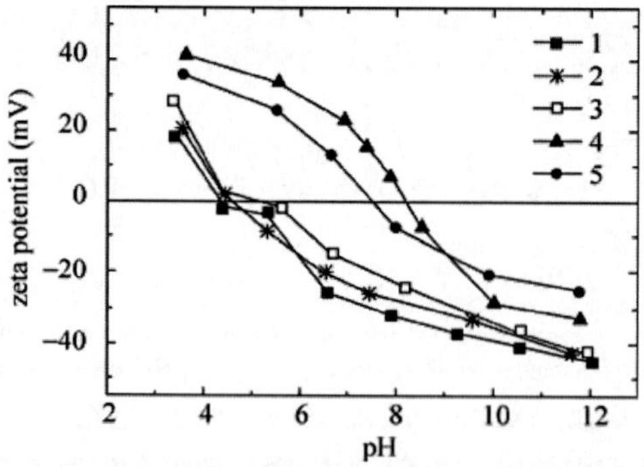

Figure 4.65 The effect of PVP on the zeta potentials of alumina powder in different aqueous solutions: 1–premix with 0 wt% PVP and 1 wt% of dispersant; 2–premix with 15 wt% PVP and 1 wt% of dispersant; 3–premix with 20 wt% PVP and 1 wt% of dispersant; 4–deionized water without PVP; 5–deionized water with 15 wt% PVP. All solutions with 10^{-2} mol/L KNO_3

solution. But, the IEP of the powder did not change so much in premix solution with the content of PVP increasing up to 20 wt%. On the whole, PVP had little influence on the colloidal characteristics of alumina powder in aqueous solutions.

The influence of PVP on rheological properties of alumina slurries and their gelation: The static rheological properties of alumina slurries with solid loading of 50 vol% and various contents of PVP in the premix solution are given in Fig. 4.66. It can be seen from the figure that, just as in other colloidal systems, the viscosity, on the whole, increases with an increase in the content of polymers. The viscosity of slurries with the same solid loading vs. the PVP contents at a shear rate of 48.1 s^{-1} is illustrated in Fig. 4.67. With an increase of PVP content in premix solution from 0 to 20 wt%, the viscosity increases from 0.101 to 1.11 Pa·s, correspondingly, which should be attributed to the adsorption of PVP on the surface of the alumina particles.

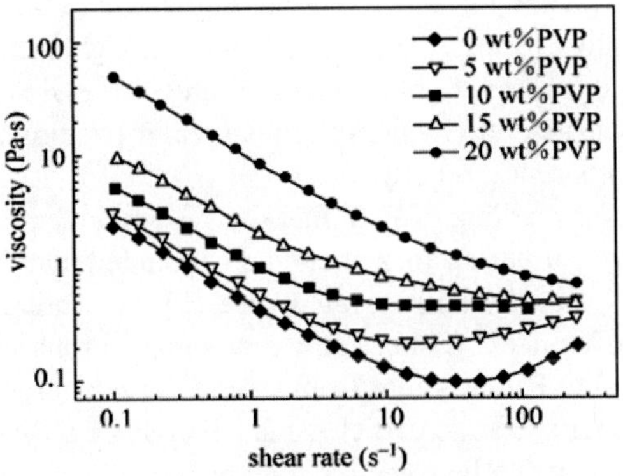

Figure 4.66 The static rheological properties of alumina slurries with solid loading of 50 vol% and various contents of PVP in the premix solution

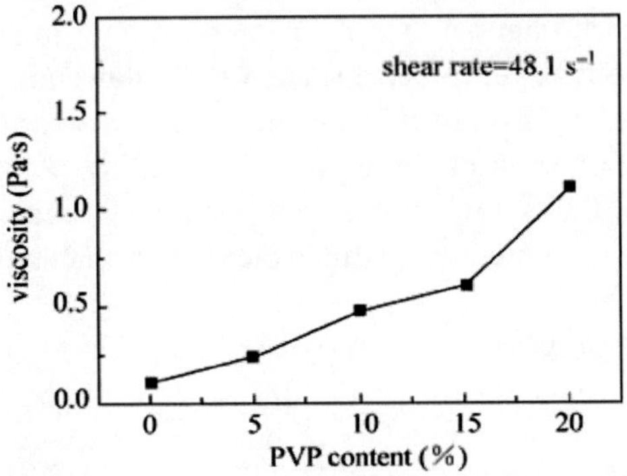

Figure 4.67 The effect of PVP content in premix solution on viscosity of 50 vol% alumina slurries

Interestingly, the slurry without PVP exhibits firstly with shear thinning and then with relatively strong shear thickening. However, with an increase in the content of PVP in the system, the shear thickening behavior becomes less obvious. For the suspension prepared from premix solution with a PVP content of 20 wt%, the slurry becomes nearly only shear thinning and belongs to almost pure pseudoplastic fluid within the measurement range.

Stable concentrated suspensions usually exhibit shear thinning behavior because of a perturbation of the suspension structure by shear (De Kruif C. G., 1990). At low shear rates, the suspension structure is similar to the equilibrium structure at rest, because thermal motion dominates over the viscous forces. At high shear rates, the viscous forces affect the suspension structure more and shear thinning occurs. However, if solid loading in the suspension is too high, the slurry may exhibit with shear thickening when the shear rate is above a critical value. Below the critical shear rate, well-stabilized particles have a tendency to form a layered structure with close packing in the layers but higher separation between layers as a result of laminar flow (Song J. H., et al., 1996). At high shear rates, the layered arrangement becomes disrupted into a random three-dimensional structure (Russel W. B., 1987). This structural change requires greater packing space and presents an increase in the apparent viscosity.

From the effect of the adding of PVP into the system on the rheological behavior of the alumina slurry for gel-casting, it can be concluded that the addition of PVP increases the stability of the suspension, especially at relatively high shear rate. The increase of differences in the static rheological properties of the alumina slurries with increasing PVP content from 0 to 20 wt% of premix solution reflect the interactions between ceramic particles and PVP. Since PVP is the only polymer in the slurries, the interaction between the polymer and alumina particles is stronger than that between the particles and other organic substances with the same content as PVP. Thus, the addition of PVP has an obvious influence on the viscosity of the slurries. As the content of PVP increases, the interactions between particles and PVP become the stronger. When the content of PVP in premix solution is up to 20 wt%, interactions between particles and PVP play such an important role in the stability of the slurry that an increase in shear stress results in destruction of polymer chains. This decreases the number of tangling points and the extended state of polymer, but the layered structure of the slurry cannot be disrupted. Even though the above speculative explanation sounds reasonable, we still need further investigations to prove it.

In the gel-casting process, the rheological properties of concentrated ceramic suspensions play an important role in controlling the shape-forming behavior and the green body properties. The results of our experiments indicate that alumina slurries with high solid loading and good fluidity can be prepared when a proper amount of PVP is added. It can be seen from Fig. 4.66 and Fig. 4.67 that when the PVP content in the premix solution is 15 wt%, the viscosity of the slurry with solid loading of 50 vol% is only 0.6 Pa·s at shear rate 48.1 s^{-1}, which is ideal for gel-casting.

Concentrated colloidal suspensions usually display viscoelastic behavior, which can be characterized by dynamic rheological measurements, namely oscillation measurements. During the measurements, a frequency-dependent shear stress or strain is applied to a suspension, and the shear modulus is obtained. The shear modulus G^* can be expressed in complex form in terms of a storage modulus G' and a loss modulus G'' as

$$G^* = G' + \mathrm{i}\, G'' \tag{4.13}$$

where $G' = G^* \cos\delta$, $G'' = G^* \sin\delta$, $\mathrm{i} = (-1)^{1/2}$, and δ is the loss angle (Collyer A. A., et al., 1988).

The storage modulus G' represents the ratio of the in-phase stress to strain and gives a measure of the elastic properties. The loss modulus G'' represents the ratio of the out-of-phase stress to strain and gives a measure of the viscous properties. G^*, G', and G'' are only a function of the angular frequency.

The variation of storage modulus vs. time of the 50 vol% alumina slurries gelated from premix solutions with varying amounts of PVP is shown in Fig. 4.68. The figure indicates that the starting time of gelation increased with an increase in PVP content. Furthermore, the rate of increase in storage modulus decreases with an increase in PVP content when gelation starts. According to the reactive mechanism of the system described in the literature (Philip Molyneux, 1982) in the presence of ammonium persulfate and MBAM, PVP can be coupled to form hydrogel at 110℃. However, after the addition of TEMED, PVP can yield hydrogel through the coupling reaction even at room temperature. So it is thought that the presence of PVP consumes initiator, catalyst, and coupling agent in the ceramic slurries; as a result, a longer time is required for AM to become a hydrogel and the rate of

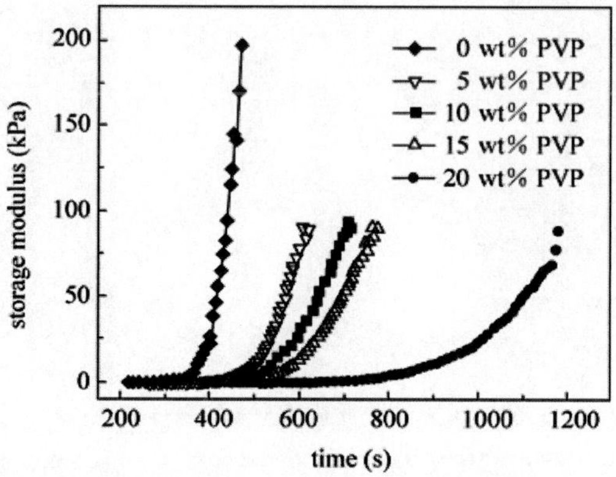

Figure 4.68 The time-dependent storage modulus of alumina slurries prepared from a premix solution with different PVP contents during gelation at 25℃

gelation also decreases. In addition, the contact among AM monomers would be hindered by an increase in the amount of PVP in the solution.

Elimination of surface spallation of green bodies with the addition of PVP and the mechanism: Cylinders of alumina green bodies (Fig. 4.69) with diameters of 2.0 cm and heights of 1.5 cm were prepared from 50 vol% alumina slurries with their premix solutions containing 0, 5, 15, and 20 wt% PVP, respectively. The upper faces of the samples, corresponding to the upper face of the mold, were exposed to air during the gelation of their slurries. The side and bottom faces, were isolated from air during the whole gelation process, therefore, no surface spallation was found on the side and bottom faces for all the samples. But for the green body without PVP, the upper face exhibits severe surface spallation (Fig. 4.69(a)). This indicates that inhibition of gelation due to the presence of oxygen can take place in the area that it is exposed to air during the gelation process. If the area is isolated from air during gelation, there will be no surface spallation in the area. Comparing with the upper faces of the green bodies, it can be found that, with an increase in PVP content, the surface spallation was gradually suppressed. When the PVP content is up to 15 wt% of premix solution, the spallation phenomenon is completely eliminated (see Fig. 4.69(d)). Considering both the rheological properties of the slurries and the elimination of surface spallation of green bodies, it is found that 15 wt% PVP content in the premix solution is optimal for the preparation of alumina slurries with low viscosity and green bodies without surface spallation. Therefore, a premix solution with 15 wt% PVP was selected for the subsequent investigations. To further explore the mechanism of eliminating surface spallation by PVP, infrared spectroscopy was used to analyze the surface and bulk of alumina green bodies prepared from the premix solutions with 15 wt% PVP and without PVP.

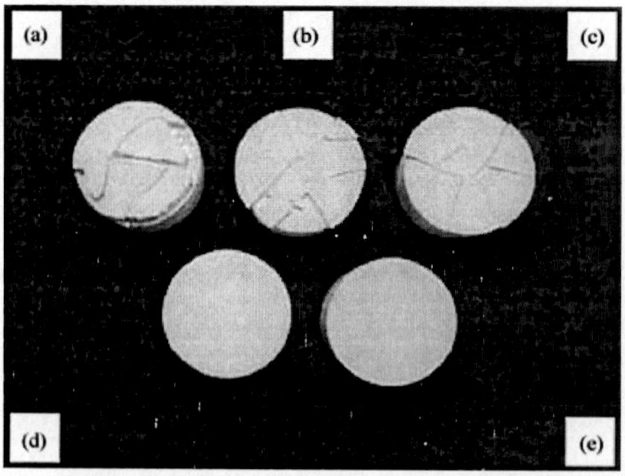

Figure 4.69 The alumina green bodies prepared from premix solution with various PVP contents (the upper face of the samples exposed to air during gelation)
(a) without PVP; (b) 5 wt% PVP; (c) 10 wt% PVP; (d) 15 wt% PVP; (e) 20 wt% PVP

Figure 4.70 shows the results of the investigation by infrared (IR) spectra to explore the mechanism for elimination of surface spallation. The IR curves for AM, PAM, PVP, and alumina powder are curves (a), (b), (c), and (d), respectively, in the figure as a reference. IR spectra for the bulk and the spalled surface layer of the sample without PVP are shown as curves (e) and (f). IR spectra of the bulk and the surface layer of the sample with PVP are also shown as curves (g) and (h).

There are a number of absorption peaks in the spectrum for AM (curve (a)). The absorption peak at 1667 cm^{-1} is the characteristic C=O stretching vibration and the absorption peak at 1610 cm^{-1} is the characteristic C=C stretching vibration for the monomers. The IR spectrum of PAM (curve (b)) differs in many aspects from that of AM (curve (a)). The major absorption peaks around 1667 cm^{-1} become much broader for the PAM sample than for AM and there is a shoulder at about 1610 cm^{-1}.The absorption peak for PAM at 1667 cm^{-1} represents the same characteristic of the C=O stretching vibration as seen in the AM sample.

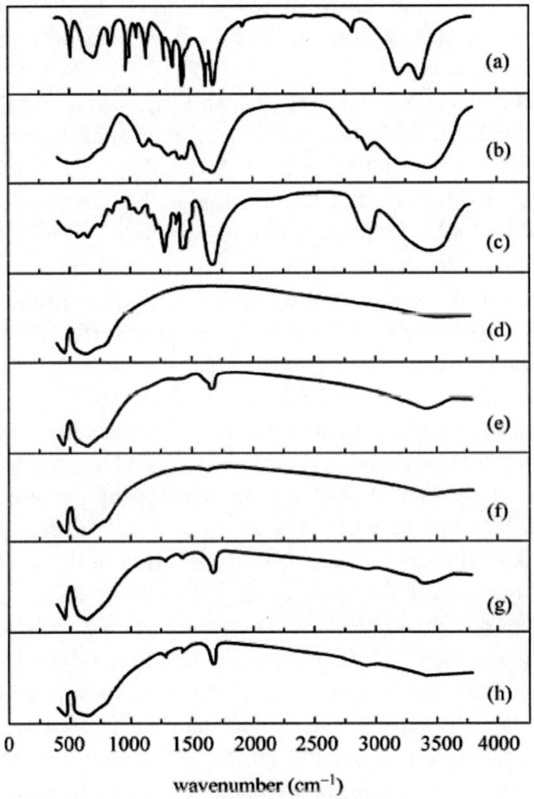

Figure 4.70 The infrared spectra for (a) AM, (b) PAM, (c) PVP, (d) alumina, (e) bulk of green body without PVP, (f) surface layer of green body without PVP, (g) bulk of green body with 15 wt% PVP, and (h) surface layer of green body with 15 wt% PVP

But the shoulder at 1610 cm^{-1} belongs to the C=C stretching vibration of the remaining monomer AM, which has not completely polymerized. In addition, the absorption peak at 3330 cm^{-1} in curves (a) and (b) corresponds to the N—H stretching vibration for both the monomer AM and polymer PAM. PVP is only an acrylic resin without N—H groups. It does not show an IR absorption at 3330 cm^{-1} due to lack of the N—H group, but has a strong characteristic absorption peak of C=O at 1667 cm^{-1} (curve (c)). Furthermore, both IR absorptions at 1450 and 1260 cm^{-1} make PVP distinct from AM and PAM. The IR analyses for AM, PAM, and PVP are in a good agreement with those reported in the literature (Santhiya D., et al., 2000; Lang L., 1974; Kumar V., et al., 1999). The broad band in the vicinity of 600 – 800 cm^{-1} as well as the strong band at 447 cm^{-1} for the α-alumina powder (curve (d)) are due to Al—O stretching vibrations. The absorption band at 1630 cm^{-1} is due to the presence of moisture in the sample.

The shapes of the IR spectra of the samples from the green bodies are similar to that of pure alumina, with weak characteristic peaks of the organic substances if they were in the samples, as shown in curves (e), (f), (g), and (h). The reason is that the dried green bodies contained more than 92 wt% alumina, but less than 8 wt% of organic substances.

Comparing curve (e) with curves (a), (b), and (d), it is found that curve (d) has the characteristic peaks of AM and PAM, i.e., the bulk of the green body without PVP contains at least either PAM or AM. From the high strength of the green bodies will be discussed below, it can be concluded that the main organic substance in the bulk should be PAM. However, by comparison of curve (f) with curves (a), (b), and (d), it can be seen that there are no characteristic peaks of AM and PAM at 1667 cm^{-1} except a minor adsorption at 1630 cm^{-1} due to moisture in curve (f). In other words, there is no AM or PAM in the spalled surface layer of the alumina green bodies without PVP according to the resolution of the IR analyzer. It could be considered that there was no PAM on the surface layer of green bodies exposed to air during its gelation without PVP. A possible explanation for the difference in the content of PAM and AM on the spalled surface layer and in the bulk of the green body is that AM on the surface of the green body did not polymerize when it was exposed to oxygen in air, and hence it may vaporize or migrate to the bulk of the sample during consolidation of the slurry.

For the green body with PVP, it is found that both the bulk (curve (g)) and the surface layer (curve (h)) demonstrate characteristics of PVP with three characteristic IR absorption peaks at 1450, 1260, and 1667 cm^{-1}. Since the intensity of the PVP absorption peak at 1667 cm^{-1} is much stronger than that of PAM and AM and there is heavy overlap between the two IR spectra, it is therefore not certain from the IR study whether PAM is also present on the surface. However, by comparison of the curves (curves (f) and (h)) of the surface layers of green bodies with and without PVP, we could still assume that the interaction of PVP chains with the alumina particles in the surface layer was responsible for the elimination of surface spallation of the green body with PVP.

From the above analysis, it can be concluded that the absence of PAM in the surface layer of a green body with the surface exposed to air during gelation causes surface spallation. But this problem can be eliminated by the addition of a proper amount of PVP to the premix solution. The improvement is mainly due to PVP and its coupled substance, which serve as binders for ceramic particles.

The influence of PVP on properties and microstructure of alumina green and sintered bodies: To investigate the influence of the addition of PVP on the properties and microstructures of gelcast alumina green and sintered bodies, premix solutions with 0, 5, 10, 15, and 20 wt% PVP were used for the whole process of gel-casting of alumina. Actually, the PVP contents based on the whole solids content in the slurries, namely, PVP contents in the dried green bodies, should be 0, 1.3, 2.8, 4.4, and 6.3 wt%, respectively.

The microstructures of alumina green bodies prepared from 50 vol% slurries containing PVP of 0 and 15 wt% after drying for 12 h at 100℃ are shown in Fig. 4.71. It can be found that there is no obvious difference in the microstructures between the green bodies with and without PVP. Both green bodies have relatively high uniformity and density. The density and flexural strength of the green bodies prepared from 50 vol% slurries containing various amounts of PVP after drying for 12 h at 100℃ are shown in Fig. 4.72. The addition of PVP has little effect on the density of the green bodies, but has a significant effect on their flexural strength. The strength of green bodies remains almost the same at 37 MPa when the PVP content in the premix varies from 0 to 10 wt%, and then declines when the PVP content is more than 10 wt%. A previous investigation has shown

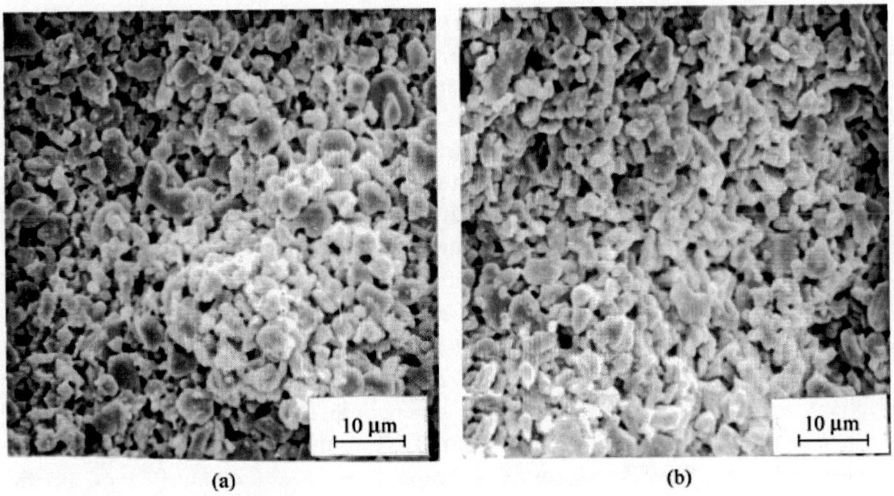

Figure 4.71 The SEM images of fracture surface of alumina green bodies prepared from premix solution with various PVP contents
(a) without PVP; (b) with 15 wt% PVP

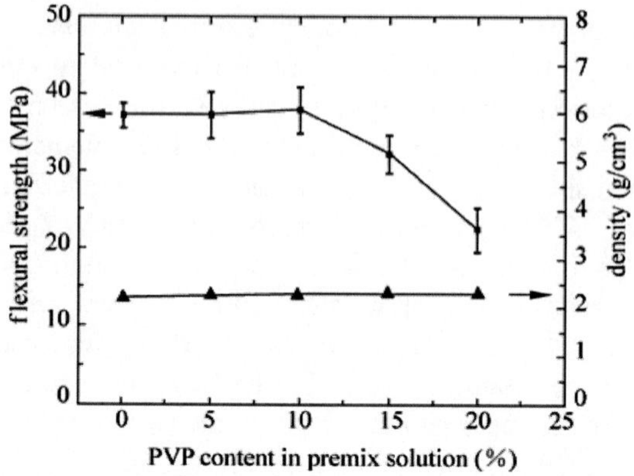

Figure 4.72 The effect of PVP content in premix solution on density and flexural strength of alumina green bodies

that the strength of alumina green bodies is determined by hydrogels that form by polymerization of acrylamide (Khan A. U., et al., 2000; Bauer W., et al., 1999). When the PVP content is relatively low, it does not significantly affect the polymerization of acrylamide. Meanwhile PVP may improve a certain amount of strength by coupling and forming hydrogels. With further increases of PVP in the premix to more than 10 wt%, the coupling of PVP can impair the polymerization of acrylamide. The strength of a hydrogel of PVP is lower than that of acrylamide; hence the strength of alumina green bodies continuously decreases with increasing PVP in the systems. However, the strength of the green bodies is still high enough for any further process even with PVP content up to 20 wt%.

After binder burnout, green bodies with different amounts of PVP were sintered at 1550℃ for 2 h, all of which attained density of about 3.6 g/cm^3, with a flexural strength of about 305 MPa. Microstructures of fractal surfaces of the materials prepared from 50 vol% slurries containing 0 and 15 wt% PVP are shown in Fig. 4.73. There is also no obvious difference between the uniform microstructures. The results above indicate that the addition of PVP, with its content up to 20 wt% in premix solutions and 6.3 wt% in the dried green bodies, has no significant effect on the final properties and microstructure of gelcast alumina ceramics.

It can be concluded that the surface spallation of gelcast alumina green bodies due to the exposure of the surface to air during gelation can be successfully eliminated by adding a proper amount of PVP in acrylamide premix solution during gel-casting. It is inferred that PVP serves as a substitute binder for polyacrylamide for the particles in the green bodies. This is important to their surface layer where the gelation is inhibited due to the presence of oxygen.

The addition of PVP to premix solution has no obvious influence on the colloidal behavior of alumina powder in the solution, but has a slightly negative effect on the rheological properties of the slurry and its gelation process. When the amount of PVP in the premix solution is up to 15 wt%, the fluidity of the

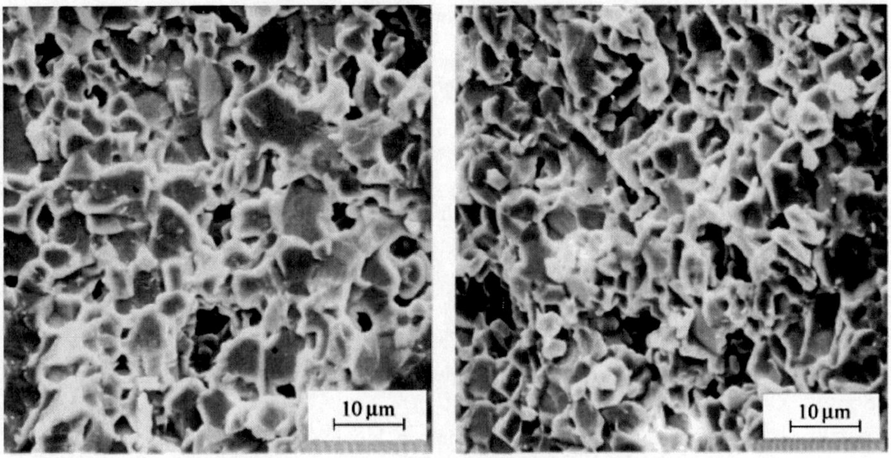

(a) (b)

Figure 4.73 The SEM of fracture surface of alumina sintered bodies prepared from premix solution with various PVP contents
(a) without PVP; (b) with 15 wt% PVP

alumina slurry with 50 vol% solid loading and the time required for its gelation are still suitable for gel-casting. The addition of PVP also has a slightly negative influence on the strength of dried green bodies when the PVP content in premix solution is more than 10 wt%, but the strength is still high enough to allow the green bodies to undergo any further processing.

It is found that for the range of PVP concentrations used in premix solution, the addition of PVP has no obvious influence on the final properties and microstructure of the alumina materials.

Overall, PVP is effective in eliminating the surface spallation of gelcast alumina green bodies with their surfaces exposed to air during gelation. The suggested mechanism implies that it should be possible to extend the use of PVP to other ceramic and monomer systems for gel-casting.

References

Abbas E. N., Harn Y. P., Draskovich B. S., et al., 1999. J. Am. Ceram. Soc., 82: 513.

Asad U. K., Brian J. B., Paul F. L., 2000. Interaction of binders with dispersant stabilized alumina suspensions. Colloid Surface A, 161: 243 – 257.

Baes C. F., Mesmer R. E., 1976. The Hydrolysis of Cations. New York: John Wiley and Sons Inc., 112 – 123.

Balzer B., Hruschka M. K. M., Gauckler L. J., 1999. J. colloid. Interf. Sci., 216: 379.

Bauer W., Ritzhaupt-Kleissl H. J., Hausselt J., 1999. Micropatterning of ceramics by slip pressing. Ceram. Int., 25: 201 – 205.

Bender J., Wagner N. J., 1996. Reversible shear thicking in monodisperse and bidisperse colloidal dispersions. J. Rheol., 40: 899 – 916.

Bergstrom L., 1994. Method for ceramic powders by temperature induced flocculation. US Pat. No. 5340532.

Bergstrom L., 1997. Hamaker constants of inorganic materials. Adv. Colloid. Interf. Sci., 70: 125 – 169.

Bergstrom L., 1998. Shear thinning and shear thickening of concentrated ceramic suspensions. Colloids Surf. A, Physicochem. Eng. Aspects, 133: 151 – 155.

Bossel C., Dutta J., Houiet R., et al., 1995. Processing of nano-scaled silicon powders to prepare slip cast structural ceramics. Mater. Sci. Eng. A, 204 (1 – 2): 107 – 112.

Brinker C. J., Scherer G. W., 1990. Sol-Gel Science. San Diego: Academic Press.

Carniglia S. C., 1981. Thermochemistry of the aluminas and aluminum trihalides. J. Am. Ceram. Soc., 64: C-62-C-63.

Cesarano Ⅲ J., Askay Ⅰ. A., 1988. Processing of highly concentrated aqueous α-alumina suspension stabilized with polyelectrolytes. J. Am. Ceram. Soc., 71(12): 1061 – 1067.

Chen Ruifeng, 2007. Generation and evolution of defects in ceramics by gel-casting:mechanisms and applications (in Chinese, dissertation). Beijing: Tsinghua University, 57 – 58.

Chong J. S., Christiansen E. B., Bear A. D., 1971. Rheology of concentrated suspensions. J. App. Polym. Sci., 15: 2007 – 2021.

Christos A., Athena T., Ioanna L., 2000. Effect of slurry rheological properties on the coating of ceramic honeycombs with yttria-stabilized-zirconia washcoats. J. Am. Ceram. Soc., 83: 1033 – 1038.

Collyer A. A., Clegg D. W., 1988. Rheological Measurement. London: Elsevier Applied Science.

De Kruif C. G., 1990. In: J. P. Hulin, A. M. Cazabat, E. Guyon et al., (eds). Hydrodynamics of Dispersed Media. Amsterdam: Elsevier Applied Science, 79 – 85.

Dewhurst D. N., Cartwright J. A., Longergan L., 1999. Marine & Petroleum Geology, 16: 793.

Gaucker L. J., Graule T. J., Baader F. H., 1999a. Mater. Chem. Phy., 61: 78.

Gaucker L. J., Graule T. J., Baader F. H., et al., 1999b. Key Eng. Mater., 159 – 160: 135.

Gilissen R., Erauw J. P., Smolders A., et al., 2000. Gel-casting—a near net shape technique. Mater. Design, 21: 251 – 257.

Giuliano T., 2003. Gel-casting ceramics: a review. Am. Ceram. Soc. Bull., 82: 43 – 47.

Graule T. J., Baader F. H., Gauckler L. J., 1994. Shaping of ceramic green body compacts direct from suspensions by enzyme catalyzed reactions. cfi/Ber.DKG, 71(6).

Guo Ruisong, Chen Yirui, Yang Zhengfang, et al., 1994. Journal of Tianjin University, 27: 665.

Helbig M. H., Hutter M., Schonholzer U. P., 2000. J. Colloid Interf. Sci., 222: 46.

Herschel H., Buckley R., 1999. Proc. Am. Soc. Test. Mater., 26: 621 – 623.

Hoffman R. L., 1972. Discontinuous and dilatants viscosity behavior in concentrated suspensions I. observation of a flow instability. Trans. Soc. Rheol., 16: 155 – 173.

Huang Y., Ma L. G., Le H. R., 2004. Improving the homogeneity and reliability of ceramic parts with complex shapes by pressure assisted gel-casting. Mater. Lett., 58: 3893 – 3897.

Huang Y., Ma L. G., Tang Q., et al., 2000. Surface oxidation to improve water-based gel-casting of silicon nitride. J. Mater. Sci., 35 (14): 3519 – 3524.

Huang Yong, Zhang Liming, Yang Jinlong, et al., 2007. Research progress of new colloidal forming processes for advanced ceramics. J. Chin. Ceram. Soc., 35(2): 1 – 10.

Jams S. Reed, 1988. Principles of Ceramics Processing. New York: John Wiley & Sons, Inc., 162 – 163.

Janney M. A., Omatete O. O., 1992. Method for molding ceramic powder using a water-based gel-casting process. US Patent, 4145908.

Khan A. U., Briscoe B. J., Luckham P. F., 2000. Interaction of binders with dispersant stabilised alumina suspensions. Colloids Surf., A: Physicochem. Eng. Aspects, 161: 243 – 257.

Kingery W. D., Bowen H. K., Uhlmann D. R., 1976. Introduction to Ceramics (2nd Editon). New York: John Wiley & Sons Inc., 785 – 787.

Kulicke W. M., Nottelmann H., 1989. Structure and swelling of some synthetic, semisynthetic and biopolymer hydrogels. In: J. E. Glass (Ed.), Polymers in Aqueous Media. Washington, DC: American Chemical Society, 15 – 44.

Kumar V., Yang T., Yang Y., 1999. Interpolymer complexation: I, preparation and characterization of a poly(vinyl acetate phthalate)-poly(vinylpyrrolidone) (PVAP-PVP) complex. Int. J. Pharm., 188: 221 – 232.

Landham R. R., Nahass P., Leung D. K., et al., 1987. Potential use of polymerizable solvents and dispersants for tape casting of ceramics. Am. Ceram. Soc. Bull., 66: 1513 – 1516.

Lang F. F., 1989. Powder processing science and technology for increased reliability. J. Am. Ceram. Soc., 72(1): 3 – 15.

Lang L., 1974. Adsorption spectra in the infrared region. Akademiai Kiado, 4.

Leong Y. K., Boger D. V., 1991. Surface chemistry and rheological properties of zirconia suspensions. J. Rheology, 35(10): 149 – 165.

Lim L. C., Wong P. M., Ma Jan, 2000. Microstructural evolution during sintering of near-monosized agglomerate-free submicron alumina powder compacts. Acta Mater., 48: 2263 – 2275.

Liu Den-Mo, 1998. Dispersion characteristic of nano-sized ceramic powder in an aqueous medium. J. Mater. Sci. Lett., 17: 207 – 210.

Liu X. L., Huang Y., Yang J. L., et al., et al., 2002. Mechanical properties of ZTA formed by gel-casting. Key Eng. Mater., 224 (2): 295 – 299.

Liu Xiaolin, Huang Yong, Yang Jinlong, 2002. Effect of rheological properties of the suspension on the mechanical strength Al_2O_3-ZrO_2 composites prepared by gelcasting. Ceram Inter., 28: 159 – 164.

Lyckfeldt O., Brandt J., Lesca S., 1999. Protein forming a novel shaping technique for ceramics. J. Euro. Ceram. Soc., 20: 2551.

Lyckfeldt O., Ferreira J. M. F., 1998. Processing of porous ceramics by starch consolidation. J. Eur. Ceram. Soc., 18: 131 – 140.

Ma J. T., Xie Z. P., Miao H. Z., et al., 2003. Elimination of surface spallation of alumina green bodies prepared by acrylamide-based gel-casting via poly(vinylpyrrolidone). J. Am. Ceram. Soc., 86(2): 266 – 272.

Ma L. G., Yang J. L., Zhao L., et al., 2002. Gel-casting of a rutile mixture applied to extrusion forming. J. Eur. Ceram. Soc., 22 (13): 2291 – 2296.

Morissette S. L., and Lewis J. A., 1999. J. Am. Ceram. Soc., 82: 521.

Nunn S. D., Kirby G. H., 1996. Green machining of gelcast ceramic materials. Ceram. Eng. Sci. Proc., 17 (3): 209 – 213.

Odian G. G., 1991. Principles of Polymerization. New York: John Wiley and Sons Inc., 262 – 266.

Omatete O. O., Bleier A., Westmoreland C. G., 1991a. Gelcast zirconia-alumina composites. Ceram. Eng. Sci. Proc., 12(9 – 10): 2084 – 2094.

Omatete O. O., Janney M. A., Strehlow R. A., 1991b. Gel-casting—a new ceramic forming process. Ceram. Bull., 70 (10): 1641 – 1649.

Omatete O. O., Janney M. A., Nunn S. D., 1997. Gel-casting from laboratory development toward industrial production. J. Eur. Ceram. Soc., 17: 407 – 413.

Omatete O. O., Tieggs T. N., Young A. C., 1991c. Gelcast reaction bonded silicon nitride composites. Ceram. Eng. Sci. Proc., 12 (7 – 8): 1257 – 1264.

Philip Molyneux, 1982. Water-Soluble Synthetic Polymers: Properties and Behavior. Boca Raton: CRC Press, 172 – 175.

Quinn G. D., 1990. Flexure strength of advanced structural ceramics: a round robin. J. Am. Ceram. Soc., 73: 2374 – 2384.

Rahaman M. N., 2003. Ceramic Processing and Sintering. New York: M. Dekker, 567 – 573.

Russel W. B., 1987. The Dynamics of Colloidal Systems. Madison: University of Wisconsin Press, 8 – 10.

Santhiya D., Subramanian S., Natarajan K. A., et al., 2000. Surface chemical studies on alumina suspensions using ammonium polymethacrylate. Colloids Surf. A: Physicochem. Eng. Aspects, 164: 143 – 154.

Sasa Novak, Tomaz Kosmac, Kristoffer Krnel et al., 2002. J. Euro. Ceram. Soc., 22: 289.

Scherer G. W., 1989. J. Non-Cryst. Solids, 108: 18.

Shih W. Y., Shih W. H., Aksay I. A., 1999. J. Am. Ceram. Soc., 82: 616 – 624.

Sigmund W. M., Bell N. S., Bergström L., 2000. Novel powder-processing method for advanced ceramics. J. Am. Ceram. Soc., 83(7): 1557 – 1574.

Si W. J., Graule T. J., Baader F. H., et al., 1999. J. Am. Ceram. Soc., 82: 1129.

Song J. H., Evans J. R. G., 1996. Ultrafine ceramic powder injection molding: the role of dispersants. J. Rheol., 40: 131 – 152.

Takai C., Tsukamoto M., Fuji M., et al., 2006. Control of high solid content yttria slurry with low viscosity for gel-casting. J. Alloys Comp., 408 – 412: 533 – 537.

Tanaka T., 1981. Gels Sci. Am., 244 (1): 124 – 138.

Tari G., Ferreira J. M., Fonseca A. T., et al., 1998. Influence of particle size distribution on colloidal processing of alumina. J. Eur. Ceram. Soc., 18: 249 – 253.

Waesche B., Steinborn G., 1997. Influence of slip viscosity on the mechanical properties of high purity alumina made by gel-casting. Key Eng. Mater., 132 – 136: 374 – 377.

Wang J., Stevens R., 1989. Zirconia-toughened alumina (ZTA) ceramics. J. Mater. Sci., 24: 3421 – 3440.

Yang J., Su L., Ma L., et al., 2002. Colloidal injection molding of ceramics. Key Eng. Mater., 224 – 226: 667.

Yang Yunpeng, Wolfgang M. S., 2003. J. Euro. Ceram. Soc., 23: 253 – 261.

Young A. C., Omatete O. O., Janney M. A., 1991. Gel-casting of alumina. J. Am. Ceram. Soc., 74 (3): 612 – 618.

Zhang Liming, Ma Tian, Yang Jinlong, et al., 2004. Rheological behavior of alumina suspensions. J. Inorg. Mater. (in Chinese), 19: 1145 – 1150.

Zhao L., Yang Jinlong, Ma Liguo, et al., 2002. Influence of minute metal ions on the idle time of acrylamide polymerization in gel-casting of ceramics. Mater. Lett., 56: 990 – 994.

Chapter 5 The Gel-Casting of Non-Oxide Ceramics

Abstract In this chapter, the non-oxide powder surface characteristics before and after treatment and the effect of surface modification on the solid loading of its aqueous suspension were studied systematically. For Si_3N_4 powders, a universal method to prepare highly concentrated suspensions has been established. The results show that the Si_3N_4 ceramics with high-reliability was obtained by surface modification of the powder and different processing strategy.

Keywords non-oxide ceramics; concentrated suspensions; Si_3N_4; SiC; surface modification; coating; oxidizing

The non-oxide ceramics, such as SiC, Si_3N_4 and Si_3N_4-bonded SiC, are important and promising advanced ceramics. The SiC ceramic materials, for example, are indispensable advanced materials in a broad range of applications, including medical biomaterials, high temperature semiconductors, synchrotron optical elements and light weight/high strength structural materials. However, the application of non-oxide ceramics has still been limited mainly owing to their low reliability and high cost.

An approach to overcome these problems is to prepare homogeneous and near-net-shaped green bodies through colloidal forming *in-situ* processes, such as temperature-induced flocculation (Bergström L.,1994), direct coagulation casting (Graule T. J., et al., 1996; Baader F. H., et al., 1996) and gel-casting (Young A. C., et al., 1991; Omatete O. O., et al., 1997). These techniques can be used not only to prepare complicated shapes, but can also be employed to decrease or eliminate defects, such as hard agglomerates, pores and cracks, which would result in relatively low reliability of structural components. Among those forming *in-situ* processes, gel-casting is one of the most potential technique for industrial applications because it is capable of producing uniform green bodies with high strength in order to machine them in green state.

Y.Huang et al., *Novel Colloidal Forming of Ceramics*
© Tsinghua University Press, Beijing and Springer-Verlag Berlin Heidelberg 2010

5.1 The Effects of Powder Surface Modification on Concentrated Suspensions Properties of Si_3N_4

5.1.1 The Contributing Factor and Elimination of Macropores in Si_3N_4 Green Bodies

In the forming process of water-based gel-casting, a complex suspension system is composed of deionized water, monomer, crosslinker, dispersant, catalyst, initiator and Si_3N_4 powder. The gas-discharging reactions, particularly those that occur during the gel-forming may form large pores in the green body if the amount of discharged gas is great enough. Thus not only the homogeneity of the green body is destroyed, but also densification becomes difficult during sintering.

Surface characteristics of Si_3N_4 powders strongly depend on the different production methods (Mezzasalma S., et al., 1996; Sanchez R. T., et al., 1996; Greil P., et al., 1991). In the process of nitridation of silicon powder, some free silicon and other impurities remained in the nitridated product before further purification, and their contents in the final powders were determined by the method of purification.

In this study, several possible gas-discharging reactions were investigated. The contributing factor of large pores in the green body was studied, and the homogeneous green body without large pores was obtained from alkali-cleaned the Si_3N_4 powder.

The two kinds of Si_3N_4 powders were used in this study. The two materials were referred as powder A (Beijing Founder High Tech. Ceram. Corp., Ltd.) and powder B (H. C. Starck, Berlin, FRG). The characteristics of them are shown in Table 5.1.

Table 5.1 The characteristics of powder A and powder B

	d_{50}(μm)	α-phase(%)	O(%)	Fe(ppm)	Al(ppm)	Ca(ppm)	$Si+SiO_2$(%)
powder A	0.98	93.1	2.2	7	1300	190	1.38
powder B	0.45	95	1.58	130	420	45	0.76
powder A after alkali-cleaning	2.55	93.1	0.83				0.09

Note: The O, Fe, Al, Ca, $Si+SiO_2$ representative the content of each component.

Concentrated Si_3N_4 suspension with 45 vol% solid loading was prepared by mixing monomer, crosslinker, dispersant and silicon nitride powders in deionized water completely. After ball-milling, catalyst and initiator were added into the slurry. After being degassed through vibration or/and vacuum-treating, the suspension was cast into non-pore molds. When heated, the slurry solidified *in-situ* because of the occurrence of gelling between monomer and crosslinker. After demolding and drying, the green body was examined by SEM. A large amount of macropores with millimeter level were found in green bodies gelcast with powder A, as shown in Fig. 5.1(a). The suspension was vibrated and vacuumed sufficiently to remove bubbles caused by air, so it can be deduced that the large pores were formed in the gel-casting process.

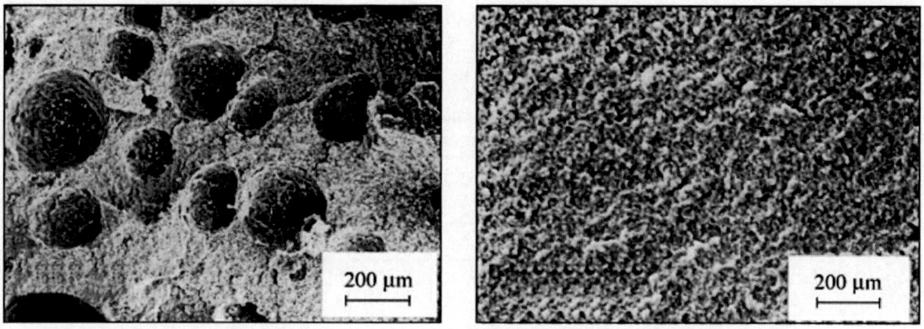

Figure 5.1 The micrographs of green bodies gelcasted with (a) as-received and (b) alkali-cleaned powder A

In the gel-casting system of powder A, there are three possible gas-discharging reactions:

① Hydrolysis of Si_3N_4:

$$Si_3N_4 + 6H_2O \longleftrightarrow 3SiO_2 + 4NH_3\uparrow \tag{5.1}$$

② Oxidation of Si_3N_4 by $(NH_4)_2S_2O_8$:

$$Si_3N_4 + 6S_2O_8^{2-} + 12OH^- \longleftrightarrow 3SiO_2 + 2N_2\uparrow + 12SO_4^{2-} + 6H_2O \tag{5.2}$$

③ Reaction between free silicon and H_2O:

$$Si + 4H_2O \longleftrightarrow H_4SiO_4 + 2H_2\uparrow \tag{5.3}$$

A series of gas-gathering experiments were designed to examine the contributing factor of pores formed during the gel-forming process. The gas-gathering set was shown in Fig. 5.2. First, the container and gas-collecting bag (0.5 L) were vacuumed after valve 1 and 2 were opened. Then valve 2 was closed and the container was put into water-bath at 80℃ for 2 h. If some amount of gas was collected in the gas bag, valve 1 was closed and the gas bag was taken off for composition analysis. Arrangement and results of the experiments were shown in Table 5.2.

It can be seen from Table 5.2 that ammonia was not detected in all experiments. This is owing to slow hydrolysis speed of Si_3N_4 and greater solubility of ammonia

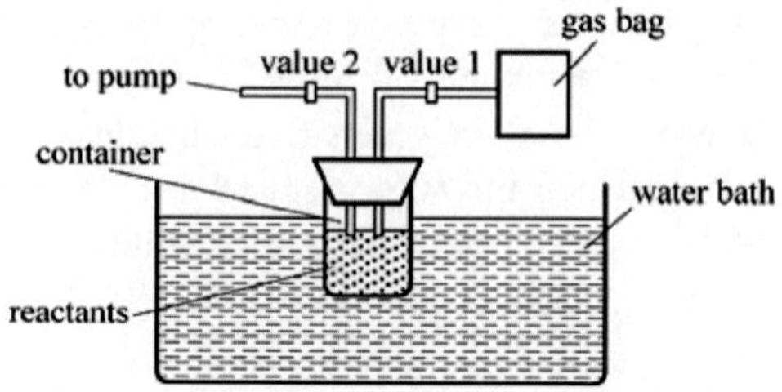

Figure 5.2 The device draft of gas-gathering experiments

Table 5.2 The arrangement and results of gas-gathering experiments

No.	Si_3N_4 powder (g)		reactants				gas composition		
			deionized water (mL)	dispersant (mL)	NaOH (g)	initiator (g)	H_2 (vol %)	N_2 (vol %)	NH_3 (vol%)
1	powder A	30	20	2	0	0.01	97.54	2.46	0
2	powder A after alkali-cleaning	30	20	2	0	0.01			
3	silicon powder	4	40	0	0	0	84.12	15.88	0
4	silicon powder	2	40	4	0	0	90.56	9.44	0
5	powder B	30	20	2	0	0.01			
6	powder B	10	40	0	10	10	11.52	88.48	0

in water. The solubility of ammonia is 89.9 mL/100 g H_2O at 0℃ and 7.4 mL/100 g H_2O at 96.6℃. For practical gel-casting process, however, Si_3N_4 suspension is ball-milled for a period, which would increase the degree of hydrolysis. Some bubbles were found in the slurry and ammonia smell was strong enough when ball-milling time reached 48 h. This indicated that solubility of ammonia was saturated in the suspension. In this case, if the slurry was cast directly without vacuum-degassing, large pores might be found in green body because of supersaturation of ammonia in the suspension at the temperature of solidification (70 – 80℃). During the vacuum-treatment process of the slurry, pH of atmosphere (measured by wet pH test paper) decreased from 11 to 9, which was close to that of slurry without ball-milling. This suggested that the ammonia yielded during ball-milling process was removed by vacuum-treatment. So it can be concluded that the ammonia formed by hydrolysis of Si_3N_4 cannot cause large amount of pores in the green body used for the present study.

It was confirmed by experiment 6 (see Table 5.2) that reaction (5.2) occurred and discharged large amount of N_2 when the amount of $(NH_4)_2S_2O_8$ was great and pH was higher (pH = 13.9). The small amount of H_2 detected at the same time will be discussed later. In the practical suspension, the initiator solution (water solution of $(NH_4)_2S_2O_8$) was diluted and the small amount was added into the slurry. For 100 mL of slurry with solid loading of 45 vol% Si_3N_4, the amount of N_2 yielded was less than 0.1 mL as $(NH_4)_2S_2O_8$ was exhausted completely. So this reaction cannot cause the large amount of pores in green body. This was confirmed by experiment 5, in which the amount of dispersant and initiator was about the same as that in practical suspension, and no gas was collected in gas bag. For other experiments, the little amount of N_2 detected may be owing to incompletely vacuum-treatment and/or leakage of air into gas bag.

To examine reaction (5.3), silicon powder was used. It can be seen from experiment 3 and 4 that Si can react with water and H_2 is obtained at 80℃. The reaction becomes stronger and more gas is collected during same period when the basic dispersant is added. The reason for the absence of gas production in experiment 5 is because of the fact that powder B has little element Si as we know from its purification process introduced above. When compared with experiment 1, the Si_3N_4 powder used in experiment 2 was powder A treated that was alkali-cleaned to eliminate free silicon, and no gas was found. The reason of a little amount of H_2 detected in experiment 6 was that little amount of element Si inside powder B was exposed to the outside layer of Si_3N_4 consumed by $(NH_4)_2S_2O_8$, or there still a little amount of free silicon on the powder surface.

According to the results and analysis presented in Table 5.2, it can be concluded that the contributing factor for the large amount of pores in the green body is H_2 produced by the reaction of free Si with basic media.

As discussed earlier, large pores in the green body prepared by powder A were formed by H_2 yielded during the reaction between free Si and water. To avoid these pores in green body, it is natural to eliminate free Si by treating the as-received powder. Alkali-cleaning was used in this study. The process is employed as follows: Powder A was suspended in NaOH solution at a concentration of 5 wt% and stirred for 6 h in a water-bath at 80℃, then the centrifugalized powder was washed by using deionized water until $pH<10$. The modified powder A was used to gas collecting experiment and no gas discharged (experiment 2 in Table 5.2). Because the content of free Si was only 0.09% (see Table 5.1).

The particle size distributions of as-received and alkali-cleaned powder A are shown in Fig. 5.3. It can be seen that the range of alkali-cleaned particle size becomes wider. Because the alkali-cleaning process caused some unreversible agglomeration of powder A and that some small particles were carried off by the deionized water. From zeta potential curves presented in Fig. 5.4, the IEP of alkali-cleaned powder shifts to right slightly and the zeta potentials of the two powders are close to each other when pH is beyond 10. The slight shift of IEP may be owing to the reduction of silica-like surface layer of powder A (GREIL P., 1989). As shown in Fig. 5.5, apparent viscosity of alkali cleaned powder A suspension was lower than that of the suspension of original powder A, and both of them were observed to behave as pseudo-plastic fluids. The lower viscosity of alkali-cleaned powder A suspension is considered to be advantageous for gel-casting. The fracture morphology of green body gelcasted with alkali-cleaned powder A is shown in Fig. 5.1. It can be seen that no large pores existed and the green body was homogeneous.

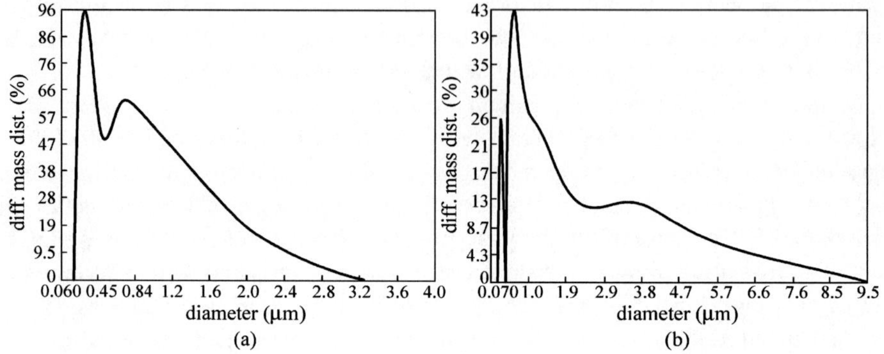

Figure 5.3 The particle size distributions of (a) as-received and (b) alkali-cleaned powder A

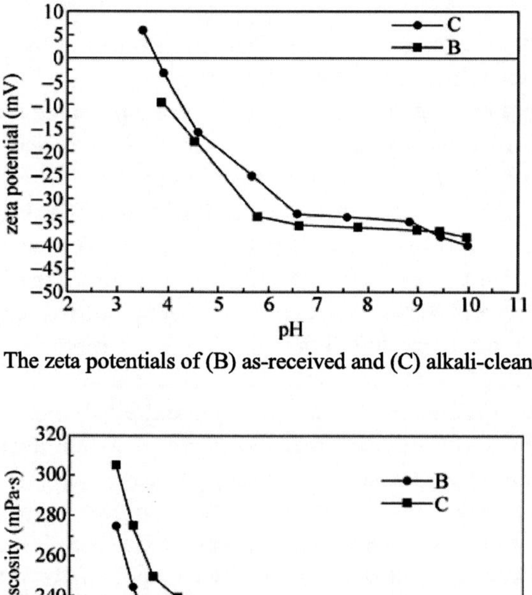

Figure 5.4 The zeta potentials of (B) as-received and (C) alkali-cleaned powder A

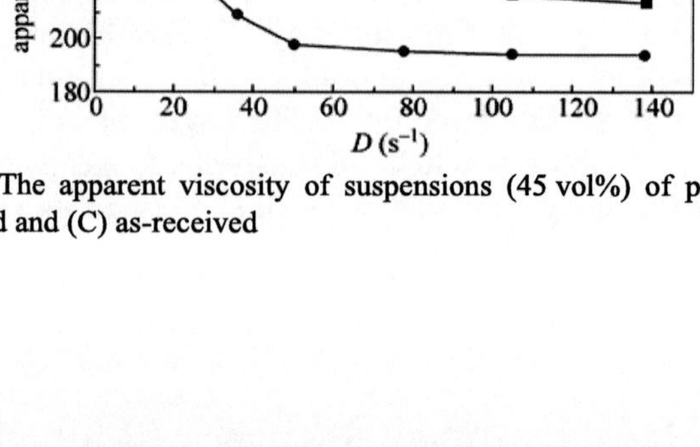

Figure 5.5 The apparent viscosity of suspensions (45 vol%) of powder A (B) alkali-cleaned and (C) as-received

It can be concluded from above discussion that ① there are three possible gas-discharging reactions in the Si_3N_4 suspension of water-based gel-casting: hydrolysis of Si_3N_4 discharging ammonia, oxidation of Si_3N_4 by $(NH_4)_2S_2O_8$ discharging nitrogen, and basic reaction of free silicon with water discharging hydrogen. ② The contributing factor of large amount of pores in green body of powder A is H_2 discharged during the reaction between free Si and water. ③ Homogeneous green body without large pores is obtained by the use of alkali-cleaned powder A in which the free Si has been eliminated.

5.1.2 The Effect of Foreign Ions on Concentrated Suspension of Si_3N_4

In this study, the influence of the concentration of ions on the colloidal behavior of concentrated suspensions of Si_3N_4 was investigated. The foreign ions in the powders of Si_3N_4 mainly come from raw materials and the preparation processes. For example, the Si_3N_4 powders of Founder Ceramics Corp. are prepared by nitridation of silicon, which is followed by attriting, wet-ball milling, and iron removing by hydrochloric acid. After washed with running water, the powders are dried for reservation. The test results indicated that the ions in silicon, a little remnant Fe^{2+}, and some amount of Fe^{3+}, Ca^{2+}, Mg^{2+} in the running water remained in the final powder of Si_3N_4.

The Si_3N_4 powders are from Shanghai Materials Research Institute with d_{50} of 2.57 μm (hereafter referred to as S0) and Founder Ceramics Corp. with d_{50} of 2.88 μm (hereafter referred to as F0). Their particle size distributions are shown in Figs. 5.6 and 5.7. The Na^+ ion exchange resin (Model Molselect CM-25) was supplied by Roanal Corp., Hungary. The deionized water was supplied by the water station of Microelectronics Institute, Tsinghua University.

The powder samples F0 and S0 are deionized by washing with deionized water and reacting with ion exchange resin respectively. The zeta potentials of the

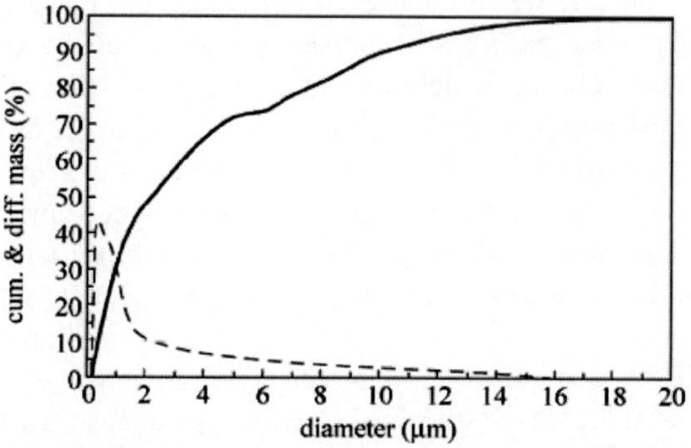

Figure 5.6 The particle size cum. and diff. mass distribution of Si_3N_4 sample S0

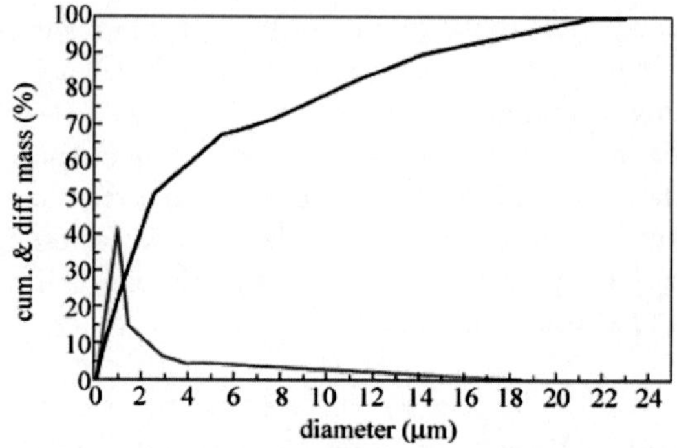

Figure 5.7 The particle size cum. and diff. mass distribution of Si_3N_4 sample F0

powders before and after deionization were measured and the conductivity and ions concentration of the supernatant of the centrifuged concentrated suspension were measured. Furthermore, the viscosity of the suspensions before and after deionization is compared.

The washing process is as follows:

(1) About 50 g of Si_3N_4 powder was mixed with 200 mL of deionized water to obtain homogeneous suspension.

(2) To centrifuge the suspension, decant the supernatant and add 200 mL deionized water again. To repeat this operation for 5 times.

(3) After drying, grinding and sieving, the final powder was obtained.

The ion exchange process is as follows:

① To mix the washed powder with an appropriate amount of deionized water and ion exchange resin, and then they are ball-milled for 24 h.

② To sieve the ball-milled suspension to remove most of the resin.

③ To calcinate the powder at 500℃ to remove the residual resin.

The calcinated powder was processed by using two different methods, referred to as method a and method b:

Method a: directly grind and sieve the calcinated powder to obtain the final powder.

Method b: wash the calcinated powder to obtain the final powder.

The interaction between particles in the slurry mainly relates to particle surface adsorption and surface charge is determined by the properties of the solid-liquid interface where the electric double layer is located (Levine S., et al., 1996; James R. O., et al., 1980; Levine S., et al., 1971). One important factor influencing the electric double layer is the surface charge of particles. According to Mezzasalma and Baldovino, the reason for the Si_3N_4-H_2O interface charge is the presence of $SiNH^{3+}$, $SiOH^{2+}$, and SiO^- under acid conditions and $Si(OH)_{n+1}^{n-1}$ and $Si(OH)_nO^{(4-n)-}$ under alkaline conditions. And the location and density of these surface groups change along with the variation of pH, which determines the charge density and charge character of surface of the Si_3N_4 (Mezzasalma S., et al., 1996). The impurities, such as C, O, etc., introduced during the production process normally make great difference in charging mechanism of different Si_3N_4 powder. Another important factor influencing the electric double layer is the species and concentration of ions in the slurry (Kitahara A., et al., 1992), which affects the height of potential barrier and distance of interaction between particles. The above mentioned two factors fully determine the interaction between particles.

The conductivity of supernatants of the suspensions prepared by powder samples F0 and powder S0 before and after washing and ion exchange is shown in Table 5.3. Washing with deionized water greatly decreased ions concentration of the powder, but it is difficult to remove the foreign ions thoroughly because there is an adsorb-resolve equilibrium of ions at the particle surface through which ions continuously resolve into water from the particle surface. After washed for 5 times, conductivity of the supernatant remained at 23 μS/cm or so. The advantage of ion exchange is that the ion exchange resin can exchange the high-valence counter-ions. This can greatly decrease the harmfulness of counter-ions according to the above-mentioned Schulze-Hardy rule. And this is also the main reason for obtaining the low conductivity by using ion exchange method b. On the contrary, the high conductivity of ion exchange method a is owing to the Na^+ introduced by calcinations of the remnant Na^+ ion exchange resin. The effect of the two deionization methods can also be seen from Table 5.4 that presents the ionography data of the supernatants.

Table 5.3 The conductivity of the supernatants of the suspensions (μS/cm)

raw materials	unprocessed (powder: H_2O =1:2 wt)	washed with deionized water (five times)	ion exchange method a	ion exchange method b
F0	175.6	23	620	10
S0	180.4	23	620	10

Table 5.4 The ion concentration of different samples (ppm)

sample	Na^+	NH_4^+	K^+	Ca^+
1	1.8	16.6	—	30.4
2	5.8	1.4	2.2	12.9
3	0.1	0.7	—	8.2
4	0.7	0.4	—	2
5	46.4	—	—	—

Notes: Sample 1 The supernatant of the suspension was obtained by mixing 20 g of unprocessed powder F0 with 40 mL of deionized water.

Sample 2 The supernatant of the suspension was obtained by mixing 20 g of unprocessed powder S0 with 40 mL of deionized water.

Sample 3 The supernatant of the suspension was obtained by mixing 20 g of washed powder F0 with 40 mL of deionized water.

Sample 4 The supernatant of the suspension was obtained by mixing 20 g of washed powder S0 with 40 mL of deionized water.

Sample 5 The supernatant of the suspension was obtained by mixing 55 g of powder F0 processed by ion exchange method a with 300 mL of deionized water.

Deionization decreased the ions concentration in the slurry, and thus changed the ions density at the particle surface through adsorb-resolve equilibrium. As a result, zeta potential of the powder was charged. Figures 5.8 and 5.9 show zeta potentials of powder samples F0 and S0 before and after deionization. It is seen that the pH_{iep} of both powder F0 and S0 drop slightly after deionization. This is owing to the decrease of the density of counter-ions cations at the particles surface. As shown in Fig. 5.10, the specific adsorbed cations exist in the Stern layer and when the concentration of OH^- in the diffusion electronic double layer increases along with the increase of pH, it results in an pH_{iep} when the potential at the slide plane-zeta potential becomes zero. When density of the cations in Stern layer decreases, the OH^- concentration at which the zeta potential equals zero, or the pH_{iep}, also decreases.

According to the formula

$$1/\kappa = (\varepsilon kT / 8\pi n e^2 z^2)^{1/2} \tag{5.4}$$

where κ—Debye parameter;

k—Boltzmann constant;

T—absolute temperature;

n—ion concentration;

z—ion valence.

When ion concentration decreases, the electric double layer will become thicker and increase the radius of slide plane. Very low ion concentration results in larger radius of slide plane and increases the resistance of the particles to becomes rearrange and slide against each other in concentrated suspensions, and therefore increases the viscosity. When dispersant J (TMAH) is added into the suspension, the viscosity of suspension of deionized powder F0 becomes less than one tenth of that of unprocessed powder F0 with the same solid loading of 49% as shown in Fig. 5.11. The reason is that more dispersant molecules are adsorbed and repulsion between particles is improved when more room is obtained in the Stern layer after deionization.

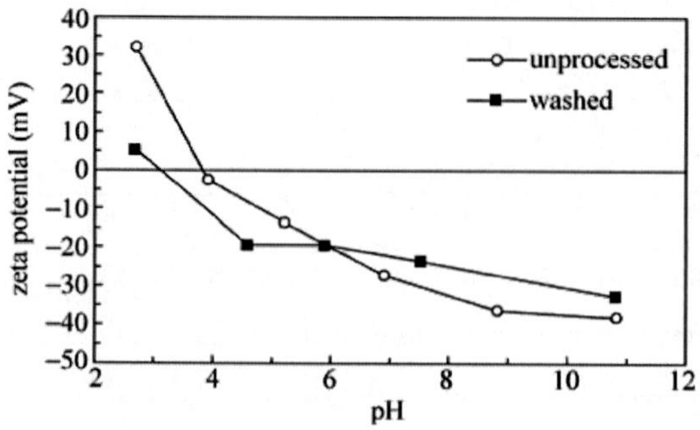

Figure 5.8 The zeta potential of powder F0 before and after deionization

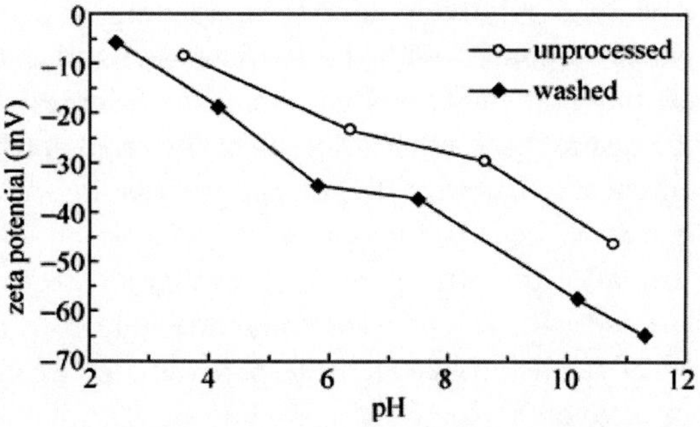

Figure 5.9 The zeta potential of powder S0 before and after deionization

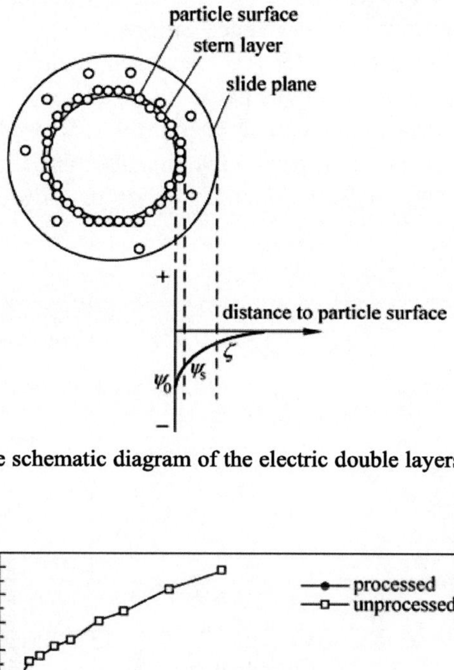

Figure 5.10 The schematic diagram of the electric double layers at the surface of a particle

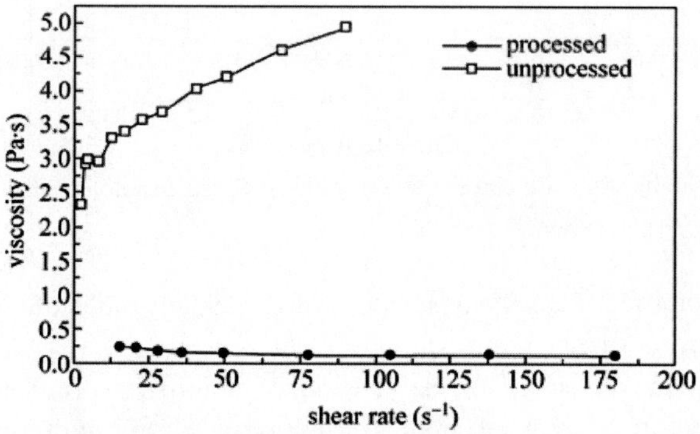

Figure 5.11 The viscosity comparison of powder F0 suspensions with solid loading of 49% before and after processing with ion exchange method b

In the suspensions with relatively low solid loading, the distance between particles is large when compared with the particle size and enlargement of the radius of slide plane may not lead to evident increase of the resistance of particles to rearrange and slide against each other. Moreover, the repulsion between particles increases after deionization, therefore the dispersity was improved and viscosity was decreased. Figure 5.12 shows the viscosity comparison of suspensions of powder S0 before and after processing with ion exchange method. It can be seen that because of the raise of zeta potential absolute value (see Fig. 5.9) and the consequent increase of repulsion between the particles, the apparent viscosity of the suspension dropped over 90%.

Both washing with deionized water and ion exchange method can effectively decrease the ion concentration in suspensions and on surface of Si_3N_4 particles effectively. And ion exchange method b can especially decrease the concentration of high-valence counterions—high-valence cations.

The pH_{iep} of the deionized powder was decreased owing to the decrease of the counter-ions—cations adsorbed on the particle surface.

The repulsion between particles in suspension is increased after deionization. Therefore the dispersity of the suspension is improved and the apparent viscosity of the suspension of powder S0 with solid loading of 20% decreases obviously after deionization.

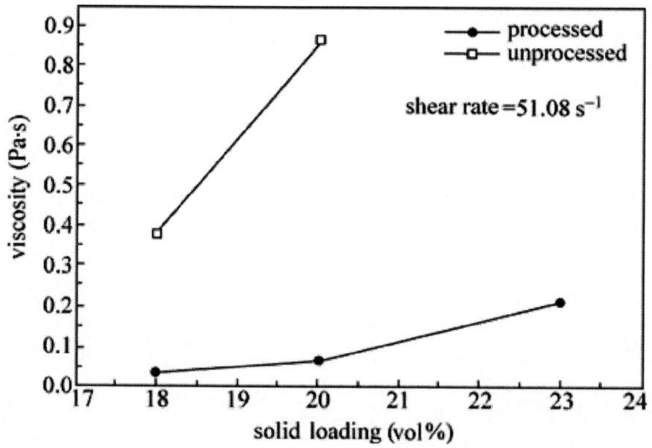

Figure 5.12 The viscosity comparison of powder S0 suspensions before and after processing with ion exchange method

Deionization can facilitate the adsorption of dispersant molecule on the particle surface and improve suspension dispersity. When dispersant TMAH is added into the suspensions, the viscosity of suspension of deionized powder F0 becomes decreases one tenth of that of unprocessed powder F0 with the same solid loading of 49%.

5.1.3 The Effect of Acid Cleaning and Calcinations on the Suspension Properties of Si_3N_4

The purpose of this study is to investigate the particle surface characteristics before and after treatment, and the effect of surface modification on the solid loading of its aqueous suspension.

The powder characteristics of two types of Si_3N_4 (referred to as FD1 and FD2), as well as the rheological properties of their aqueous suspensions, were studied. Both Si_3N_4 powders used in this study (Beijing Founder High Tech. Ceramic Corp., Ltd., Beijing, China) were obtained by direct nitridation of Si. The only difference between them is in the pulverization process. One powder is pulverized through vibration with an aqueous medium (FD1), and the other is pulverized through ball milling with an ethanol medium (FD2). The final products are obtained after drying without further purification.

The FD1 and FD2 were acid-cleaned as follows: the mixture of Si_3N_4 powder with 5% HCl (100 g:200 mL) was put into a ball-milling jar (without milling-balls) and trundled for the first 24 h. After the mixture was centrifuged, the supernatant was removed and fresh acid was added to the jar. Then, the suspension was trundled for another 24 h. The process was repeated for 4 times. Finally, the acid-cleaned powder was leached with de-ionized water for 4 or 5 times to remove residual acid. The powder was then dried.

To form a silica layer on the particle surface, the Si_3N_4 powder was calcined in air at 600℃ for 6 h. A standard muffle furnace with a temperature controller was used with a static air atmosphere. The depth of the powder layer was 10 mm.

(1) Powder Characteristics

Figure 5.13 shows the particle size distribution of the powders FD1 and FD2. Both these powders have a wide distribution and a size at the micrometer level. The specific surface area (SSA) of the powder FD1 is larger than that of the FD2

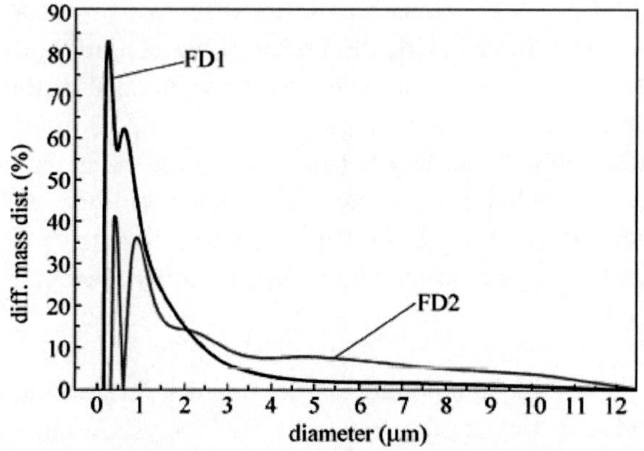

Figure 5.13 The particle-size distribution of the powders FD1 and FD2

powder, and that of the SSAs of the treated samples are almost the same as the corresponding as-received powders (see Table 5.5). The difference in size distribution may be owing to the pulverization process. The introduced energy of vibration milling for FD1 was larger than that of ball milling for powder FD2. Both bulk and surface pulverization mechanisms had an equivalent impact during the vibration-milling process, while the surface pulverization mechanism played an important role during ball milling.

Table 5.5 The surface oxygen content and specific surface area of powders FD1 and FD2

powder	oxygen content (mg/m^2)	specific surface area (m^2/g)
FD1 (as-received)	2.98	12.4
FD1 (acid-leached)	2.91	12.5
FD1 (acid-leached and calcined)	5.97	12.2
FD2 (as-received)	0.56	8.2
FD2 (acid-leached)	0.53	8.0
FD2 (calcined)	11.93	8.3

Table 5.5 shows the surface oxygen content of the samples. For the calcined samples, the surface oxygen content increased remarkably when compared with the as-received powders, which implies a silica-like layer formed on the particle surface after calcination.

Table 5.6 shows ion conductivity and ($Mg^{2+} + Ca^{2+}$) concentration of the supernatants, and Table 5.7 shows the chemical composition of the two as-received powders. The ion conductivity and concentration of ($Mg^{2+} + Ca^{2+}$) of the as-received powder FD1 are much higher than those of the as-received powder FD2. The considerable difference resulted mainly from the pulverization process. For powder FD1, the liquid medium used during pulverization process was tap water, which contains many kinds of high-valence cations, such as Mg^{2+}, Ca^{2+}, and $Fe^{3+/2+}$. Because there was no further purification after pulverization, both ions from the tap water and soluble ions of the powder itself remained. For the powder FD2, however, the liquid medium was ethanol and no foreign ions were introduced during pulverization, which can be confirmed from the chemical composition of the two as-received powders (see Table 5.7). For the acid-cleaned FD1 sample with small ion conductivity and ($Mg^{2+} + Ca^{2+}$) concentration, the majority of soluble ions were eliminated during acid-cleaning and subsequent leaching by de-ionized water. For the calcined FD1 powder, ion conductivity and ($Mg^{2+} + Ca^{2+}$) concentration are even lower than the acid-cleaned sample, which may be because of the formation of nonsoluble compounds on the powder surface and/or diffusion of soluble ions from the particle surface inwards. Table 5.7 shows the surface oxygen content of the samples. For the calcined samples, the surface oxygen content increased remarkably compared to the as-received powders, which implies a silica-like layer formed on the particle surface after calcination.

Table 5.6 The ion conductivity and ($Mg^{2+} + Ca^{2+}$) concentration of supernatants of powders FD1 and FD2

powder	ion conductivity (mS/cm)	$Mg^{2+} + Ca^{2+}$ concentration (ppm)
FD1 (as-received)	0.716	74.7
FD1 (acid-cleaned)	0.186	11.8
FD1 (calcined)	0.015	6.8
FD2 (as-received)	0.136	21.1
FD2 (acid-cleaned)	0.180	6.8
FD2 (calcined)	0.047	4.2

Table 5.7 The characteristics of the as-received powders

powder	chemical composition (%)†						
	Si	N	O	Fe	Al	Ca	Cl
FD1	48.4	37.3	6.4	0.58	0.78	0.27	0.14
FD2	60	40	2.8	0.50	0.28	0.13	

† The chemical composition was obtained by sequential X-ray fluorescence (XRF), and the SSAs were measured by the BET method. The oxygen content determined by XRF agreed well with the data measured by the inert gas fusion method (shown in Table 5.5), which were 6.43% and 2.28% for the as-received powders FD1 and FD2, respectively.

Figure 5.14 gives the zeta potential versus pH curves of the powders FD1 and FD2. According to Fig. 5.14, the as-received powder FD1 has a higher diffuse layer potential than the as-received powder FD2, which indicates that it would be easier to disperse FD1 than FD2. For treated samples, however, the difference in zeta potentials is not distinctive at a high pH. These powders also seem to have extremely low isoelectric points (IEP) (pH<2), which conflicts with results presented in the literature (Greil P., et al., 1991; Bergström L., et al., 1989). In the case of this investigation, the powders FD1 and FD2 have more surface oxygen (see Table 5.5), which indicates the low IEP, according to Hackley et al. (1993).

The diffuse reflectance infrared Fourier transform (DRIFT) spectra, from 2400 to 4000 cm^{-1} of the as-received and treated samples, are shown in Figs. 5.15, 5.16, 5.17. According to the literature (Nilsen K., et al., 1987; Ramis G., et al., 1989; Busca G., et al., 1986a,b) the peak ranging from 3400 to 3200 cm^{-1}, corresponds to the amine structure, and the one at 3000 cm^{-1} is caused by the stretching vibration of C—H bonding. The spectrum of the as-received powder FD1 (see Fig. 5.15) has a peak at 3000 cm^{-1}, but it is difficult to discern the peak from 3400 to 3200 cm^{-1}. The spectrum of the as-received powder FD2 (see Fig. 5.16) demonstrates the existence of the amine structure and C—H bonding,

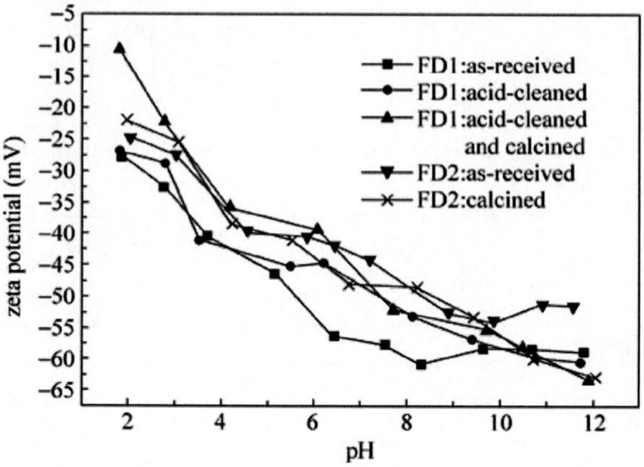

Figure 5.14 The zeta potential versus pH of powders FD1 and FD2

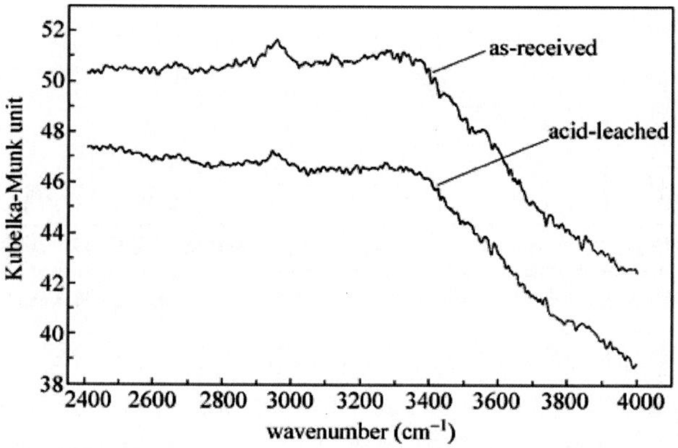

Figure 5.15 The DRIFT spectra of as-received and acid-leached powder FD1

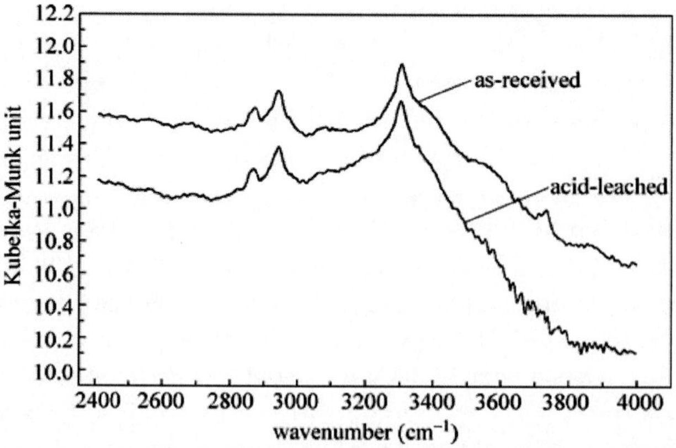

Figure 5.16 The DRIFT spectra of as-received and acid-leached powder FD2

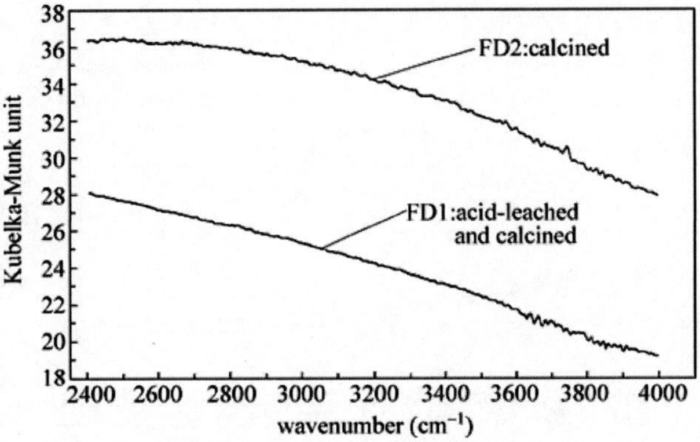

Figure 5.17 The DRIFT spectra of calcined powders FD1 and FD2

and the peak at 2900 cm^{-1} is also caused by C—H stretching. The two peaks, at 3400 to 3200 cm^{-1} and 3000 cm^{-1} of the as-received powder FD2, are more pronounced than those of the as-received powder FD1, which implies that the amount of amine structure and C—H bonding in FD2 are more developed than in FD1. The spectra of acid-cleaned FD1 and FD2 are similar to those of their as-received powders. As shown in Fig. 5.17, the patterns of calcined FD1 and FD2 are very similar and considerably different from their as-received samples.

(2) Rheological Properties of Concentrated Si_3N_4 Slurries

Rheological Properties of Acid-Cleaned Si_3N_4 Powders: For the as-received and treated powders, the viscosity versus shear rate curves of the FD1 and FD2 are shown in Figs. 5.18 and 5.19, respectively. It is clear that concentrated suspensions cannot be directly prepared from the as-received powders. As is shown in Fig. 5.18, acid cleaning can improve the aqueous dispersibility of FD1 dramatically, and the slurry viscosity of the acid-leached FD1 decreased greatly when compared with its as-received sample. Although acid-leaching can also improve the dispersibility of FD2 (see Fig. 5.19), the apparent viscosity of a 40 vol% suspension of acid-leached FD2 is still beyond 1500 mPa·s.

Rheological Properties of Calcined Powders: The rheological curves of calcined FD1 and FD2 are also shown in Figs. 5.18 and 5.19. For the direct-calcined powder FD1, the viscosity of a 40 vol% slurry decreased greatly when compared with the as-received sample, but it was still much higher than that of the acid-cleaned powder (see Fig. 5.18). This implies that although calcination can improve the aqueous dispersibility of FD1, the degree of improvement is not as pronounced as with acid cleaning. For the acid-cleaned and calcined FD1 sample, rheological curves of 40 and 50 vol% slurries are shown in Figs. 5.18 and 5.20, respectively. The aqueous dispersibility of this sample is better than the acid-leached one, and the viscosity of the 50 vol% slurry is lower than 300 mPa·s at low shear rates. It shows shear thickening behavior when the shear rate is beyond 50 s^{-1}.

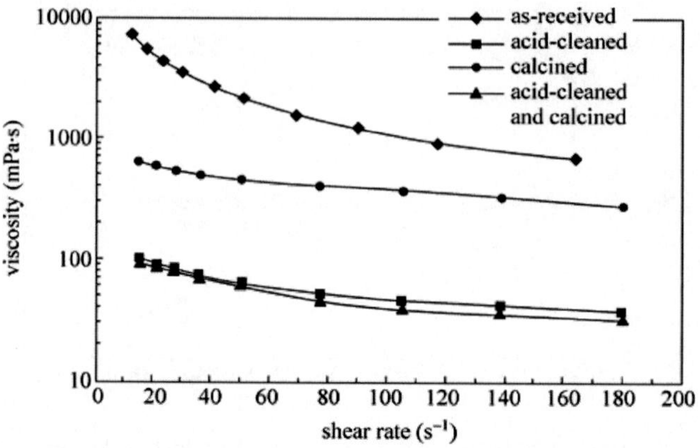

Figure 5.18 The rheological curve of solid loading 40 vol% suspension of powder FD1 (pH = 11.5)

For the calcined powder FD2, Figs. 5.19 and 5.20 show rheological curves of the slurries with solid loading of 40 and 50 vol%, respectively. The slurry with solid loading of 40 vol% behaves almost as a Newtonian liquid and its viscosity remains at 50 mPa·s. The suspension with solid loading of 50 vol% can be found to show pseudoplastic characteristics at low shear rates, and shear-thickening when the shear rate is beyond 35 s^{-1}. From rheological curves of the as-received and treated FD2 slurries, calcinations enhances the solid loading dramatically to 50 vol%, while the improvement from acid-leaching is inferior.

High concentration slurries (⩾ 50 vol%) cannot be prepared from the as-received powders of FD1 and FD2. Both acid-leaching and calcination can improve their aqueous dispersibility, but the effect of the same method varies with different powders. The solid loading of the FD1 slurry can reach 50 vol% through acid-cleaning and calcination, while the effect of direct calcinations is not equal to

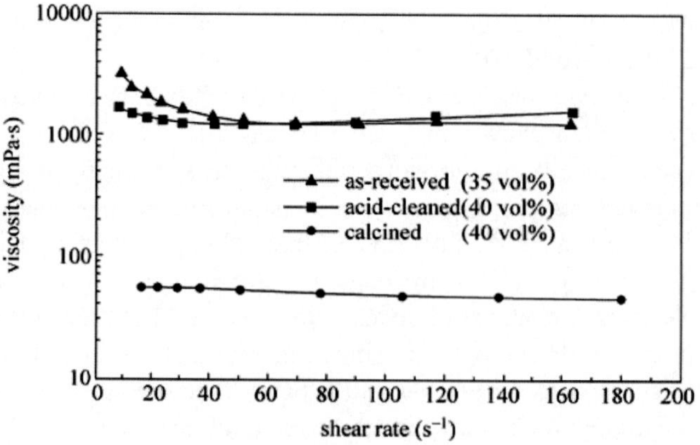

Figure 5.19 The rheological properties of FD2 suspensions (pH = 11.5)

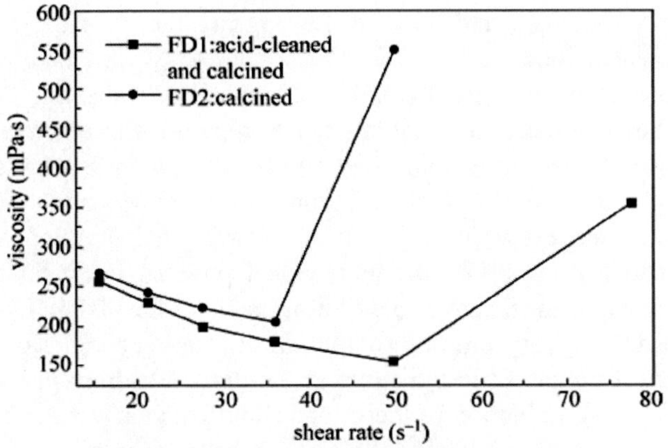

Figure 5.20 The rheological properties of solid loading 50 vol% suspension of treated powders FD1 and FD2 (pH = 11.5)

acid-cleaning. For the FD2 slurry, however, calcination can enhance the solid loading to 50 vol%, while acid-leaching does not produce a similar effect. With regard to the slurries of FD1 and FD2 with 50 vol% solid loading (Fig. 5.20), although both exhibit similar rheological behaviors, FD2 is more strongly dilatant than FD1 at high shear rates. The different dilatancy of the slurries may be owing to the differences in particle size and size distribution.

(3) Discussion

The flocculation of the slurry is affected by the concentration of electrolytes. According to the Schulze-Hardy rule, in an aqueous suspension at 25℃ (Kitahara A., et al., 1992)

$$n_c = 8\times10^{-22}\left(\frac{\gamma^4}{A^2 z^6}\right) \tag{5.5}$$

where, n_c is the critical flocculation concentration, $\gamma = [\exp(Z/2)-1]/[\exp(Z/2)+1]$ (where $Z = ze\varphi_0/(kT)$), z is the valency of the counter-ion, e is the electron charge, k is the Boltzmann constant, and T is the absolute temperature), and A is the Hamaker constant. The critical flocculation concentration n_c is inversely proportional to the sixth power of the counter-ion valency. The higher the counterion valency, the lower the critical flocculation concentration. When compared with mono-valence counter-ions, high-valence counter-ions will lead to coagulation at lower molar concentrations. The presence of specific surface impurities on dry Si_3N_4 powder, which are primarily imparted from the processing steps (milling, leaching, etc.), has been reported (Bergström, L., et al., 1989; Peuckert M., et al., 1987; Rahaman M. N., et al., 1986). According to Malghan (1992), cationic (Na^+, Al^{3+}, Ca^{2+}, Fe^{3+} etc.) and anionic (Cl^-, F^-, etc.) species can get dissolved from the surface impurities and form surface complexes by reacting with surface silanols. Thus, the soluble counter-ions, especially the counter-ions with high-valence, will affect the slurry rheology, and may be the limiting factor in the preparation of concentrated suspensions.

For the as-received powder FD1, ion conductivity and ($Mg^{2+}+Ca^{2+}$) concentration of the supernatant are higher when compared with other samples, while the rheological property of its solid loading 40 vol% slurry is very poor. For the acid-leached powder FD1, however, ion conductivity and (Mg^{2+} + Ca^{2+}) concentration are far lesser than those in the as-received sample, indicating that most of the soluble species in the as-received powder, as well as the soluble high-valence counter-ions, have been eliminated. The DRIFT spectrum of acid-leached FD1 is very similar to that of the as-received sample, and the difference of zeta potentials is indistinct at the discussed high pH. According to Hackley et al.(1994), reduction of these ions should also affect the zeta potential, but it did not occur in this study. So, it can be concluded that the dramatic improvement of dispersibility is caused by the elimination of soluble high-valence counter-ions. For the powder FD1, which was acid-leached and calcined, the zeta potential is almost the same as that of the acid-cleaned sample, while the DRIFT spectrum is different. The viscosity of the solid loading 40 vol% slurry has a slight decrease when compared with the acid-leached powder, implying that oxidation has an indistinctive effect on aqueous dispersibility of the acid-leached sample. For the direct calcined powder FD1, however, ion conductivity and ($Mg^{2+}+Ca^{2+}$) concentration is lesser than those of the acid-leached sample. The remarkable reduction of soluble species may be caused by the formation of insoluble species or the diffusion of these species from the particle surface inwards. From Fig. 5.18, it is clear that the effect of the direct calcination of FD1 on dispersibility is inferior to acid leaching. These insoluble species may be unfavorable for dispersibility, so the improvement is restricted when compared with acid leaching. Thus, it can be concluded that for the as-received powder FD1, the crucial factor limiting the aqueous dispersibility is the presence of soluble species, especially the high-valency counter-ions. The results of acid-leaching correlate with those obtained by Laarz et al. (1997), while the effects of oxidation are conflicting. They attributed the improvement of dispersibility to a characteristic physical change in the powder, and presumed that the effect of oxidation was unexpected.

For the as-received powder FD2, the dispersibility is so poor that the viscosity of a slurry with 35 vol% solid loading is beyond 1000 mPa·s (see Fig. 5.19). Soluble species from the powder are not responsible for the poor dispersibility because the ion conductivity and ($Mg^{2+}+Ca^{2+}$) concentration of the as-received powder are very small.

The dispersant used in this study is an aqueous solution of $(CH_3)_4NOH$, which is an organic alkali, so the slurry of Si_3N_4 is stabilized via electrostatic repulsion at pH=11. Surface charging of Si_3N_4 particles would contribute to the dissolution of surface groups as follows:

$$[SiOH_2]^+ \xleftarrow{H^+} [SiOH] \xrightarrow{OH^-} [SiO]^- + H_2O \tag{5.6}$$

$$[Si_2NH_2]^+ \xleftarrow{H^+} [Si_2NH] \xrightarrow{OH^-} [Si_2N]^- + H_2O \tag{5.7}$$

$$[SiNH_3]^+ \xleftarrow{H^+} [SiNH_2] \xrightarrow{OH^-} [SiNH]^- + H_2O \tag{5.8}$$

These reactions are pH-dependent and essentially equivalent to the specific adsorption of protons or hydroxyls. Although the charging mechanism of Si_3N_4 in basic solution has not been well-established yet (Kulig M., et al., 1991; Harame D. L., 1984; Greil P., 1989). The same viewpoint is that the silanol group Si—OH is favorable for surface charging. If other conditions remain constant, the more silanols, the larger the electrostatic repulsion between the particles. From the viewpoint of preparing highly concentrated slurries, the substitution of Si—OH for amine structures is beneficial to improve the dispersibility of Si_3N_4 powder. On the other hand, it is difficult to prepare highly concentrated aqueous slurries if the amount of amine structures or other hydrophobic groups is dominant on the particle surface. According to Schwelm et al. (1993) stable esters, which are hydrophobic with the configuration Si—O—C—R, would be formed on the Si_3N_4 surface by the reaction of alcohol with silanol groups. The CH_3 groups of the alcohol are believed to shield the negative charge of the electron pairs at the oxygen ion, which are the active sites for deflocculant coupling. When temperature is up to 500℃ the esters can be removed from the powder surface.

The DRIFT spectrum of the as-received powder FD2 shows that the dominant surface groups are amine structures (Si_2NH and $SiNH_2$) and C—H bonds, which are indistinct for the sample FD1. From the pulverization process of FD2, it can be concluded that the Si_3N_4 powder would react with alcohol to form esters on the particle surface, and that the distinctive C—H bonding is caused by the ester Si—O—C—R. Hydrophobic groups are dominant on the particle surface of the as-received powder FD2, so it is easy to understand the poor aqueous dispersibility. The DRIFT pattern of acid-leached FD2 is similar to that of the as-received one, and the rheological improvement of the slurry is limited. For the calcined powder FD2, however, both amine structures and C—H bonding disappeared completely, and the DRIFT spectrum of calcined FD2 is similar to that of the acid-cleaned and calcined sample FD1. The solid loading of slurries for these two samples has reached 50 vol%, while the viscosity was below 300 mPa·s at low shear rates. For the as-received powder FD2, the results indicate that the crucial factor limiting the preparation of the highly concentrated slurry is the hydrophobic groups on the particle surface.

Two types of commercial Si_3N_4 powder were studied in this section, and only their pulverization processes were found to be different. Several differences in size distribution, soluble species, and rheological properties of the aqueous slurries were found. Both acid leaching and calcination were found to improve their aqueous dispersibility, but the degree for the same method varied with different powders. For the as-received FD1 powder, the crucial factor limiting dispersibility is high-valence counter-ions, which can be eliminated almost completely through acid leaching. Because of the diffusion or reaction of the soluble species, improvement occurs with the use of direct calcination, but is inferior to acid leaching. For the as-received powder FD2, the surface hydrophobic groups of Si—O—C—R were found to limit the solid loading. For FD2, the effect of calcination on dispersibility is remarkble, while the effect of acid cleaning is not similar to that of calcination. For acid-cleaned and calcined FD1 and FD2, the solid loading of their aqueous suspensions has reached 50 vol% with a viscosity below 300 mPa·s.

5.1.4 The Effect of Liquid Medium and Surface Group on Dispersibility of Si_3N_4 Powder

The purpose of this section is to investigate the effects of liquid media and ball-milling process on the particle surface group and the aqueous dispersibility of Si_3N_4 powders.

The commercial α-Si_3N_4 powder used in this study was prepared by direct nitridation of Si powder (grade M11, H. C. Starck, Berlin, FRG) and powder properties of the as-received product was shown in Table 5.8. The liquid media used to modify the particle surface group of Si_3N_4 powder are deionized water (about $18\times10^6\Omega$), analytically pure ethanol and isopropanol, respectively.

To study the influence of liquid media on particle surface group and aqueous dispersibility, the Si_3N_4 powder was treated as follows. Mixtures of the as-received Si_3N_4 powders with different liquids (100 g∶200 mL) were ball-milled for 3 days at the rotary speed of 100 r/min, and then dried at 100℃ for 5 days. The weight ratio of ZrO_2 milling-balls to Si_3N_4 powder is 1∶1.

Table 5.8 The composition and properties of the as-received Si_3N_4 powder

chemical composition						specific surface area (m^2/g)	α phase (%)
N (%)	O (%)	C (%)	Fe (ppm)	Al (ppm)	Ca (ppm)		
38.69	1.44	0.20	15	500	61	12.9	91.7

To compare influence of ball-milling process on the particle surface group and aqueous dispersibility, for the mixture with ethanol as liquid medium, the weight ratio of milling-balls to Si_3N_4 powder and the rotary speed are increased to 3∶1 and 400 r/min, respectively. Other parameters such as ratio of powder weight to liquid volume and drying process are invariant.

All the treated powders are divided into two portions: one part is used directly for rheology measurement and DRIFT spectra test, the other portion is calcined at 600℃ for 6 h first and then is used for the same examinations. A standard muffle furnace with a temperature controller, having a hearth size of $25\times15\times12$ cm^3, was used with an air static atmosphere, and with the powder layer of about 10 mm.

(1) Effect of Liquid Media Refering Water, Ethanol and Isopropanol

The DRIFT spectra of the as-received and liquid-treated Si_3N_4 powders are shown in Fig. 5.21. According to refs.(Stadelmann H., et al., 1989; Nilsen K., et al., 1987; Ramis G., et al., 1989; Busca G., et al., 1986a,b), the wide peak from 3200 to 3500 cm^{-1} corresponds to amine structures (Si_2—NH and Si—NH_2) on the particle surface, while the peak at 3740 cm^{-1} is resulted from the isolated Si—OH surface group. In the range of 800 – 1500 cm^{-1}, previous investigations of DRIFT spectra for undiluted Si_3N_4 samples have shown (Busca,1986; Dai, J. Q, 2001) that the strong sharp peak at about 1200 cm^{-1} is owing to anti-symmetry stretching vibration of siloxane Si—O—Si surface group. It can be seen from Fig. 5.21(a) that the amine structures (3200 – 3500 cm^{-1}) have obscure change and the weak peak corresponding to isolated Si—OH group disappeared after liquid-treatment. For powders ball-milled in ethanol and isopropanol (referred to M11-ethanol and M11-isopropanol, respectively), there are three peaks emerged during 2800 – 3000 cm^{-1}, which correspond to the stretching vibration of C—H bonding, while these peaks are nonexistent in DRIFT spectrum of sample treated in deionized water (referred to M11—H_2O). When compared with powder M11-ethanol, the C—H peak of powder M11—isopropanol is found to be more intensive. At the wave number range of 800 – 1500 cm^{-1} (Fig. 5.21(b)), compared to the as-received powder (referred to M11—raw), the peak intensity of Si—O—Si group for powder M11—H_2O decreases and that of sample M11-isopropanol increases, while the peak intensity of M11-ethanol keeps invariant.

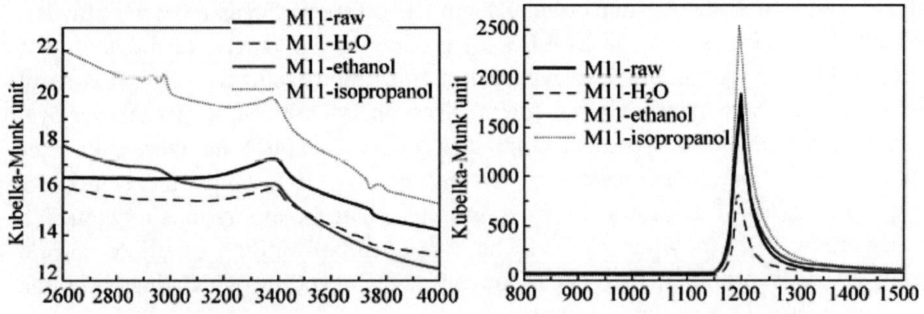

Figure 5.21 The DRIFT spectra of the silicon nitride powders ball-milled in different liquids

(a) 4000 – 2600 cm^{-1}; (b) 1500 – 800 cm^{-1}

As previously investigated (Dai J. Q., et al., 2001), the surface oxygen content of the as-received Si_3N_4 powder is 0.72 wt%, and there are excess of Si—O—Si groups on the particles surface, which can be hydrolyzed gradually in aqueous solution as follows:

$$Si—O—Si + 2H_2O \longrightarrow 2Si—OH \tag{5.9}$$

Thus, the number of Si—O—Si surface groups will decrease owing to the hydrolysis reaction, which is consistent with observations in the DRIFT spectrum of powder M11—H_2O.

Both the powders ball-milled in ethanol and isopropanol are quite volatilizable, the emergence of C—H bonding implies the reaction of the liquids with particles surface. The reaction may occur as follows:

$$\text{Si—O—H} + \text{HO—CH}_2\text{CH}_3 \longrightarrow \text{Si—O—CH}_2\text{CH}_3 + \text{H}_2\text{O} \tag{5.10}$$

$$\text{Si—OH} + \text{HO—CH (CH}_3)_2 \longrightarrow \text{Si—O—CH(CH}_3)_2 + \text{H}_2\text{O} \tag{5.11}$$

The number of the Si—OH groups (isolated and adjacent) is observed to decrease and that of the Si—O—Si surface species is found to remain invariant according to these two reaction equations. Emergence of the C—H bonding and disappearance of the isolated Si—OH groups in their DRIFT spectra lend support to the postulate. This finding is also consistent with the results obtained by Schwelm et al. (1993), who observed that the Si—OH groups at the Si_3N_4 particle surface react with isopropanol and form esters of Si—O—C—R. Their results also demonstrated that the Si—O—C—R surface groups are very stable and can be burned out at temperatures above 450℃. The peaks of amine structures for powders M11—ethanol and M11—isopropanol are the same as those of the as-received sample, implying that there is no reaction between amine structures and liquids of ethanol and isopropanol. For powder M11—ethanol, the number of the Si—O—Si surface groups remains similar to that of the as-received powder, which demonstrates that the Si—O—Si groups do not react with ethanol. For sample M11—isopropanol, however, the obvious increase of Si—O—Si amount indicates that there exists other reaction of isopropanol with Si_3N_4 particle surface. It has been proposed in a previous literature (Dai J. Q., 2001) that there is another kind of Si—O bonding (neither Si—O—Si nor Si—OH), which is unstable and easy to form Si—OH groups in aqueous solution, at the amorphous oxygen-rich layer on the Si_3N_4 powder surface. The long-term ball-milling of silicon nitride powders in isopropanol may cause the following reaction, which increases the amount of Si—O—Si surface groups:

$$\begin{aligned}&2\text{Si—O} + \text{Si—H} + \text{HO—CH—(CH}_3)_2 \longrightarrow \\ &\qquad \text{Si—O—Si} + \text{Si—O—CH—(CH}_3)_2 + \text{H}_2\text{O}\end{aligned} \tag{5.12}$$

The rheological properties of solid loading 40 vol% aqueous slurries of the as-received and liquid-treated powders are shown in Fig. 5.22(a). In contrast to rheological property of slurry M11—raw, apparent viscosity of slurry M11-H_2O is reduced, while that of slurries M11-ethanol and M11-isopropanol are increased. As mentioned above, hydrophobic Si—O—C—R groups have been formed on particle surface of the powders ball-milled in ethanol and isopropanol. This hydrophobic surface group deteriorates the aqueous dispersibility (Dai J. Q., 2001). For the liquid-treated powders after thermal oxidation at 600℃ for 6 h, as exemplified in Fig. 5.22(b) their 40 vol% suspensions show the same rheological behaviors. It can be clearly seen from Fig. 5.23 that the DRIFT spectra of these liquid-treated samples are very similar after calcination, which indicates that the hydrophobic Si—O—C—R groups are removed and the surface characteristics of the liquid-treated powders are recovered, which are identical after calcination.

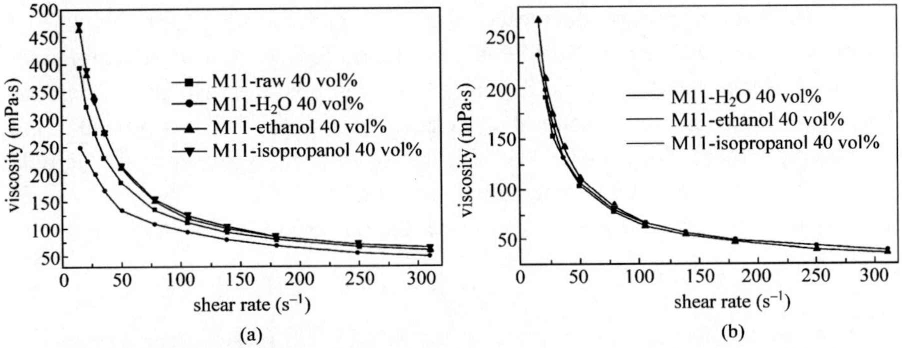

Figure 5.22 The rheological properties of the slurries of the liquid-treated Si_3N_4 powders (a) before and (b) after calcination

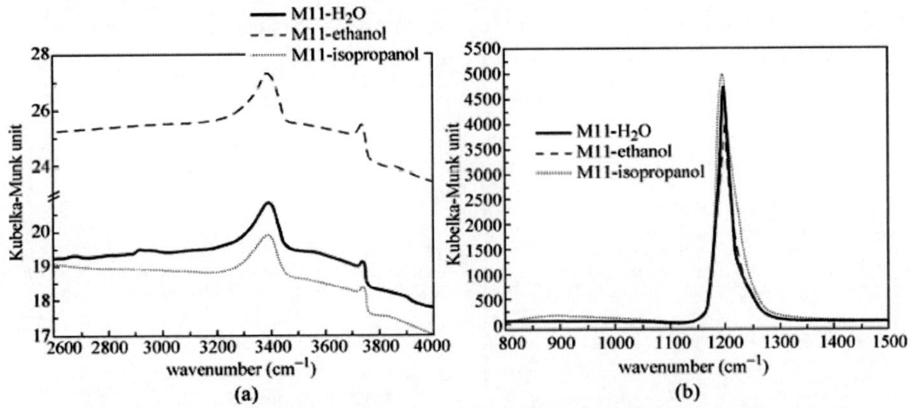

Figure 5.23 The DRIFT spectra of the liquid-treated Si_3N_4 powders after calcinations (a) 4000 – 2600 cm^{-1}; (b) 1500 – 800 cm^{-1}

(2) The Effect of Ball-Milling Process

Though aqueous dispersibility of the liquid-treated powders was modified, the change was not very notable, which may be owing to the weak reaction of liquid media with particle surface of Si_3N_4 powder. To study influence of ball-milling process on the aqueous dispersibility and particle surface group, the weight ratio of milling-balls to powders ball-milled in ethanol was increased from 1 : 1 to 3 : 1, and the rotary speed was increased from 100 to 400 r/min at the same time. The treated samples were referred to as powders M11-11* and M11-34*, respectively. As shown in Fig. 5.24, the DRIFT spectra of powder M11-34 have remarkable difference when compared with those of sample M11-11. Firstly, amount of the amine structures and C—H bonding increases dramatically, and the peak positions also have deviation (shown in Fig. 5.24(a)). Secondly, peak of the adjacent Si—OH surface groups at 3500 – 3600 cm^{-1} emerges. Thirdly, amount of the Si—O—Si surface groups decreased obviously and the peak position shifted visibly to lower frequency. Besides the condensation reaction of ethanol with Si—OH surface species to form Si—O—C—R groups, the results also demonstrated that other reactions occurred at the particle surface, which is related to the ball-milling process. For powder M11-34, during ball-milling process, the particle surface endured intensively mechanical impact, which broke the Si—O—Si bonding and caused reaction of ethanol with the ruptured Si—O—Si groups:

$$\text{Si—O—Si} + \text{HO—CH}_2\text{CH}_3 \longrightarrow \text{Si—O—CH}_2\text{CH}_3 + \text{Si—OH} \qquad (5.13)$$

According to this reaction, amount of the Si—O—Si groups decreases, while that of Si—O—C—R and Si—OH group simultaneously increases, which is confirmed by the DRIFT spectra of powder M11-34. Further more, peak shift of the Si—O—Si groups from 1190 to 1175 cm^{-1} also demonstrates that the presence of large tensile stress in the cyclic Si—O—Si structure is owing to the intensive impact. The DRIFT spectra of powder M11-34 show conclusively that

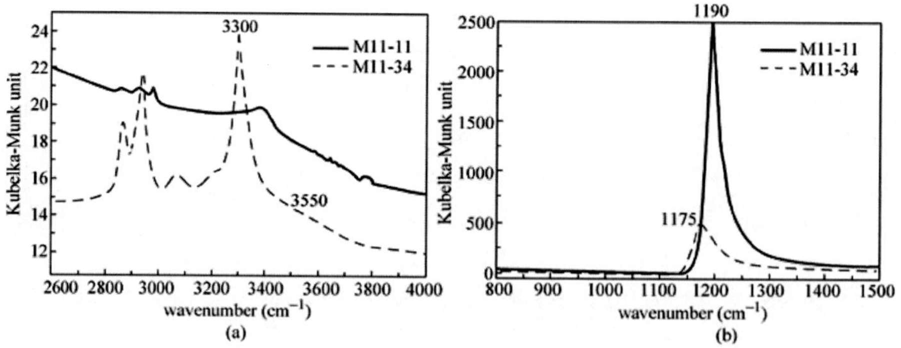

Figure 5.24 The DRIFT spectra of Si_3N_4 powders after different ball-milling process (a) 4000 – 2600 cm^{-1}; (b) 1500 – 800 cm^{-1}

* Note: M11-11: milling-balls to powders = 1 : 1; the rotary speed 100 r/min;
M11-34: milling-balls to powders = 3 : 1; the rotary speed 400 r/min.

the dominant process during ball-milling is the reaction of ethanol with the ruptured Si—O—Si groups, which generates large amount of Si—O—C—R groups and obviously consumes the Si—O—Si structures.

The slurry rheological properties of the slurries of powders M11-11 and M11-34 are plotted in Fig. 5.25. For the ball-milled powders before calcination (Fig. 5.25(a)), the aqueous dispersibility of sample M11-34 is much inferior to that of M11-11. It is the large amount of hydrophobic Si—O—C—R groups on the particles surface that deteriorates the dispersibility. For the thermal-oxidized powders (calcination, see Fig. 5.25(b)). However, the apparent viscosity of slurry M11-34 is lower than that of slurry M11-11, owing to the small amount of Si—O—Si groups of sample M11-34 after calcination, as shown in Fig. 5.26.

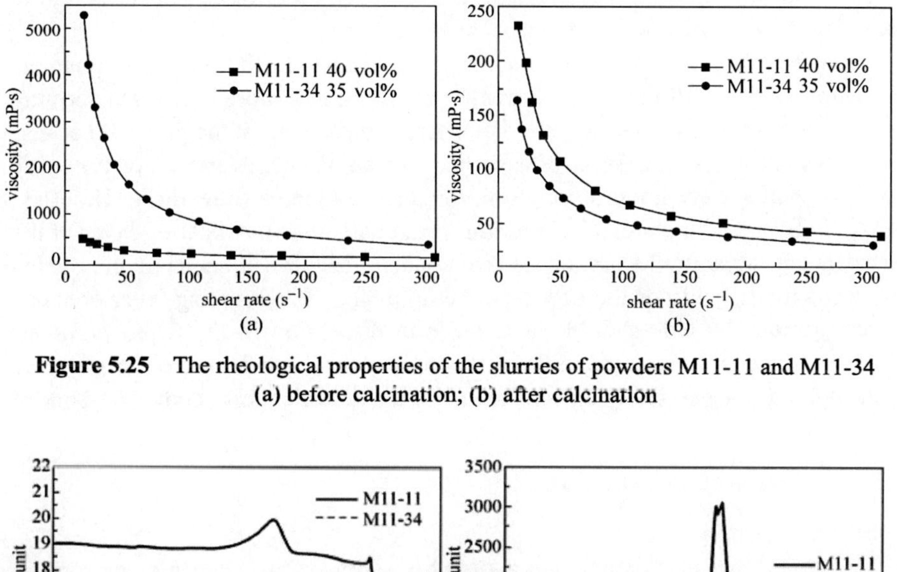

Figure 5.25 The rheological properties of the slurries of powders M11-11 and M11-34 (a) before calcination; (b) after calcination

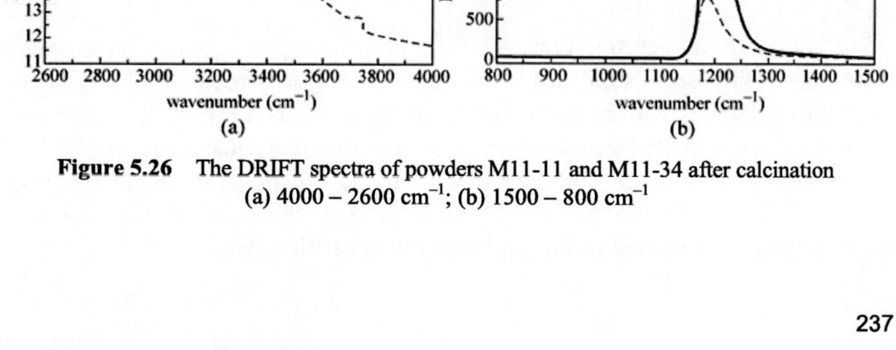

Figure 5.26 The DRIFT spectra of powders M11-11 and M11-34 after calcination (a) 4000 – 2600 cm^{-1}; (b) 1500 – 800 cm^{-1}

5.2 The Gel-Casting of Si_3N_4 Ceramics

5.2.1 The Preparation of Si_3N_4 Ceramics with Surface-Coated Si_3N_4 Powder

One of the key steps of gel-casting is the preparation of a highly concentrated ceramic suspension with low viscosity, a characteristic which is strongly dependent on the surface chemistry of the ceramic powder. The surface status and characteristics of Si_3N_4 particles are relatively complex (Mezzasalma S., et al., 1996; Greil P., et al., 1991; Sanchez R. T., et al., 1996). To obtain consistent rheological behavior in slurries from different batches of Si_3N_4 powders, the technical parameters must be continuously modified, which makes the process very difficult to control during industrial production and increases production costs.

By coating Si_3N_4 with Al_2O_3 and Y_2O_3, we expected different Si_3N_4 powders to exhibit similar colloidal characteristics, resulting in a more consistent forming process. In addition, we also expected to improve dispersion of the powder, because the surface of the coated Si_3N_4 powder takes on some characteristics of the oxide coatings, which have good dispersibility in aqueous media (Fagerholm H., et al., 1994; Shih W. H., et al., 1996; Fagerholm H., et al., 1996). A coating layer of the powders can separate the surface of the powders from the aqueous media, which improves the stability of the powders. Additionally, if the coating layer contains sintering aids, the aids should cause uniform dispersion in the Si_3N_4 particles. Uniform dispersion can reduce the diffusion distance and improve the sintering rate, and, thus, be especially advantageous for the densification of covalent-bonded compounds.

5.2.1.1 Preparation Procedure

1) Surface Coating

Y_2O_3 and Al_2O_3 usually are coated onto Si_3N_4 powder as sintering aids.

Two main methods for coating Al_2O_3 and Y_2O_3 onto Si_3N_4 powder are by the hydrolysis of metal alkoxides (Joshi R. N., et al., 1994) and by the precipitation of inorganic salts (Kim J. S., et al., 1997; Garg A. K., et al., 1990; Yang J., et al., 1999).

For inorganic salt precipitation, $Al(NO_3)_3$ and $Y(NO_3)_3$ are made to react with $CO(NH_2)_2$ to produce insoluble basic salts. The basic salts are then physically or chemically adsorbed onto the surface of the Si_3N_4 particles, forming a coating layer. After calcination, the basic salts decompose, and the oxide coating forms.

The advantages of the precipitation process are the low price of the salt precursor and the feasibility of upscaling the technique. Owing to those advantages, inorganic salt precipitation was used as the coating method in this study.

The Si_3N_4 used for coating was a commercially available powder (Founder High Technology Ceramics Co., Ltd., Beijing, China). All the Si_3N_4 powders mentioned in this chapter are obtained from this source, unless otherwise cited.

The process of coating Si_3N_4 particles with Y_2O_3-Al_2O_3 is illustrated in Fig. 5.27. An aqueous solution containing 0.200 mol/L of $Al(NO_3)_3$ (purity 98.0%, Yili Fine Chemical Corp., Yili, China), 0.276 mol/L of $Y(NO_3)_3$ (purity 99.0%, Lanzhouhuaxin Chemical Corp., Lanzhou, China), and 50 g/L of urea (purity 99.0%, Beijing Chemical Reagents Co., Beijing, China) was heated at 95℃ for 100 min and then cooled quickly to room temperature. The resulting clear liquid is termed as the coating solution. An aqueous dispersion containing 73.4 g of Si_3N_4 powder, etched previously by a solution of 5 wt% HF, 33 wt% HNO_3 and 62 wt% H_2O, to remove surface SiO_2, in 800 mL of deionized water, was agitated ultrasonically for 10 min and then added to 200 mL of the coating concentrate. The mixture, with $pH = 5.5 \pm 0.2$, was kept at 50℃ and vigorously stirred in a closed flask for up to 48 h, and then dried on a hot plate to obtain it in the form of a powder. The Si_3N_4 powder coated with yttrium-aluminum salts was passed through a sieve (100 mesh) and calcined at 1000℃ in N_2 gas for 2 h, to produce a Si_3N_4 powder coated with 7.5% Y_2O_3 and 2.5% Al_2O_3. A pure precursor of the Y_2O_3-Al_2O_3 coating was also prepared similarly, for comparison.

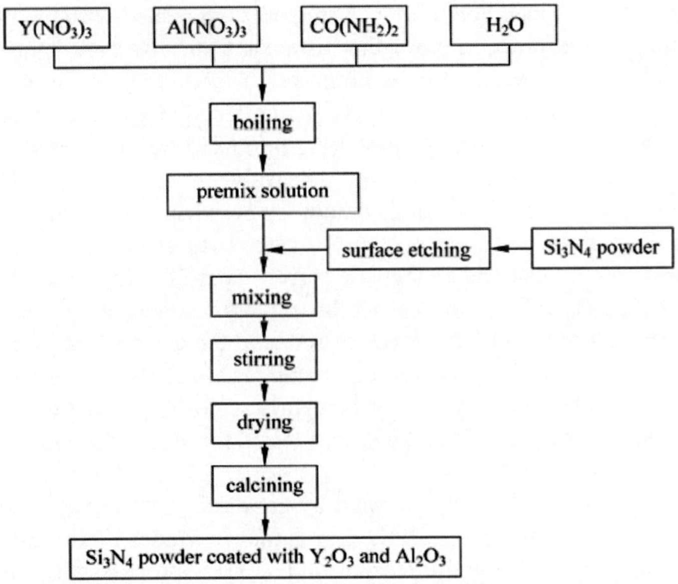

Figure 5.27 The flow chart of the Y_2O_3-Al_2O_3 coating process

The suspensions were rolled to create a good dispersion of ceramic powders, rather than for milling purposes. Our previous study showed no obvious change on the surface coating of Si_3N_4 whiskers after 24 h of rolling, and hence we will not discuss the effect of rolling on the surface coating in this study.

2) Gel-Casting

The gel-casting process was similar to that of previous studies (Young A. C., et al., 1991; Omatete O. O., et al., 1997). First, the original Si_3N_4, with 7.5 wt% Y_2O_3 and 2.5 wt% Al_2O_3 powder (Jiangsu Wuxian Special Ceramics Plant, Wuxian, China), or a coated Si_3N_4 powder was suspended, with TMAH as a dispersant, in the premixed solution. The slurry, with a solid loading of 45 vol%, was degassed for 10 min after rolling for 24 h in polyethylene bottles, using alumina balls. The suspension was degassed for another 3 min when the initiator and catalyst were added. All the above mentioned operations were conducted at room temperature. Subsequently, the slurry was cast into a nonporous mold, which was then kept at 70℃. After the monomers had been polymerized, the green bodies were demolded and dried at room temperature, under controlled humidity, to avoid cracking and nonuniform shrinkage caused by rapid drying. The microscopic morphology of the green bodies was observed by scanning electron microscopy (SEM; Model CSM950, OPTON Co., Ltd., Munchen, Germany).

3) The Gas Chromatography

The Si_3N_4 suspensions with the same contents of coated or mechanically mixed sintering aids for gel-casting, used for comparison, were all at a solid loading of 45 vol%. After the gel-casting process, large amounts of macropores were observed in the original Si_3N_4 green bodies, which might be associated with the gas evolved from reaction of the powders with the dispersant and the monomer-containing premixed solution. To verify this assumption, a laboratory-made gas-gathering device, shown in Fig. 5.2, was designed to collect the gas formed during the reactions. The reagents (ceramic powders, premixed solution, and dispersant) were contained in a gas-tight container. The valves 1 and 2 of the device were initially opened, and the gas bag and container were vacuumed before valve 2 was closed. The container was then put into a water bath at 80℃, and the reagents were agitated continuously by a magnetic stirrer. After 30 min, water was breathed into the container via valve 2, until all of the gas in the container had been squeezed into the gas bag. The valve 1 then was closed, and the gas bag was removed from the instrument, for analysis by gas chromatography (Model 3400, Beijing Analytical Instrument Plant, Beijing, China). For comparison, Si_3N_4 powder (SN-E10, UBE Industries, Ltd., Yamaguchi, Japan) was also tested for its gas-discharging reaction.

4) Sintering and Testing

The coated Si_3N_4 green bodies formed by gel-casting were subjected to binder out and gas-pressure sintering (GPS) in a furnace (Model FPW 180/250-2200-100-SP, KCE Corp., Rödental, Germany). The GPS process was conducted at 1750℃ for 2.5 h, under a N_2 pressure of 3 atm (1 atm = 9.8066×10^4 Pa), then at 1900℃ for 2.5 h, under a N_2 pressure of 60 atm. The heating and cooling rates were 10℃/min.

Because the original Si_3N_4 green bodies prepared by gel-casting contained numerous macropores, investigating the effect of the surface coating on the mechanical properties of the Si_3N_4 ceramics by comparing ceramics gelcast from the original and the coated Si_3N_4 powders was meaningless. The original Si_3N_4 samples were prepared through cold isostatically pressing, dry-pressed at 0.1 MPa, and then cold isostatically pressed at 2 MPa, for comparison with the gel-casting samples and to allow discussion of the effect of surface coating and gel-casting on the performances of the Si_3N_4 ceramics. The original Si_3N_4 green bodies, with densities of 55% of the theoretical value, were sintered by the same procedure used for the coated and gelcast bodies.

5.2.1.2 The Characteristics of Coated Si_3N_4 Powder

The micrographs of Si_3N_4 particles coated with Al_2O_3 by the same coating process used in this work have already been illustrated in a previous paper (Tang Q., et al., 1998).

The TEM micrographs of the Y_2O_3-Al_2O_3-coated Si_3N_4 particles in the present work are shown in Fig. 5.28. A layer of precipitates, with a thickness varying from 5 to 40 nm, is closely attached to the particle surfaces. However, a more careful observation of the figure shows that the surface of the Si_3N_4 powder is not completely covered by precipitates. In addition, some precipitates make no contact with the Si_3N_4 particles.

The XRD curves (see Fig. 5.29) demonstrate that the coated layer is a Y_2O_3-Al_2O_3 layer consisting of Y_2O_3, δ-Al_2O_3, $Y_3Al_2(AlO_4)_3$, and $Y_4Al_2O_9$ resulting from decomposition of the Y_2O_3-Al_2O_3 precursor.

The zeta potentials of the coated Si_3N_4 particles are shown in Fig. 5.30. Unlike any curve of the pH-dependent zeta potential for Al_2O_3, Y_2O_3, or Si_3N_4, the zeta-potential curve for the coated Si_3N_4 powder exhibited three isoelectric points (IEPs), i.e., three "charge" reversals in the pH range. A previous study (Hackley V. A., et al., 1994) has shown that up to three reversals of the electrokinetic potential can occur when polyvalent metal ions (Al^{3+} and Y^{3+} ions in our work) are present

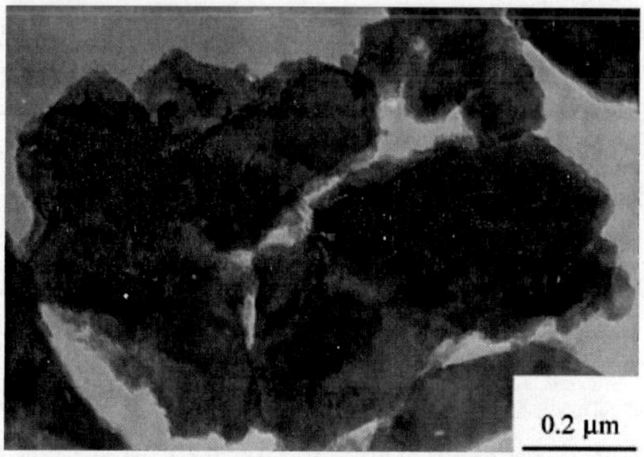

Figure 5.28 The TEM micrographs of Y_2O_3-Al_2O_3-coated Si_3N_4 particles

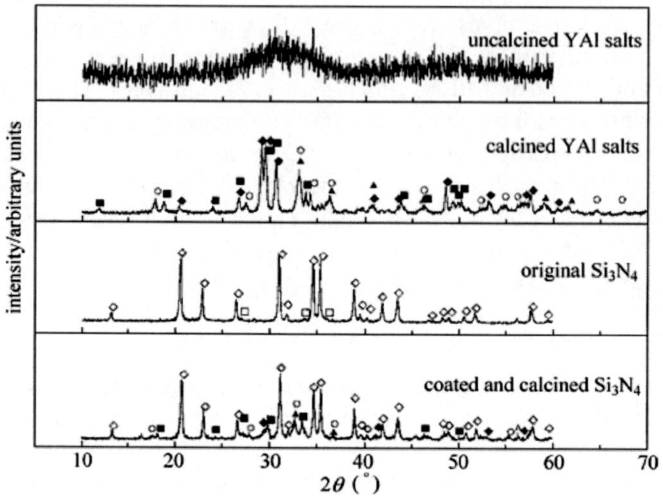

Figure 5.29 The XRD curves of Y_2O_3-Al_2O_3 precursor salts and uncoated/coated Si_3N_4 powders (♦–Y_2O_3, ▲–δ-Al_2O_3, ○–Y_3Al_2 $(AlO_4)_3$), ◊–α-Si_3N_4, □–β-Si_3N_4, ■–$Y_4Al_2O_9$)

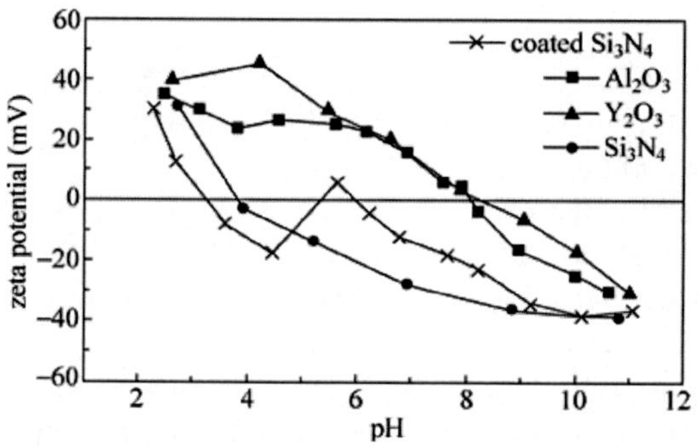

Figure 5.30 The zeta potentials of some ceramic powders

during the acid-base titration of oxides. Furthermore, the study attributed these three "charge" reversals, with an increase in pH, to the native substrate (the surface of the Si_3N_4 in the present work), the onset of surface hydroxide precipitation (Y_2O_3-Al_2O_3 in the present work), and the mixed substrate/precipitate surface, and theorized that the appearance of three IEPs indicates inadequate coverage of Y_2O_3-Al_2O_3 on the surface of the Si_3N_4 particles, as observed by using the TEM.

At any rate, when the original or the coated Si_3N_4 was suspended in an aqueous solution, the absolute values of its zeta potentials were higher at a high pH, inducing higher electrostatic interparticle repulsion, which is important for establishing a colloidally stable suspension. Thus, both the original and the coated powders easily got dispersed in an alkaline solution. Figure 5.31 shows the flow curves of the slurries prepared from the two powders. At pH = 10.5 ± 0.2, the viscosity of the coated Si_3N_4 slurries with the same solid loading as that of the uncoated Si_3N_4 slurry decreased significantly. When the solid loading were increased to 50.6 vol%, the viscosity of the coated Si_3N_4 slurry was still much lower than that of the uncoated Si_3N_4 slurry with a solid loading of 45 vol%.

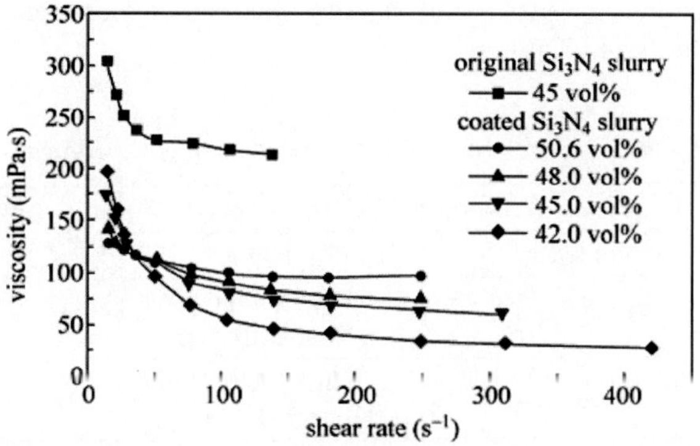

Figure 5.31 The viscosity comparison of suspensions of original and coated Si_3N_4 powders

However, the increase in the rheological properties of the coated Si_3N_4 slurries has not been adequately explained by the zeta-potential data presented in Fig. 5.30, where the zeta-potential for the coated and original Si_3N_4 powders are almost identical at $pH = 10.5 \pm 0.2$. The electrokinetic behavior of a powder is only one of the contributors to the colloidal characteristics of the powder. In the present work, changes in the particle-size distribution and shape of the Si_3N_4 powder may contribute to the increase in the rheological properties of the coated Si_3N_4 slurry; however, confirmation of the exact contributors requires further investigation.

5.2.1.3 The Structure and Characteristics of Gelcast Si_3N_4 Green Bodies

Following the solidification of the slurry prepared from the uncoated Si_3N_4 powder, numerous millimeter-sized macropores were found in the green bodies, as shown in Fig. 5.1. As the slurry had been vibrated and vacuumed before it got solidified, there were no macroscopic bubbles remained. Therefore, the macropores in the green bodies were not introduced by the bubbles in the slurry. Thus, the macropores must have been formed during the solidification of the slurry. The relative quantity of the ingredients in the Si_3N_4 slurry indicates that the three gas-discharging reactions may occur in the system (see reaction (5.1) – (5.3)).

During gel-casting of the original Si_3N_4, a high content of the basic dispersant, TMAH, in the Si_3N_4 suspensions produced large amounts of bubbles in the slurries, even after milling only for several hours. A previous study (Raider S. I., et al., 1976) reported that Si_3N_4 can be hydrolyzed in an aqueous solution, even at room temperature and neutral pH, which confirms reaction (5.1). This hydrolysis is an intrinsic difficulty for gel-casting Si_3N_4-containing aqueous suspensions. During our experiments, we continuously vacuumed the slurry-containing flask, to remove the bubbles from the suspensions before casting, but this exercise failed to create a macropore-free original Si_3N_4 green body after solidification. To determine the macropore-forming reaction during the solidification of the slurry system, we conducted a series of gas-gathering experiments. The experimental arrangement and results are shown in Table 5.9.

Table 5.9 The results of gas chromatography experiments

experiment No.	reactant (reacting at 80℃ for 30 min)		pH	H_2 (mL)	N_2 (mL)
	powder	solution			
1	Si_3N_4, 40 g	H_2O, 250 mL, $(NH_4)_2S_2O_8$, 5 g	3.36	0	30.8
2	Si_3N_4, 40 g	H_2O, 250 mL, $(NH_4)_2S_2O_8$, 0.04 g	6.60	0	0
3	Si_3N_4, 40 g	H_2O, 250 mL, TMAH①, 12.5 mL	12.50	96.5	0
4	Si, 0.25 g	premix②, 250 mL	4.02	20.2	0
5	Si, 0.25 g	H_2O, 250 mL, TMAH, 6.0 mL	12.20	146.4	0
6	UBE Si_3N_4, 40 g	H_2O, 250 mL, TMAH, 12.5 mL	12.50	0	0
7	oxidized $Si_3N_4$③, 40 g	H_2O, 250 mL, TMAH, 12.5 mL	12.50	0	0
8	coated $Si_3N_4$④, 40 g	H_2O, 250 mL, TMAH, 12.5 mL	12.50	0	0

Note: ① TMAH: tetramethyl ammonium hydroxide ($(CH_3)_4NOH$), 10 wt%.
② Premix solution: $C_2H_3CONH_2$ (AM), $(C_2H_3CONH)_2CH_2$ (MBAM), and deionized water.
③ Oxidized at 800℃ for 2 h.
④ Coated Si_3N_4 coated by Al_2O_3 and Y_2O_3.

No ammonia was detected in any of the experiments, possibly because of its relatively high solubility in water (Zhou Ninghuai, 1980) and the limited resolution of the analytical instrument used, but the appearance of ammonia during the experiments was confirmed by its smell.

Large amounts of N_2 gas were detected in experiment 1, suggesting that the reaction (5.2) could occur in the system at a high concentration of initiator $(NH_4)_2S_2O_8$. However, during the practical gel-casting process, the actual amount of the initiator added to the slurry is much lower than that used in experiment 1, which would generate a negligible amount of discharged nitrogen gas, as certified by experiment 2. Thus, the reaction (5.2) cannot be considered as a major cause for the high amount of macropores in the gelcast Si_3N_4 green bodies. Moreover, experiment 3 illustrated that Si_3N_4 powder can also react with the TMAH ($(CH_3)_4NOH$) water solution and discharge H_2 gas at 80℃. A small amount of free silicon was present in the Si_3N_4 powder used in our experiments. Therefore, the free silicon logically could be expected to hydrolyze the subsequent reaction (5.3), releasing H_2 gas.

Experiment 4 indicated that silicon powder could get hydrolyzed in an aqueous solution at 80℃, discharging a small amount of H_2 gas. Experiment 5 further indicated that more H_2 gas was produced in the system when a basic dispersant was added. The Si_3N_4 powder used for gel-casting was prepared by nitrification of the silicon powder. The extent of the reaction did not reach 100%, and thus, more or less free silicon remained in Si_3N_4 powder. The content of free silicon contained in the powder provided by the manufacturer is <1 wt%. The UBE Si_3N_4 powder was prepared by the decomposition of imide, thus contained no free silicon. The absence of H_2 gas in experiment 6 certified the difference between the two powders. Finally, no H_2 gas was detected in experiments 7 and 8 when the Y_2O_3-Al_2O_3-coated powder or oxidized Si_3N_4 powders were heated (800℃ for 2 h in air) in the basic aqueous solution, which indicates that surface modification caused by the coating and surface oxidation of Si_3N_4 could effectively restrain the deleterious gas-discharging reactions in the slurry system.

Several early investigations (Hackley V. A., et al., 1997; Popper P., et al., 1960; Ezis A., 1974) have reported that silicon has a tendency to hydrolyze in water, which is an exothermic reaction, and that difficulties are associated with silicon processing in an aqueous environment. When silicon was mixed with an aqueous solution in those studies, hydrogen-gas evolution was observed. The degree of reactivity increased as the particle size decreased and the alkalinity increased, and severe gassing was reported for $pH>10$. According to the present authors' knowledge, no early study has reported the reaction between free silicon contained in Si_3N_4 powder and an alkaline solution, partly because of the fact that the free silicon in Si_3N_4 powder is commonly much lesser, and thus, the reaction is slow and inconspicuous and partly because Si_3N_4 slurries with $pH>10$ were not heated to temperatures as high as 70 – 80℃ in early researches.

The results presented in Table 5.9 indicate that the hydrolysis of free silicon in Si_3N_4 powder in the basic dispersant-containing aqueous solution was one of the major causes of macropores in the present Si_3N_4 green bodies. However, it is not clear whether the reaction between Si_3N_4 and TMAH in the solution was the other major cause responsible for macropores, because the amount of NH_3 in our aqueous system was difficult to detect, owing to its high solubility in water.

As the coating layer of Y_2O_3-Al_2O_3 effectively restrained the deleterious gas-discharging reactions in the slurry system, macropore-free green bodies were obtained via gel-casting, as shown in Fig. 5.32(b). According to our analysis, three mechanisms contributed to the elimination of macropores in the green bodies:

① The coated Y_2O_3-Al_2O_3 layer acted as a barrier, retarding contact and reaction between the basic solution and the free silicon contained in the Si_3N_4 powder.

② The content of free silicon was decreased by hydrolysis before gel-casting, when the Si_3N_4 powder was treated in the aqueous coating concentrate at 50℃ for 48 h.

③ The content of free silicon was decreased by nitridation before gel-casting, when the yttrium-aluminum salt-coated Si_3N_4 powder was calcined in N_2 gas at 1000℃ for 2 h.

Further investigation is necessary to identify the exact roles of the three mechanisms.

5.2.1.4 The Improvement of Mechanical Properties

Both Al_2O_3 and Y_2O_3 can be simultaneously coated onto the surface of Si_3N_4 particles in the same proportion and quantity as used for sintering aids. With this type of coated powder, complex shaped green bodies can be gelcasted and then sintered into final components.

The relative densities of the sintering samples prepared by either gel-casting coated Si_3N_4 or by cold isostatically pressing of the original Si_3N_4 were 98% of the theoretical density. The mechanical characteristics of the Si_3N_4 ceramic bodies prepared by gel-casting the coated Si_3N_4 powder were much better than those of the bodies prepared by cold isostatically pressing of the original powder. The room-temperature bending strength and Weibull modulus of the gas-pressure-sintered gelcast samples produced from the Y_2O_3-Al_2O_3-coated Si_3N_4 powder, via gel-casting, were 840.57 MPa and 15.55, respectively, whereas those of materials prepared from the uncoated powder, by conventional cold isostatic pressing, were 582.70 MPa and 6.04, respectively.

The micrographs of the polished and etched surfaces of the two samples, as shown in Fig. 5.32, may explain the discrepancies in their mechanical performances. Figure 5.32 also shows that the grain sizes of the cold isostatically pressed samples were nonuniform and that some of the particles were very large and obviously spherical, which may be the main reason for its low bending strength. The grains in the coated and gelcast sample, on the other hand, were spindle-shaped, which indicates that the coating promoted liquid formation and grain rearrangement but not excessive grain growth. Apparently, homogeneous distribution of the sintering aids is considered to be beneficial for the formation of green bodies with uniform microstructures by coating and gel-casting and may improve the mechanical properties of the sintered bodies.

Differences in grain shape can also influence the rupture mode of the samples, as shown in Fig. 5.33. The large grains in Fig. 5.33(a) can be easily removed, because they have large interface areas and high interfacial strengths. Thus, when

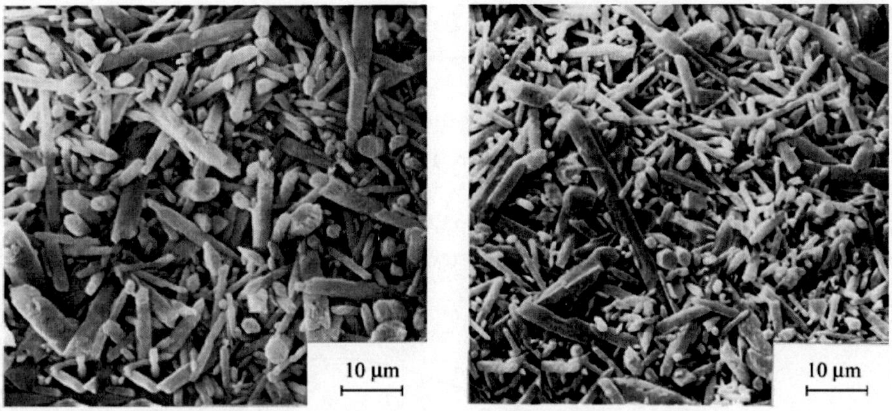

(a) (b)

Figure 5.32 The micrographs of polished and etched planes of different GPS Si_3N_4 samples
(a) uncoated powder, by dry and isostatic pressing; (b) coated powder, by gel-casting

the angle between the interface and the crack plane is relatively large, the crack will propagate mostly through the interface. When the angle is small, the crack will spread mostly along the grain boundary, and, as a result, the grain will flake off. The grains in Fig. 5.33(b) are more spindle-shaped, and thus, more grains are found to flake off and pull out. All these phenomena increase distance and energy dissipation during crack spreading, thus increasing the toughness of the sample. Therefore, the spindle shape of the grains may be another reason, besides grain uniformity, for the high Weibull modulus of the coated samples.

(a) (b)

Figure 5.33 The micrographs of fracture surfaces of different GPS Si_3N_4 samples
(a) uncoated powder and produced by dry and isostatic pressing; (b) coated powder and produced by gel-casting

5.2.2 The Preparation of Si_3N_4 Ceramics with Surface-Oxidized Si_3N_4 Powder

A method of oxidation to improve water-based gel-casting of Si_3N_4 was studied in this section. This method is not only effective but also easy for utilization. Because an effective coating layer of silica (SiO_2) was formed on the surface of Si_3N_4 powder.

The Si_3N_4 powder (Beijing Founder High Technology Ceramics Corporation) was used in this study with an average size $d_{50} = 0.6$ μm (see Fig. 5.34).

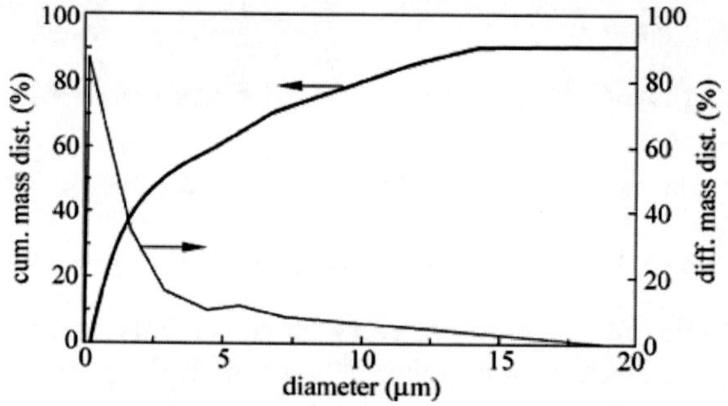

Figure 5.34 The particle size of cum. and diff. mass distribution of Si_3N_4

The photoelectron spectra were obtained by using an ESCALAB220i-XL electron spectrometer with a base pressure of 1×10^{-9} mbar and an EP300 electron spectrometer with base pressure of 1×10^{-8} mbar. The apparent viscosity of Si_3N_4 suspensions was measured by rotating viscosimeter (model NSX-11, Chengdu Instrument Plant, China). The ion concentration of the suspension was measured by IONPACDX-100 ion chromatography instrument. The ion conductivity of suspensions was measured by ion conductivity meter (model DDS-II A_T, Shanghai Leici Instrument Plant, China).

5.2.2.1 The Influence of Surface Oxidation on the Composition and Surface Status of Si_3N_4 Powder

Under the conditions of high temperature and high oxygen pressure, the following reaction takes place:

$$Si_3N_4 + 3O_2 = 3SiO_2 + 2N_2 \uparrow \tag{5.14}$$

After SiO_2 appears, the reaction on the interface between SiO_2 layer and Si_3N_4 powder is controlled by diffusion of oxygen across the SiO_2 layer. In the meantime, the following reactions occur:

$$Si_3N_4 + SiO_2 = 2Si_2N_2O \tag{5.15}$$

$$2Si_2N_2O + 3O_2 = 4SiO_2 + 2N_2 \uparrow \tag{5.16}$$

The Si_3N_4 powders were oxidized in the molybdenum silicide furnace, by maintaining the temperature for 2 h at 600, 800, 850 and 900℃ respectively. The weight of oxidized powders was compared with that of the raw powders, and the results were shown in Fig. 5.35. The thermal gravity analysis (TGA) curves of Si_3N_4 at 600, 800, 850, and 900℃ respectively were shown in Fig. 5.36.

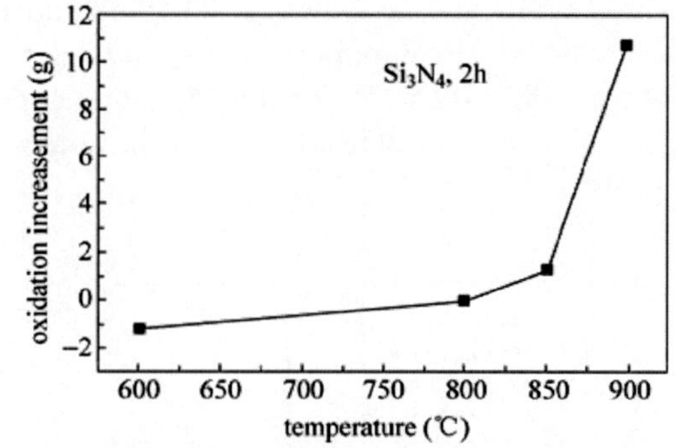

Figure 5.35 The oxidation increasement of Si_3N_4 powder oxidized in the molybdenum silicide furnace for 2 h

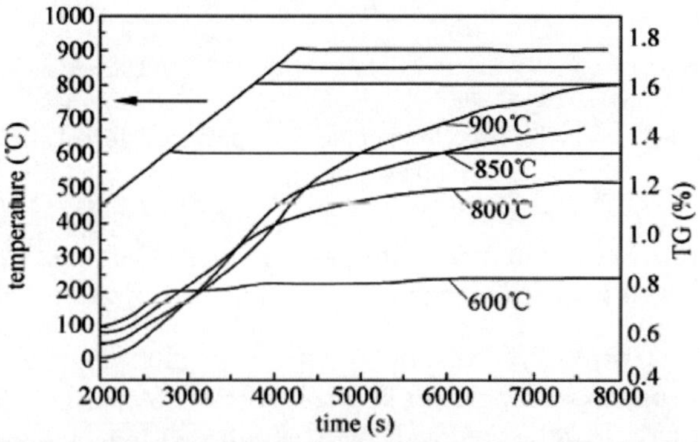

Figure 5.36 The TGA curves of oxidized Si_3N_4 at different temperature

It can be observed through Fig. 5.36 that the oxygen content from 600 to 850℃ increases slowly, but it rises sharply when the oxidizing temperature reaches 900℃. Furthermore Fig. 5.37 indicates that the gain in the oxidation speed at 900℃ is relatively high. This is because that the transformation from α-quartz ($\rho = 2.533$ g/cm^3) to α-tridymite ($\rho = 2.228$ g/cm^3) at 870℃ takes place in the SiO_2 layer. Owing to the change of density, SiO_2 film layer cracks, and the oxygen infiltrates through the crack, leading to more active reactions. At 900℃ the sharp oxidation reaction results in the large increase of SiO_2 content in the Si_3N_4 powder. In fact, the real increase of SiO_2 is more than that measured in our study, because the weight loss during the experiment, which can be attributed to the combustion of the oxygen and carbon in the Si_3N_4 powder (Castanho M., et al., 1997a, b), is not considered. As is well known, too much of glass phase derived from excess of SiO_2 will degrade mechanical properties of Si_3N_4 at elevated temperature remarkably. Obviously, it is necessary that the Si_3N_4 powder should be oxidized below 900℃.

Figure 5.37 presents the X-ray photoelectron spectrometer (XPS) resulted from the oxidation powder at 850℃. In Fig. 5.37(a), Si_{2p} peak divides into two, one at 102.30 eV representing Si_2N_2O and the other at 103.49 eV representing SiO_2. In Fig. 5.37(b), N_{1s} peak at 397.91 eV represents Si_2N_2O (Castanho M., et al., 1997a,b; Brow K., et al., 1986). It can be concluded that powder surface consists of Si_2N_2O and SiO_2 mainly, and that the surface modification based on oxidation can yield a layer of coating.

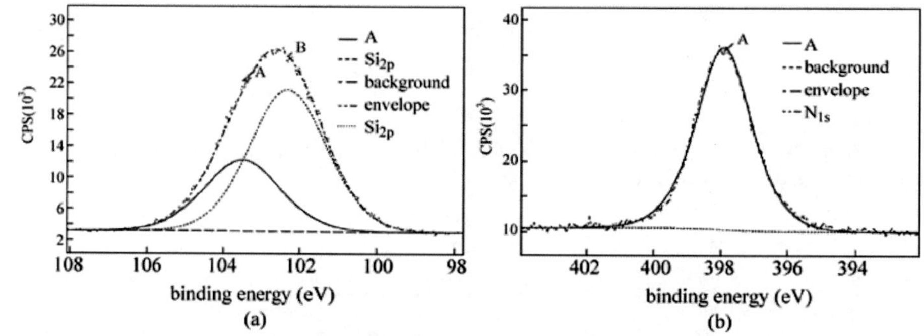

Figure 5.37 The XPS spectrum of the Si_3N_4 powder oxidized at 850℃
(a) Si_{2p}; (b) N_{1s}

5.2.2.2 The Influence of Surface Oxidation on Slurry Colloidal Characteristics and Green Body Structure

In the gel-casting process of Si_3N_4 some cations in the slurry, namely anti-ions, are very disadvantageous to rheologies of slurries (Kitahara A., et al., 1992). Impurity ions of Si_3N_4 powder can be removed by washing and ions exchange resin, but these methods are expensive and time-consuming. It has been found in our study that oxidizing at high temperature could easily and effectively reduce the ion concentration in the slurry. Original Si_3N_4 powder and several kinds of oxidized Si_3N_4 powders were dispersed into deionized water respectively and corresponding homogenous suspensions were acquired. After a few days, they were centrifuged and the ion conductivity of the watery solution on top of the suspensions was measured. Figure 5.38 shows the curves of the ion conductivity versus time. With the rise of the oxidizing temperature, the ion conductivity of the watery solutions decreased greatly, indicating that the ion concentration in the slurry decreased remarkably. The decrease may be attributed to the isolation of the SiO_2 layer and the change of the ions' formation on the surface of Si_3N_4. Table 5.10 presents the variation of K^+ and NH_4^+ concentration versus time. It can be seen from the table that the concentration of both K^+ and NH_4^+ decreased after the powder has been oxidized. The higher the oxidizing temperature, the more greatly the ion concentration decreased. The decreasing of K^+ concentration may be owing to the volatilization of K salts and the formation of insoluble solid solution. The concentration of NH_4^+ decreased obviously because of the decomposition of ammonium salts and the isolating function of the oxide layer on Si_3N_4 particles. In addition, Fig. 5.38 indicates that ion conductivity of the oxidized powder varies relatively smoothly, while that of the raw powder varies rather sharply. Thus, it could be concluded that a homogeneous coating was formed on the surface of the oxidized powder retarding the deleterious reaction on the surface of the Si_3N_4 powder effectively, and that after oxidizing the improvements of rheologies of Si_3N_4 slurries should be partly attributed to the decrease of impurity ions.

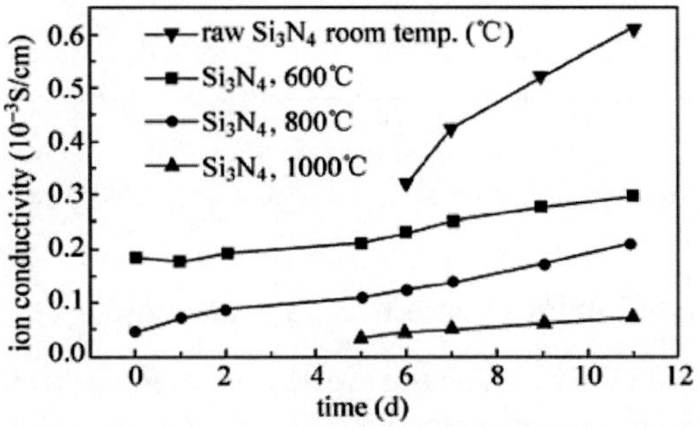

Figure 5.38 The curves of the ion conductivity of the Si_3N_4 suspensions versus time

Table 5.10 The variation of K^+ and NH_4^+ concentration versus time

	original Si_3N_4 powder (0 d)	original Si_3N_4 powder (7 d)	Si_3N_4 powder oxidized at 600℃ (11 d)	Si_3N_4 powder oxidized at 800℃ (11 d)
K^+(ppm)	3.47	3.43	1.13	0.59
NH_4^+(ppm)	24.8	92.73	10.57	1.21

Figure 5.39 shows the viscosity at different shear rates of the slurries of the Si_3N_4 powders oxidized at different temperature and with the same solid loading of 42 vol% (the solid density is considered as 3.18 g/cm^3). It indicates that below 850℃ the suspension viscosity decreased along with the increase of oxidizing temperature, and that above 850℃ the suspension viscosity increased again because of the sharp increase of SiO_2 content and the resulting increase of solid loading. To obtain slurries with low viscosity, 850℃ was determined as oxidizing temperature.

As has been stated before, some gas-discharging reactions between Si_3N_4 particle and medium occur when the raw Si_3N_4 powder comes into contact with the medium. In the gas-gathering experiment (Dai J. Q., et al., 1999), the harmful reactions were detected in the slurry of the unprocessed Si_3N_4 powder. And no reaction was detected between oxidized Si_3N_4 powder and the basic medium, which indicated that surface oxidizing could effectively restrain the deleterious gas-discharging reactions in the slurry system.

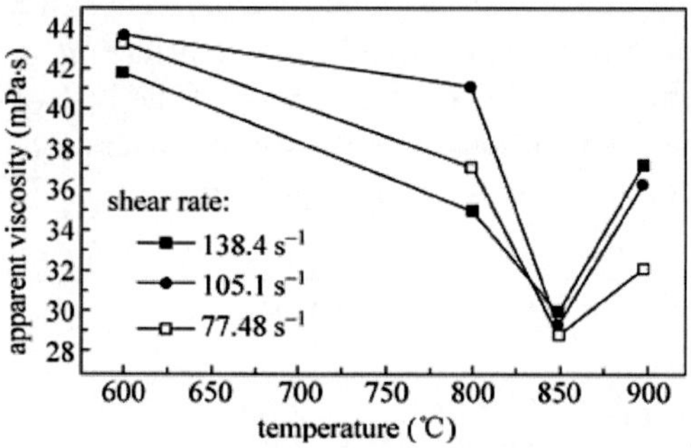

Figure 5.39 The viscosity of suspensions of oxidized Si_3N_4 powders at different temperatures

5.2.2.3 Influence of Surface Oxidation on Sintering

Because oxygen content in Si_3N_4 powder increased after oxidizing, the liquid phase in the process of sintering also increased. This influences the mechanical property and microstructure of samples. The mechanical properties of the gelcast samples of the oxidized powder as well as the dry-pressed and then isostatic-pressed samples are shown in Table 5.11.

Table 5.11 The comparison of mechanical property of different samples

	GPS	
	bending strength at room temperature $\sigma_{f,RT}$ (MPa)	Weibull modulus m
gel-casting of oxidized powder	785.16	11.78
dry press and isostatic pressed sample	582.7	6.04

It is shown in Table 5.11 that mechanical property of gelcast samples is much better than that of dry-pressed and then isostatic pressed samples. The difference in this mechanical property may be owing to the defects caused by inhomogeneity in the dry-pressed and then isostatic pressed samples, as well as their different grain boundary phases.

5.2.3 The Preparation of Si_3N_4 Ceramics with Combination Processing

This section aims to produce high-reliability silicon nitride ceramics via processing strategy using a combination of coating, oxidizing, gel-casting, cold isostatic pressing and gas pressure sintering.

5.2.3.1 Experimental Procedure

1) Materials and Reagents

The Si_3N_4 powder used in this study (H. C. Starck, M-11, Germany) consists of 91.7% β-phase. It exhibits a specific surface area of 12.85 m^2/g and a mean particle size of 430 nm. The major impurities in the powder are C and O with 0.20 and 1.44 wt%, respectively, and a total metal content (Fe + Al + Ca) of 0.058 wt%. $Y(NO_3)_3 \cdot 6H_2O$ and $AlCl_3 \cdot 6H_2O$, coprecipitated with $NH_3 \cdot H_2O$, were applied as precursors of Y_2O_3-Al_2O_3 sintering aids coated on the surface of Si_3N_4 particles. Deionized water was employed as a reacting medium and, followed by alcohol, as a washing liquid in the coating process. Powders of yttria and alumina (Jiangsu Wuxian Special Ceramics Plant, China) with a purity of 99%, were used as sintering aids for uncoated Si_3N_4 powder.

2) Surface Modification of Si_3N_4 Powder

To adjust the colloidal characteristics of the Si_3N_4 powder and the properties of the resultant ceramics, the as-received powder was coated with its sintering aids and then oxidized in air.

The coating process of Si_3N_4 particles with Y_2O_3-Al_2O_3 sintering aids is shown in Fig. 5.40. The precursors $Y(NO_3)_3 \cdot 6H_2O$ and $AlCl_3 \cdot 6H_2O$ were firstly dissolved in deionized water at a concentration of 0.4 mol/L followed by 2 h of stirring. The resulting clear liquid is called the "coating solution". Secondly, as-received Si_3N_4 powder was dispersed in deionized water to prepare an aqueous Si_3N_4 suspension with solid loading of 10 wt% and with a basicity of pH = 10 adjusted by addition of $NH_3 \cdot H_2O$. The suspension was then stirred and ultrasonically agitated for 4 h to achieve high homogeneity. Afterwards, a certain amount of coating solution was dripped into the suspension at a rate of 10 drops/min, which was vigorously stirred and kept constant at pH = 10 by simultaneous addition of concentrated $NH_3 \cdot H_2O$. The mixture was kept stirred for another 30 min after the coating solution was completely added. Nitrate and chlorine ions as well as deionized water were totally eliminated by washing and filtering for 4 times with deionized water and for another 4 times with alcohol. The resultant Si_3N_4 powder coated with Y-Al hydroxides was then dried at 80℃ for 24 h and passed through a sieve (100 mesh) followed by calcination in N_2 atmosphere for 2 h. Pure Y-Al hydroxides were also prepared, analyzed by DTA, calcined at different temperatures and detected by using XRD to determine a proper calcinations temperature for the Si_3N_4 powder. The content of sintering aids was adjusted to 7.5 wt% Y_2O_3 and 2.5 wt% Al_2O_3. After calcination, the coated Si_3N_4 powder was sieved with a 100 mesh and oxidized at 600℃ for 3 or 6 h in air to reach a certain oxygen content on the surface of the powder which corresponds to a silica layer. For comparison, as-received Si_3N_4 powder was also oxidized at 600℃ for 3 h in air.

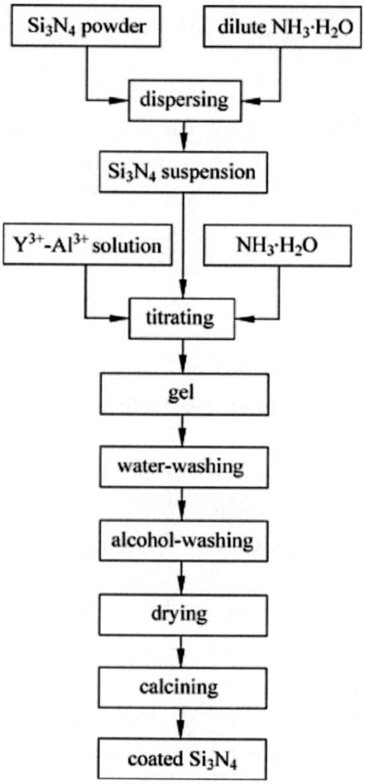

Figure 5.40 The process for coating of Si_3N_4 powder with sintering aids

3) Forming and Sintering

Gel-casting was employed to prepare Si_3N_4 green bodies with high homogeneity, which can be further improved by subsequent cold isostatic pressing. The gel-casting process used was similar to that used in previous studies (Young A. C., et al., 1991; Omatete O. O., et al., 1997). First, Si_3N_4 powder with 7.5 wt% Y_2O_3 and 2.5 wt% Al_2O_3 was suspended with TMAH as dispersant in the premixed solution of AM and MBAM. The suspensions at $pH = 10.5 \pm 0.2$ were degassed for 5 min after rolling for 24 h in polyethylene bottles using alumina balls as milling media. The slurry was degassed for another 3 min, when the initiator and catalyst were added. All the above mentioned operations were carried out at room temperature. Afterwards, the slurry was cast into a nonporous mold, which was then kept at

60 – 80℃. After consolidation, the green bodies were demolded and dried under controlled humidity to avoid cracking and non-uniform shrinkage owing to rapid drying. Subsequent binder burnout was carried out at 600℃ for 3 h in air followed, if necessary, by cold isostatic pressing for 1 min at a pressure of 200 MPa.

GPS of the green bodies was carried out in a KCE furnace (FPW 180/250-2200-100-SP, KCE Corp., Germany). The GPS schemes were as follows: 1750℃ for 1.5 h under the N_2 pressure of 0.3 MPa and then 1900℃ for 1.0 h (sample SN5), 1.5 h (samples SN1–SN4), 2.0 h (sample SN6) or 2.5 h (sample SN7) under the N_2 pressure of 6 MPa with both heating and cooling rates of 10℃/min.

5.2.3.2 The Influence of Surface Modification on the Colloidal Properties of Si_3N_4 Powder

The coating followed by oxidation is applied to modify the surface characteristics of Si_3N_4 powder, and its rheological performance. For the coating process, a proper calcination procedure should be determined for the transformation of Y-Al hydroxide to Y-Al oxide on the surface of the powder. DTA and XRD were selected for this aim. The DTA curve of pure Y-Al hydroxides reveals an exothermic peak at 892℃, which corresponds to the transformation temperature (see Fig. 5.41). However, calcination of the hydroxides at 900℃ for 2 h in N_2 atmosphere fail to produce crystalline oxides; only when calcined at 1000℃ for 2 h, the resulting products are clearly crystalline (see Fig. 5.42).

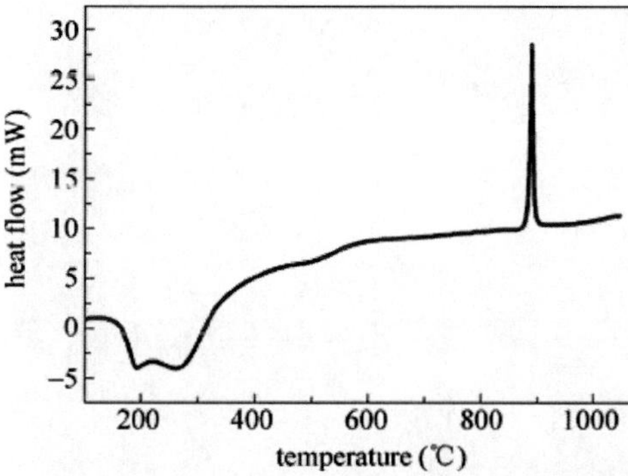

Figure 5.41 The TGA curve of transformation from Y-Al hydroxide to Y-Al oxide

The TEM micrographs of Y-Al oxide coated Si_3N_4 particles are shown in Fig. 5.43. It can be observed in this figure that a layer of nano-sized precipitates is closely attached to the surface of large Si_3N_4 particles and the thickness of precipitates varies from about 5 to 40 nm. A more careful observation of the figure shows that the surface of small Si_3N_4 particles is not completely covered by precipitates. This is a typical feature for the coating of sub-micron powder with nano-sized precipitates, since the smallest Si_3N_4 particles are also in the nano-size regime. However, colloidal characteristics and rheological properties of the surface-modified powder are still obviously adjusted.

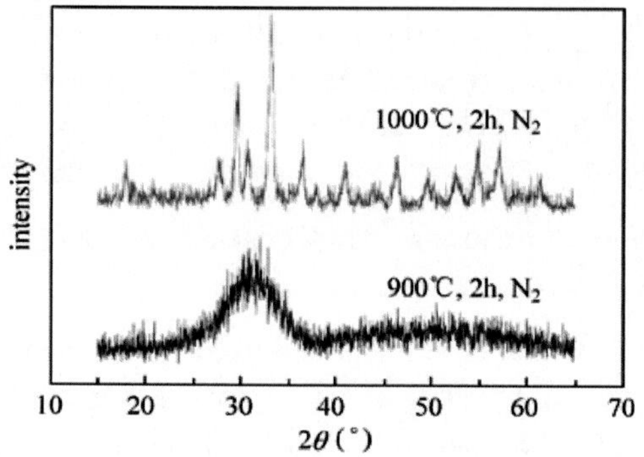

Figure 5.42 The XRD patterns of Y-Al hydroxides calcined at different temperatures

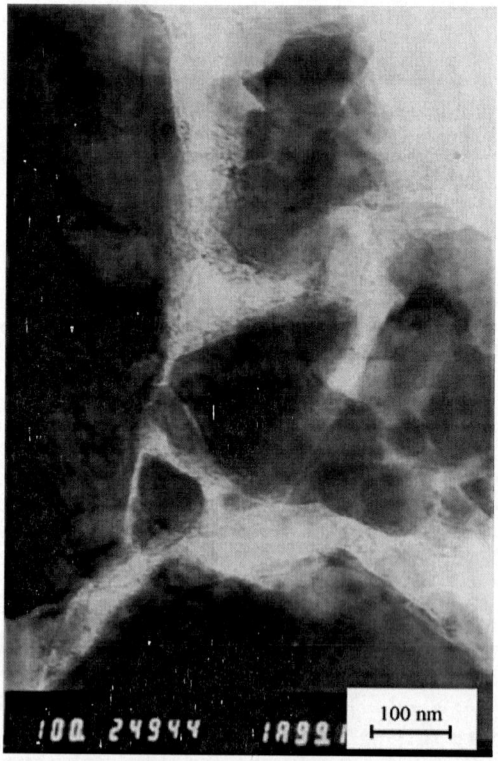

Figure 5.43 The TEM micrograph of coated Si_3N_4 particles

As reported in previous studies (Huang Y., et al., 2001; Tang Q., et al., 1998), the zeta potentials of coated Si_3N_4 particles are rather complex. Unlike any curve of pH-dependent zeta potentials for Al_2O_3, Y_2O_3 or Si_3N_4, the zeta potential for coated Si_3N_4 powder possesses three IEPs, i.e. three "charge" reversals in the pH range, which indicates an inadequate coverage of Y-Al oxides on the surface of Si_3N_4 particles, as observed by TEM. However, the dispersibility of the powder could be improved by coating. Earlier studies also demonstrate that the dispersion of Si_3N_4 powder is enhanced after being oxidized below 850℃ in air. In the current study, it is found that the dispersibility of the coated and oxidized (600℃, 3 h in air) powder is improved when compared with that of the as-received powder (see Fig. 5.44). At pH $= 10.5 \pm 0.2$, the viscosity of slurries from coated and oxidized Si_3N_4 powder with the same solid loading is greatly decreased when compared with the original Si_3N_4 slurry. When the solid loading is increased up to 50 vol%, the viscosity of the coated and oxidized Si_3N_4 slurry is still much lower than that of the as-received Si_3N_4 slurry although it has a lower solid loading of 45 vol%. The reason is complex, but it is commonly believed that oxides on the surface of non-oxides such as Si_3N_4 lead to surface properties similar to those of oxides, which exhibit good dispersibility in aqueous media (Fagerholm H., et al., 1996).

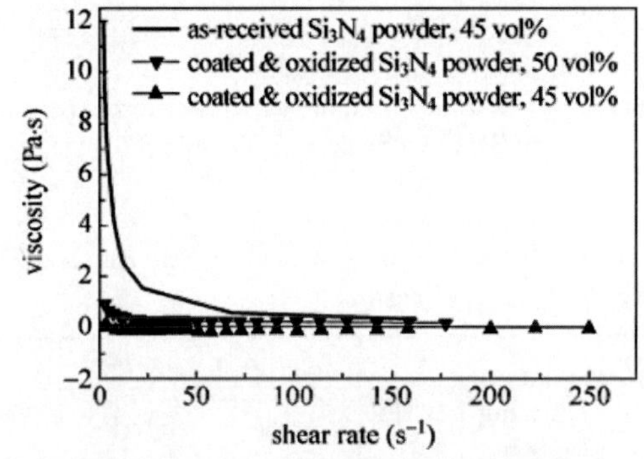

Figure 5.44 The rheological curves for the suspensions of as-received and surface modified (coated and oxidized) Si_3N_4 powders

5.2.3.3 The Influences of Coating, CIP and Oxidation on the Properties of Si_3N_4 Ceramics

In previous studies, it was found that relative to dry pressing of original Si_3N_4 powder with mixed sintering aids, both gel-casting of Si_3N_4 powder with coated sintering aids and gel-casting of oxidized powder with mixed sintering aids were able to produce Si_3N_4 ceramics with higher bending strength and reliability. Hence, it was concluded that coating and oxidizing as well as gel-casting were beneficial for the resulting properties of the material. In this study, we combined coating, oxidation, gel-casting and CIP to prepare Si_3N_4 ceramics and to investigate their influences on the final properties of the material. There were four batches of Si_3N_4 specimens, each one corresponding to a processing strategy listed as follows:

SN1: mixing sintering aids + oxidation (600℃, 3 h) + gel-casting + GPS
SN2: coating sintering aids + oxidation (600℃, 3 h) + gel-casting + GPS
SN3: coating sintering aids + oxidation (600℃, 3 h) + gel-casting + CIP + GPS
SN4: coating sintering aids + oxidation (600℃, 6 h) + gel-casting + CIP + GPS

The GPS scheme was 1750℃ for 1.5 h under the N_2 pressure of 0.3 MPa and then 1900℃ for 1.5 h under the N_2 pressure of 6 MPa. The processing strategies and their results are shown in Fig. 5.45 and Table 5.12.

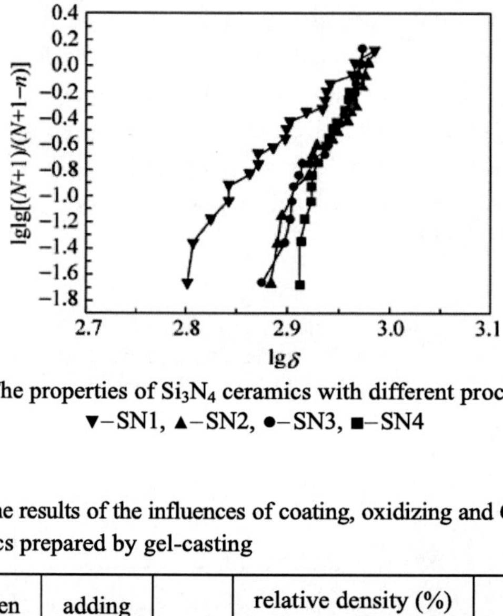

Figure 5.45 The properties of Si_3N_4 ceramics with different processing schedules ▼–SN1, ▲–SN2, ●–SN3, ■–SN4

Table 5.12 The results of the influences of coating, oxidizing and CIP on properties of Si_3N_4 ceramics prepared by gel-casting

No.	surface oxygen content of Si_3N_4 powder (wt%)	adding sintering aids by	CIP	relative density (%)		bending strength (MPa)	Weibull modulus
				green bodies	sintering bodies		
SN1	2.94 (600℃, 3 h)	mixing	No	47.7	99.0	773.5 ± 95.1	8.6
SN2	2.73 (600℃, 3 h)	coating	No	47.7	98.9	814.5 ± 68.5	12.4
SN3	2.73 (600℃, 3 h)	coating	Yes	64.2	98.8	868.9 ± 59.8	15.1
SN4	3.64 (600℃, 6 h)	coating	Yes	64.2	99.2	877.2 ± 40.5	22.8

The surface oxygen contents of the Si_3N_4 powders presented in Table 5.12 do not include those of Y-Al oxides coated on the surface of the powder. The original Si_3N_4 powder contains 1.44 wt% oxygen, with 0.72 wt% in the bulk and

0.72 wt% on the surface. When oxidized at 600℃ for 3 h in air, the oxygen content on the surface of the original powder is up to 2.94 wt%, while that of the coated powder is up to 2.73 wt%, which is slightly lesser, indicating that the coated layer on the surface of Si_3N_4 powder inhibited to a certain extent by the oxidation of the powder. The restraint of existing layers on the further oxidation of Si_3N_4 powder is also illustrated by the influence of the oxidation time at 600℃: for 3 and 6 h, the increments are not proportional to the oxidation times.

Even though there remains a small difference in the oxygen contents of the Si_3N_4 powders for SN1 and SN2 samples, it can be assumed that the main contributor to higher reliability of SN2 relative to SN1 is the addition of sintering aids via coating. There are three advantages of coating Si_3N_4 powder with its sintering aids in aqueous gel-casting: ① modification of the surface characteristics of Si_3N_4 particles to improve their dispersibility and, thus, the homogeneity of its slurry; ② uniform distribution of sintering aids in green bodies to promote sintering processes; ③ partial separation of the surface of Si_3N_4 powder from the aqueous media with high basicity to avoid the formation of macro-pores in green bodies. These advantages result in uniform Si_3N_4 green bodies, homogeneous sintered bodies and consequently high reliability of the material.

In Table 5.12, the influence of CIP on bending strength and reliability of Si_3N_4 ceramics gel-casted from coated and oxidized Si_3N_4 powder becomes evident by comparing SN2 and SN3. Bending strength and Weibull modulus of the former are (814.5 ± 68.5) MPa and 12.4, while those of the latter are measured to be (868.9 ± 59.8) MPa and 15.1, which indicates that CIP obviously improves the reliability of the material. This can be explained by analyzing the microstructures of green bodies obtained with and without CIP, as shown in Fig. 5.46. The Si_3N_4 powder is chemically unstable in aqueous media, especially in alkaline solution. There were millimeter-sized macro-pores in Si_3N_4 green bodies formed during gel-casting if the processing condition was not properly controlled. By decreasing the pH of the slurry and coating of the Si_3N_4 powder with its sintering aids, green bodies without macropores were prepared (Zhou L. J., et al., 2000). However, there are still some micron-sized pores in the green body, as shown in Fig. 5.46(a). After CIP, not only the micro-pores in the bodies disappears, but also the relative density of the green bodies increases, as can be seen from Fig. 5.46(b), both of which further optimize the homogeneity of the green bodies and the final ceramics. Accordingly, the reliability of the material is increased.

Comparison of SN3 and SN4 samples reveals the influence of the oxygen content of Si_3N_4 powder on bending strength and reliability of Si_3N_4 ceramics, as shown in Table 5.12. The bending strength of sample SN3 and SN4 are almost similar, while the Weibull modulus of the latter is 22.8, which is much more than that of the former with a value of 15.1. Similar to coating and CIP, oxidization was

(a) (b)

Figure 5.46 The micrographs of the coated and oxidized Si_3N_4 green bodies obtained by gel-casting

(a) without CIP; (b) with CIP

found to have more influence on the reliability rather than the strength of the Si_3N_4 based ceramics. The coated Si_3N_4 powders of SN3 and SN4 were oxidized at 600℃ for 3 and 6 h in air, which leads to a surface oxygen content of 2.73 and 3.64 wt%, respectively. The surface oxygen content of Si_3N_4 powder is owing to silica, which also acts as a sintering aid and plays an important role on sintering and densification of green bodies and the properties of the ceramic. Natansohn et al. (1993) investigated the role of surface oxygen and found that there was an optimal oxygen content at which the fracture strength of Si_3N_4 reaches its maximum. They found that the approach was only effective when the oxygen content was adjusted by thermal treatment, i.e. oxidation of Si_3N_4 powder. Other methods of oxygen adjustment, such as silica additions or chemical and physical treatment, did not result in Si_3N_4 based ceramics of equivalent properties. Nevertheless, whether this is the case for the material under investigation remains to be clarified in future. Further investigations are also needed to ascertain the optimal surface oxygen content of the coated Si_3N_4 related to the maximum strength and reliability of Si_3N_4 ceramics. However, the sintering scheme was optimized in this regard, as discussed in the following section.

5.2.3.4 The Influence of GPS Scheme on the Properties of Gelcast Si_3N_4 Ceramics

To determine an optimal GPS scheme for Si_3N_4, three additional schemes with various second holding time were selected beside that of SN4, which are as follows: 1750℃, 0.3 MPa (N_2), 1.5 h(SN4)+1900℃, 6 MPa (N_2), 1.0 h (SN5), 2.0 h (SN6) and 2.5 h (SN7). The resulting properties of Si_3N_4 sintered with those schemes are shown in Table 5.13 and Fig. 5.47.

Table 5.13 The properties of Si_3N_4 ceramics with various second holding time in two-stage GPS

No.	second holding time (h)	relative density (%)	bending strength (MPa)	Weibull modulus
SN5	1.0	98.9	915.9 ± 52.0	18.7
SN4	1.5	99.2	877.2 ± 40.5	22.8
SN6	2.0	99.0	944.7 ± 29.5	33.9
SN7	2.5	98.0	804.2 ± 77.9	10.8

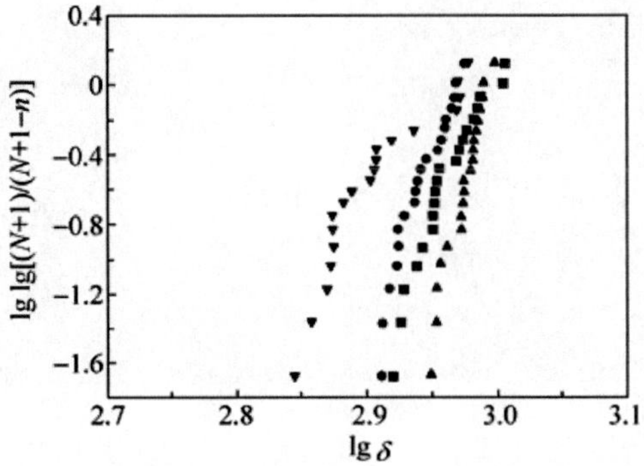

Figure 5.47 The properties of Si_3N_4 ceramics with different holding time in two-stage GPS

■–SN4, ●–SN5, ▲–SN6, ▼–SN7

As can be seen from Table 5.13, a maximal bending strength as high as 944.7 MPa (batch SN6) could be achieved, and that the differences in bending strength for SN4, SN5 and SN6 are relatively small. There is also a maximum of the Weibull modulus as high as 33.9 (batch SN6), but the differences in Weibull modulus are relatively large. This means that in the current investigation the sintering scheme has much influence on the Weibull modulus, but little effect on bending strength of the material. Micrographs of fracture surfaces of the ceramics are shown in Fig. 5.48. It can be seen from (a) to (d) in Fig. 5.48 that there is a more obvious evidence of roughness and pull-out on the fracture surface of the sample SN6 with the highest Weibull modulus. According to previous studies, high Weibull modulus is theoretically and experimentally connected to a strong R-curve effect (Cook R. F., et al., 1988; Tuan W. H., et al., 1994). Previous studies also revealed that bridging and pull-out was one of major reasons for the R-curve behavior (Fang Y., et al., 2000; Yang J. F., et al., 1999). Therefore, both toughening mechanisms are proposed to be also dominant factors for the high reliability of Si_3N_4 ceramics observed in our investigation. However, further investigations have been carried out to explore the mechanism responsible for the high reliability of this material.

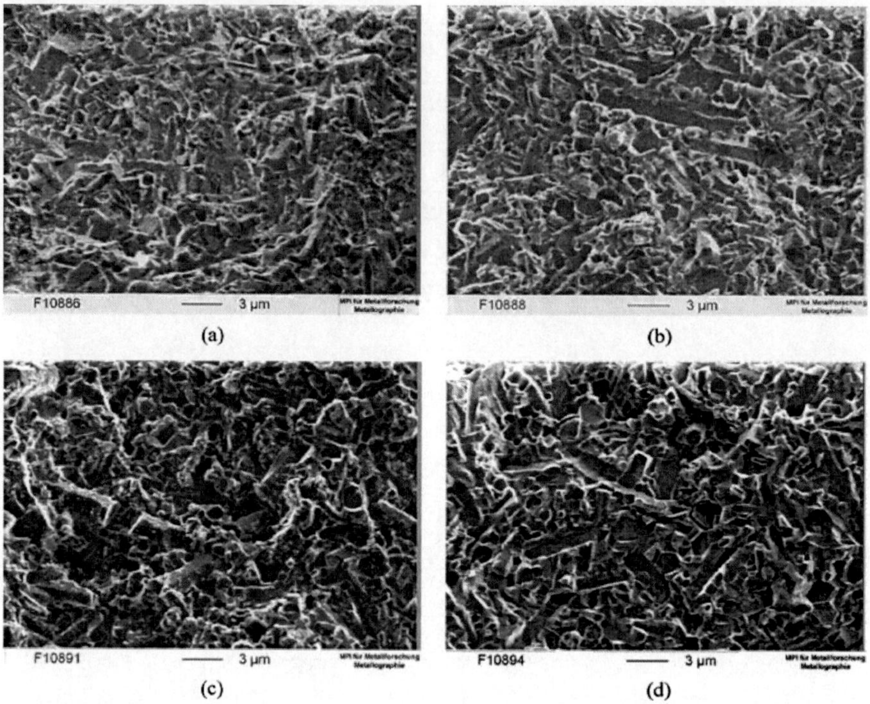

Figure 5.48 The micrographs of Si_3N_4 ceramics with various second holding time in two-stage GPS

(a) 1.0 h; (b) 1.5 h; (c) 2.0 h; (d) 2.5 h

5.3 The Gel-Casting of SiC Ceramic and Si_3N_4 Bonded SiC Ceramic

5.3.1 The Gel-Casting of Concentrated Aqueous SiC Ceramic

This section focuses on investigating colloidal behavior of SiC powder in aqueous solution, preparing high concentrated aqueous SiC slurries and casting complicated shapes of uniform SiC green bodies.

5.3.1.1 The Colloidal Behavior of SiC Powder

The pH-dependent zeta potentials of the SiC powder in aqueous solutions are shown in Fig. 5.49. As shown in the figure, the IEP of SiC in deionized water is acidic, below which zeta potentials are positive, and above which zeta potentials are negative. At strongly pH, the absolute values of the zeta potentials are relatively high.

During synthesis, transportation and storage, SiC powder is inevitably oxidized, forming an oxygen rich layer on its surface. The oxidized layer is chemically similar to the surface of silica (Hackly V. A., et al., 1997). When the powder is treated with water vapor or dispersed in deionized water, the layer is hydrated, causing silanols (Si—OH) on the surface of SiC powder. Silanols are amphoteric, relatively strongly acidic and weakly alkalescent (Crimp M. J., et al., 1986). At low pH range (pH<3.3 in Fig. 5.49), silanols react with OH^-, leaving $Si—O^-$ with negative charges on the surface of the powder, resulting in negative zeta potentials. In low pH range, silanols react with H^+, yielding $Si—OH_2^+$ with positive charges on the surface of SiC powders, resulting in positive zeta potentials. The reactions can be expressed as follows:

$$Si—OH + OH^- \rightleftharpoons Si—O^- + H_2O \tag{5.17}$$

$$Si—OH + H^+ \rightleftharpoons Si—OH_2^+ \tag{5.18}$$

Since silanols are of relatively strongly acidic, they can easily react with OH^-, the absolute values of zeta potential for SiC powder are higher at strongly basic pH than at strongly acidic pH.

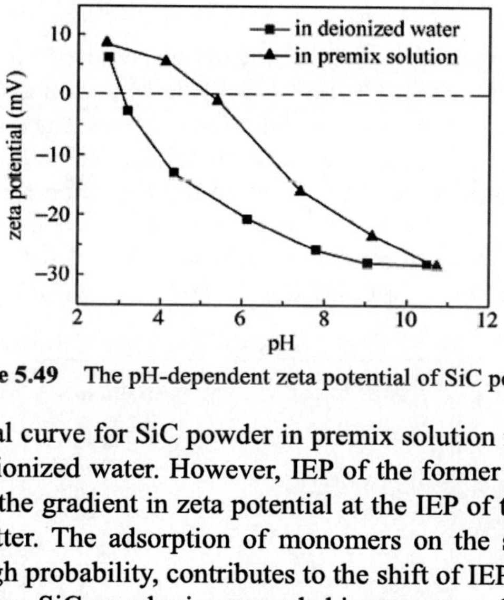

Figure 5.49 The pH-dependent zeta potential of SiC powder

The zeta potential curve for SiC powder in premix solution is similar to that of SiC particles in deionized water. However, IEP of the former is higher than that of the latter, while the gradient in zeta potential at the IEP of the former is lower than that of the latter. The adsorption of monomers on the surface of the SiC particles, with a high probability, contributes to the shift of IEP.

Nevertheless, when SiC powder is suspended in aqueous solution, the absolute values of its zeta potentials are larger at high pH, inducing larger electrostatic inter-particle repulsion, which is important to establish a colloidally stable suspension. So SiC powder is easy to disperse in alkaline solution.

Figure 5.50 shows curves for rheological property of slurries (at 50 vol% solid loading) with various pH. The slurry at pH = 10.60 or 11.38 is found to possess good rheological property. When pH is lower than 10.10, the zeta potential of SiC particle is not sufficient to establish a stable suspension, so the slurry is of shear-thickening and increases sharply in viscosity at high shear rate. When pH is higher than 11.92, ionic strength in the suspension is too high, causing a viscous slurry. Therefore, the suspensions need a proper pH range (10.5 – 11.5), i.e., the dispersant of TMAH should be added to the slurries in an appropriate quantity.

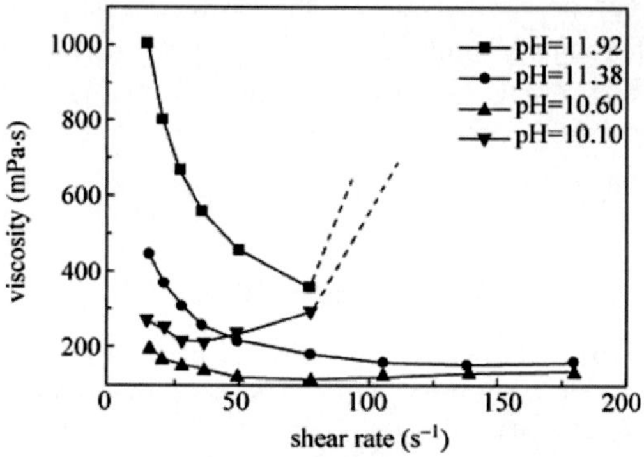

Figure 5.50 The rheological behavior of SiC suspensions at different pH (at 50 vol% solid loading)

Figure 5.51 includes curves for rheological behavior of SiC slurries (at pH = 10.5) at different solid loadings. As shown in the figure, when their solid loading are less than 50 vol%, the viscosity of the SiC slurry increases gently with increase in solid loading, but when they are more than 50 vol%, the fluidity decreases sharply with increase in solid loading. The operation in gel-casting process is responded to shear rate at 100 s^{-1}, so the suspension at 51 or 52 vol% does not fit gel-casting.

The suspensions, with a formulation of 91 wt% SiC, 6 wt% Al_2O_3 and 3 wt% Y_2O_3, were developed for consolidation in this study. The viscosities of these

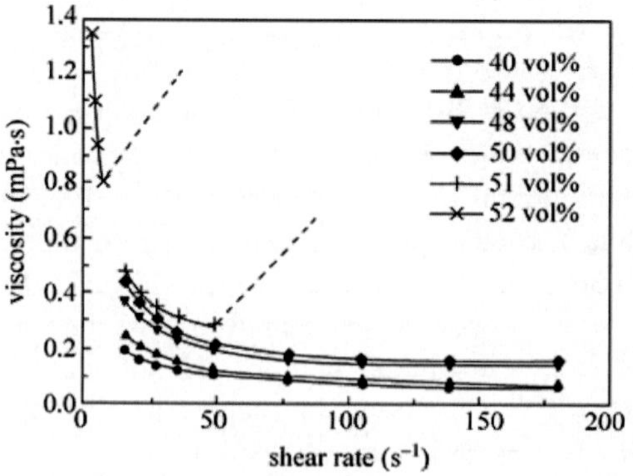

Figure 5.51 The rheological behavior of SiC suspensions at different solid loadings (at pH = 10.5)

slurries are not increased because alumina and yttria powders are also well dispersed in the basic solution, as demonstrated in Fig. 5.52.

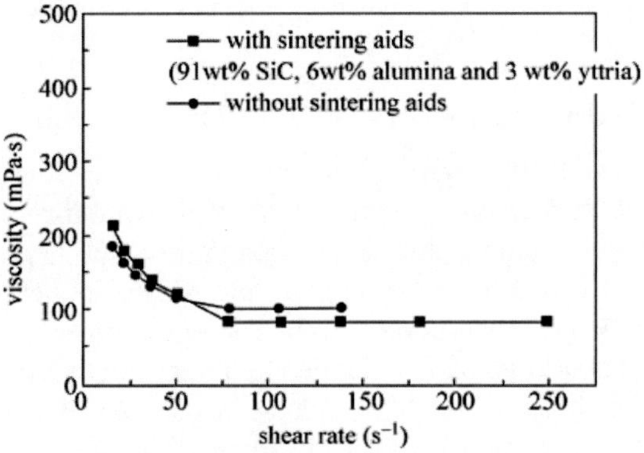

Figure 5.52 The effect of sintering aids on rheological behavior of SiC slip (at 50 vol% solid loading and pH = 10.5)

5.3.1.2 Consolidation of SiC Suspension

When cast into a mold at constant temperature of 70℃, SiC suspensions were consolidated within 10 min. Figure 5.53 is the SEM micrographs of the fracture surfaces of the SiC green bodies, which were cast from slurries, at 50 vol% solid loading and pH = 11.52, degassed by the two methods. Sample A was prepared by mechanical vibration, and sample B by vacuum pumping. As shown in Fig. 5.53, there are more and larger pores in sample A; fewer and smaller pores in sample B.

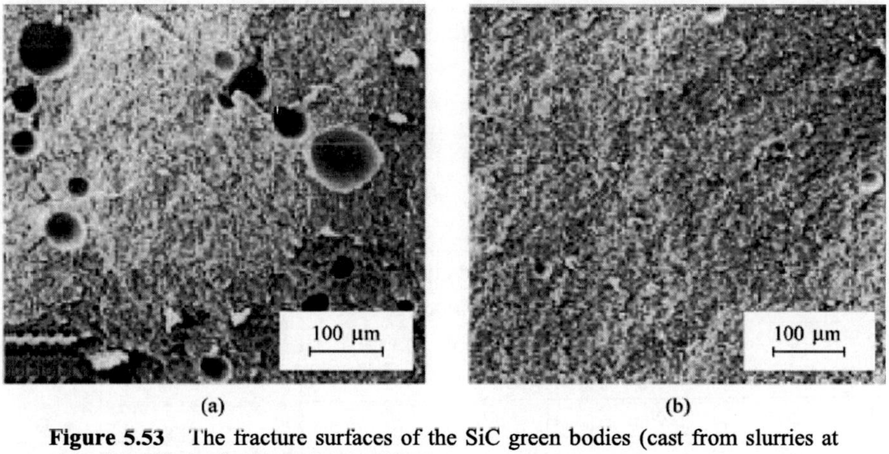

Figure 5.53 The fracture surfaces of the SiC green bodies (cast from slurries at 50 vol% solid loading and pH = 11.52)
(a) Sample A, by mechanical vibrating; (b) Sample B, by vacuum pumping

When milled or when initiator and catalyst are added, the SiC slurries are inevitably puffed, leaving air bubbles in them. If not excluded, the air bubbles will be inflated and yield pores in green bodies when suspensions were consolidated at 70℃. On the other hand, AM reacts with TMAH and discharges NH_3, the reaction is expressed as:

$$C_2H_3CONH_2 + OH^- = C_2H_3COO^- + NH_3(g) \tag{5.19}$$

At room temperature, the degree of the reactivity is very low and the NH_3 gas is dissolved in the suspensions owing to its high solubility (Zhou Ninghuai, 1980). But at 70 – 80℃, the reaction is strong and the solubility of NH_3 decreases greatly. If the rate at which the reaction gives out NH_3 is higher than the rate that slurries discharge the gas, then pores will be formed in green bodies. Because of these two causes, degassing must be applied before casting slurries.

Comparing the sample A and B, it can be concluded that vacuum pumping plays a better role on degassing, but it still cannot eliminate holes in green bodies. The pores in sample B may be propably owing to the resultant gas discharged from an unexpected reaction due to pH = 11.52 of the slurry.

To confirm the reaction discharging the resultant gas in suspensions, a series of experiments were designed for gathering the gas. As it is difficult to collect the gas during consolidation, owing to the gas being trapped in the green bodies, the experiments for gathering the gas produced by Si-hydrolysis in suspensions are operated in suspensions with lower solid loading and for longer duration without initiator and catalyst. Table 5.14 shows the results. The solution formulation in experiment 1 is that of slurry with pH = 10.5 after milling. Its result suggests that the suspensions do give off extra H_2 at 70 – 80℃. The NH_3 could not be detected because of its high solubility in water. Experiments 2 and 3 clarify that the H_2 is the result of SiC powder reacting with TMAH. According to chemical compositions of SiC powder, free Si is the only possible substance that is capable of reacting with TMAH, which is authenticated by experiment 4.

Table 5.14 The results of experiments for gas chromatography

No.	reactants (reacting at 80℃ for 30 min)		H_2 (mL)
	powder	solution	
1	SiC/200 g	premix solution[①]/250 mL, TMAH[②]/12.5 mL	195.7
2	SiC/200 g	premix solution/250 mL	none
3	SiC/200 g	H_2O/250 mL, TMAH/12.5 mL	291.3
4	Si/0.25 g	H_2O/250 mL, TMAH/12.5 mL	280.9
5	oxidized SiC/200 g	H_2O/250 mL, TMAH/12.5 mL	4.08
6	SiC/200 g	premix solution/250 mL, TMAH/25 mL	380.4

Note: ① premix solution: AM, MBAM and deionized water.
② TMAH: 10 wt%.

Several early investigations (Hackly V. A., et al., 1997; Popper P., et al., 1960; Ezis A., et al., 1974) reported that Si had a tendency to hydrolyze in water, which is an exothermic reaction

$$Si(s) + 2H_2O = SiO_2(s) + 2H_2(g) \tag{5.20}$$

and there were difficulties associated with Si processing in an aqueous environment. When mixing Si with aqueous solution, H_2 evolution was observed. The degree of reactivity increased with decreasing particle size and increasing alkalinity, and severe gassing was reported for pH>10. According to the authors' knowledge, there is no early study on the reaction between SiC powder and alkaline solution, partly because of the fact that free Si in SiC powder is commonly less than 1 wt%, and thus, the reaction is slow and inconspicuous; partly because in early researches, SiC slurries were not heated to as high as 70 – 80℃ with pH>10.

Experiment 5 illustrates that when free Si on the surface of the powder is reduced by oxidization in air at 800℃ for 2 h, there is less resultant H_2. However, when compared with experiment 2, experiment 6 shows that the rate of the reaction, and thus, the volume of H_2 are proportional to the concentration of TMAH. The two experiments indicate that there are at least two methods to decrease or even eliminate the resultant H_2 leading to pores in green bodies. One is to reduce the content of free Si on the surface of SiC powder by oxidization or other methods; the other is to decrease concentration of TMAH, namely, pH of SiC suspensions.

Figure 5.54 displays the SEM of the fracture surfaces of the SiC green bodies cast from slurries degassed by vacuum pumping at 50 vol% solid loading and

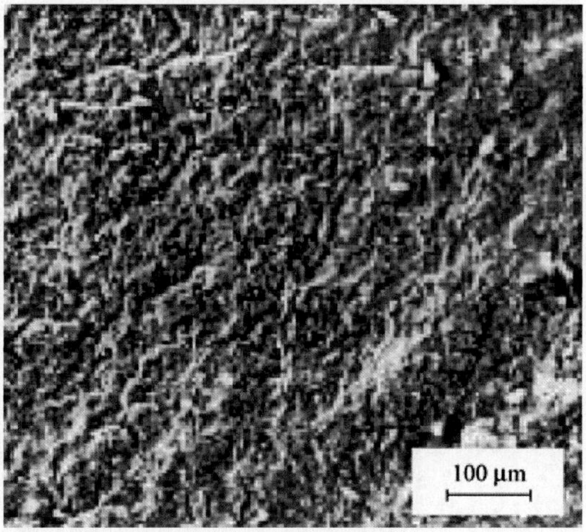

Figure 5.54 The fracture surface of the SiC green body prepared by vacuum pumping (cast from slurries at 50 vol% solid loading and pH = 10.5)

pH = 10.5. It shows a uniform green body without pores. Figure 5.55 shows some complex shapes gelcasted from the suspensions. The density of these bodies is 54%, while their strength is 31.6 MPa ($3 \times 4 \times 30$ mm^3), which indicates that the green bodies are machinable. After binder burnout, there were no shrinkage, crack and distortion in them.

Figure 5.55 The complex shapes of SiC green bodies obtained by gel-casting

5.3.2 The Gel-Casting of Aqueous Slurry with Si_3N_4 Bonded SiC

In this section, gel-casting of aqueous slurry with coarse SiC (about 1 mm in size) and fine silicon particles was investigated to fabricate Si_3N_4 bonded SiC materials.

5.3.2.1 Experimental

1) Materials Preparation

Three grades of SiC powders, 18 – 34 mesh (0.75 – 1.4 mm), 34 – 70 mesh (0.36 – 0.75 mm) and >70 mesh (<0.36 mm), and two commercial metallurgical grades of silicon powders, 200 mesh (127 μm) and 600 mesh (42 μm), were used. A typical chemical analysis for silicon and SiC powder is given in Table 5.15. The main impurities in both powders are Fe, Ca, and Al. A surface oxide layer represents another common impurity in the silicon powders. The weight ratio of SiC:Si in the samples is 79 : 21 unless specified.

The green bodies formed by gel-casting were dried and fired at 1450℃ for 10 h in an alumina tube furnace in N_2 atmosphere. The heating and cooling rates used were 3℃/min.

2) Specimen Analysis

The apparent solid density measurements of the specimens at different stages, namely after drying, debinding and sintering, were determined by the Hg immersion method. Flexural strength of both green body and sintered body were examined with an Instron 4505 instrument. Three-point bending method was employed for the test, with a crosshead speed of 3.0 mm/min and a span of 90 mm. Bars of green body for the strength test were directly gelcasted into a mold of 11 mm in width and 10 mm in height. A sheet sample was gelcasted and sintered at 1450℃, from which bars were cut by a diamond saw into $12 \times 10 \times 150\ mm^3$ in size for flexural strength test. Three bars were measured for each specimen.

Table 5.15 The chemical composition of starting powders

	Si	SiC	Fe	Al	Ca	other
Si (%)	98.2		0.8	0.5	0.3	2.1
SiC (%)	0.5	98.1	0.9	0.2	0.2	0.1

5.3.2.2 Gel-Casting of SiC/Si Mixtures

The gel-casting process is schematically shown in Fig. 5.56. A mixture of starting SiC and silicon powders was dispersed in a pre-mixed solution that had been prepared by dissolving suitable amounts of AM and MBAM in water. Stable SiC/Si suspensions with 50 – 72 vol% solid loading could be obtained by adding a commercial dispersant (DURAMAX D-3005) which is mainly composed of acrylic polymer, and mixing the particles with the solution for around 20 min. Then, an initiator solution was added. After homogeneous mixing, the suspension

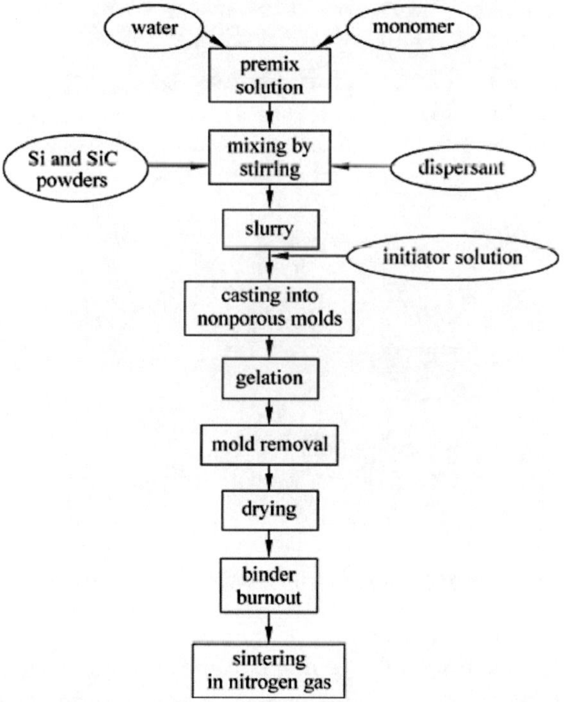

Figure 5.56 The flowchart of SiC/Si gel-casting process

was cast into a nonporous metal mold. All above mentioned operations were carried out at room temperature. The slurry-containing mold was then placed into an oven at 80℃. Gelation or consolidation of the suspension occured through polymerization of monomers and was completed in 10 – 30 min to form a wet green part that was strong enough to be demolded owing to the formation of a 3D polymer network. After drying at about 100℃, strength of the green body was further increased as a result of the strengthened interaction between the polymer network and particles (see Fig. 5.57). This forming mechanism makes it possible to produce complex shape parts with different sizes. Examples of green parts of different shapes prepared by this process are shown in Fig. 5.58.

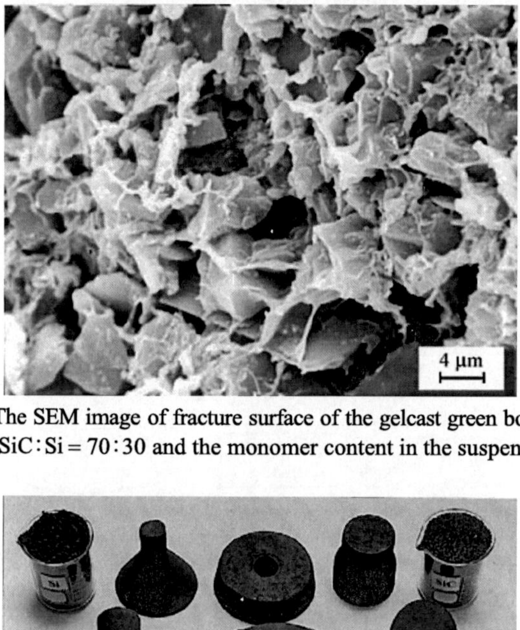

Figure 5.57 The SEM image of fracture surface of the gelcast green body in which the weight ratio of SiC : Si = 70 : 30 and the monomer content in the suspension is 2.2 wt%

Figure 5.58 The gelcast green parts with complex shapes and high green strength

It was found that the concentration of initiator and the temperature used for gelation had great influence on the gelling rate and appearance of the green body. When the volume ratio between initiator and monomer varied from 3 to 10, the gelling process was accelerated and completed in a short time. The gelling process was also accelerated by increasing the temperature. At room temperature (25℃) the gelling process required more than 2 h to get completed, and it resulted in green bodies that could be easily deformed. On the other hand, it took less than 30 min for completion of gelling when the samples were heated at 60 – 80℃ and it produced good green bodies with little defects.

5.3.2.3 The Effect of Solid Loading and Monomer Content on Density

High density and low porosity are important for silicon nitride bonded silicon carbide (SNBSC) refractory materials because properties, especially fracture strength, of the materials depend on density after firing. For fine ceramics such as Al_2O_3, ZrO_2 and Si_3N_4, the green density of the gelcast samples may range from 50 – 65% of theoretical density, but the final density of 90 – 99% of theoretical value can be readily achieved after sintering owing to large shrinkage occurring through liquid phase or solid state sintering mechanism (Prabhakaran K., et al., 2000; Huang Y., et al., 2000; Omatate O. O., et al., 1991a,b). For reaction bonded SNBSC materials, however, there is very little shrinkage during firing and the voids are partially filled by the silicon nitridation products. Therefore, the density of the final product is critically dependent on that of the green body. Aqueous gel-casting is carried out in nonporous molds through which water cannot be removed during the gelling process. This is fundamentally different from slip casting where water in slurry is absorbed by plaster molds through capillary action. Consequently, the green density produced by gel-casting strongly relies on the solid loading of the suspension. A high green density generally requires a high loading of solid particles in the slurry; but if the solids content is excessively high, then the slurry could have poor flow ability owing to high viscosity, resulting in inadequate particle compaction and low ultimate density.

The effect of solid loading on density was examined and the results are shown in Fig. 5.59, where various solids content were dispersed in the same monomer solution in which the weight ratio between monomer and water was kept at 4:25. It was observed that both the green density and final density (after sintering) increased significantly with increasing solid loading in slurries. We noticed that the density values of dried samples at 110℃ were higher than those of the burnout samples at 600℃. This is because sample mass was reduced owing to the burnout of the organic binder and the sample volume did not change at the burnout temperature.

Slurry with higher solid loading, such as 70 vol%, comprising a mixture of coarse SiC and fine Si particles showed a better stability when compared with those with lower solid loading of 50 – 60 vol%. It was observed that sedimentation of coarse particles occurred for the slurry with lower solid loading of 50 – 60 vol% and the sedimentation of coarse particles, however, almost was absent in the slurry with higher solid loading more than 70 vol% owing to higher parking density and viscosity. In addition, less drying shrinkage was observed for the green body with higher solid loading because of reduction of water content in the suspension.

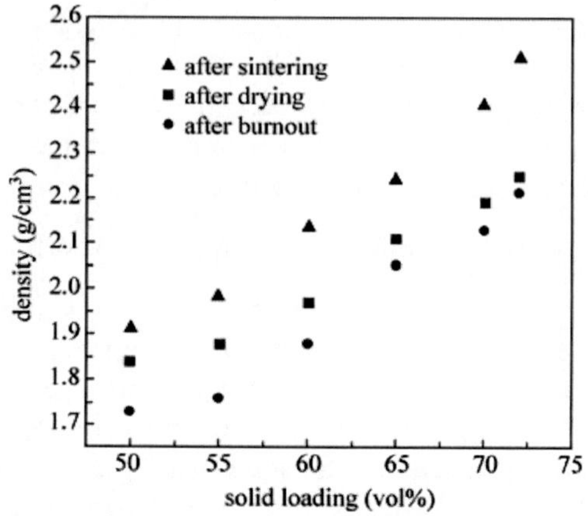

Figure 5.59 The effect of solid loading on density of the gel-cast samples at different stages (after drying at 110℃, polymer burnout at 600℃ and after sintering at 1450℃ for 10 h in N_2 atmosphere)

The sample density was also affected by the amount of organic monomer dissolved in the slurry. In the present work, the organic component is mainly introduced from the monomer in the initial solution, which polymerizes and produces a polymer network to bind the solid particles after the addition of the initiator. Figure 5.60 shows the influence of various content of organic monomer on densities of the samples in different stages, where the ratio between ceramic powder volume and monomer solution volume was kept identical for all samples (that is, all samples have the same solid loading of 72 vol% for their slurries), but dissolved monomer content in monomer solution is different, the weight ratios of monomer to water are 3:100, 6:100, 10:100, 16:100, 20:100, 30:100, 40:100, respectively. The percentage of the organic binder based on ceramic powder weight is 0.38, 0.72, 1.16, 1.77, 2.20, 2.95 and 3.66 wt%, respectively.

With regard to the samples containing less than 1.16 wt% organic binder, there was little difference in density before and after organic burnout at 600℃. However, as the organic content increased to 3.66 wt% (relative to the weight of solid particles) or 40 wt% (based on the water weight) the density significantly decreased from 2.304 g/cm^3 in the dried sample to 2.138 g/cm^3 after organic burnout. This is because of the fact that some space of air pores were occupied by the organic binder mass and higher density was reached with increased organic binder mass for the dried samples. However, more air pores would be left after organic burnout, which resulted in lower density when compared with drying stage for the respective samples, especially those with high organic monomer content of 3.66 wt%. Thus, the difference in density before and after burnout increased with the organic monomer content in the samples as shown in Fig. 5.60.

As discussed earlier, lower organic content in the samples with the same solids content is helpful for the density increase after either polymer burnout or nitridation

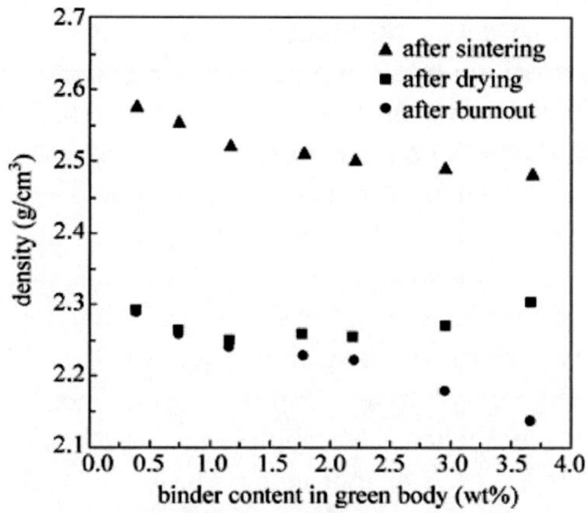

Figure 5.60 The influence of organic monomer content on density of the gelcast samples with same solid loading of 72 vol% (SiC : Si = 79 : 21 in weight ratio) at different stages (after drying at 110℃, polymer burnout at 600℃ and after sintering at 1450℃ for 10 h in N_2 atmosphere)

sintering. For example, the samples containing 0.38 wt% and 3.66 wt% organic polymer showed a density of 2.575 g/cm^3 and 2.480 g/cm^3, respectively after nitridation at 1450℃ for 10 h. It can be observed from Fig. 5.60 that final density of the samples after sintering strongly depends on the density after organic burnout stage and not after the green density in drying stage. This is because the magnitude of mass increase is almost the same during sintering stage for these samples with same silicon content.

The results suggest that smaller amount of monomer was selected for these ceramic systems consisting of coarse particles for gel-casting. This is in sharp contrast to that observed in gel-casting of ultra-fine particle systems, where the organic monomer content is often around 2 to 4 wt%. There is a major difference in the size of starting powders between the SNBSC material and fine ceramics. The particle size of SiC used for making SNBSC ranges from a few hundred microns to above 1 mm, which is much greater than that used for fine structural ceramics (usually d<1 μm). The coarse powders have a smaller surface area than the fine ones, and consequently, a green body made by coarse powders requires lesser polymer for binding the particles when compared with those made by fine powders. The above results suggest that solid loading and binder content are the two major factors that have strong influence on the density of SNBSC materials formed by the gel-casting process. To achieve high final density of the material, the solid loading in slurry should be high and the amount of monomer used should be low, to provide reasonable strength to the green body.

5.3.2.4 The Binder Removal and Silicon Oxidation During the Burnout Process

Polymer in the gelcast green body must be carefully removed before nitridation to suppress sample bloating at nitridation temperature and to avoid excessive amounts of carbon remaining in the system. For fabrication of fine ceramics such as Al_2O_3 and ZrO_2, binder removal is carried out between 600 and 800℃ at which there would be little change in chemistry of the oxide powders. For the SNBSC system, however, silicon oxidation could occur at the polymer burnout temperature, producing a silica layer on the surface of silicon powder, although SiC particles would be less affected at this temperature range. It is expected that the oxidation of silicon would have some effect on the subsequent silicon nitridation. To study this effect, TGA was performed between 500 and 800℃ in air for a gel sample that was formed through polymerization of the organic monomer. Residue of the polymer gel was obtained when the sample was fired below 550℃ in air, but above this temperature the gel was thoroughly burned off in 2 h (see Table 5.16). Weight changes of silicon powder samples were determined at the same temperature range by using TGA. According to XRD analysis, silica was the sole product of silicon oxidation between 600 and 800℃, and the reaction can be represented as follows (Moulson A. J., 1979):

$$Si + O_2 \longrightarrow SiO_2 \tag{5.21}$$

This reaction is accompanied by a weight gain and thus the quantity of silica produced by oxidation of silicon could be calculated from this equation when the weight gain is measured from TGA. Assuming that the starting silicon powders are in a spherical shape and homogeneously covered by a silica layer after oxidation, the thickness of the silica layer could be calculated based on the quantity of silica produced and the silicon particle size. The average silicon particle size is 43.5 μm in the present work. The mass gain measured by TGA at different temperatures and the calculated results of silica layer thickness are summarized in Table 5.16. As shown in the table, when the temperature increased from 500 to 700℃ the mass gain of silicon increased from 0.058 to 0.498 wt%, corresponding to an increase in the silica layer thickness from 0.014 to 0.122 μm. At 800℃, the mass gain was too high to be employed in the present study. From these results, a suitable burnout temperature of 600℃ was selected, at which binder removal could be completed in 2 h with only a very small amount of silicon oxidation being produced.

The influence of silicon oxidation on the final density of the material was studied on the gelcast SiC/Si samples containing 1.77 wt% of polymer. The samples were first fired at temperature between 500 and 800℃ for 2 h in air for polymer burnout and then at 1450℃ for 10 h in N_2 atmosphere. The results are shown in Table 5.16. Samples that had polymer burnout below 700℃ yielded a high density of 2.544 g/cm^3 after nitridation at 1450℃. Those that underwent initial firing at 700℃ or above in air showed a relatively lower density around 2.54 g/cm^3. The result indicates that a small amount of silica on the silicon surface does not markedly hinder nitridation or decrease final density. It is known that nitridation of a silicon

Table 5.16 The results of gelcast samples fired at different temperatures in air for polymer burnout, then fired at 1450℃ for 10 h in N_2 atmosphere

heat treatment conditions	500℃, 2 h in air	550℃, 2 h in air	600℃, 2 h in air	650℃, 2 h in air	700℃, 2 h in air	800℃, 2 h in air
residue for gel (%)	8.080	0.003	0.000	0.000	0.000	0.000
mass gain for silicon (%)	0.058	0.114	0.202	0.363	0.498	1.828
silica layer thickness (μm)	0.014	0.028	0.049	0.082	0.122	0.448
density of SNBSC after firing at 1450℃ (g/cm^3)	2.536	2.562	2.564	2.560	2.544	2.541

Note: The sample contained 1.77 wt% polymer (relative to the weight of solid particles).

powder compact occurs in three stages: induction, growth and termination (Sheldon B. W., et al., 1988). At the induction stage, the first step is the decomposition of the native vitreous silica layer covering the silicon particles (Sheldon B. W., et al., 1988; Giridhar R. V., et al., 1998), which exposes the underlying silicon to the nitrogen gas. This can take place when the partial pressure of oxygen is below a certain value. It was reported (Giridhar R. V., et al., 1998) that when the partial pressure of air was below about 4 torr in 1 atm of nitrogen at 1370℃, a small amount of iron (as low as 55 ppm by weight) could drastically raise the removal rate of vitreous silica with respect to pure oxide. However, if the thickness of the surface oxide layer on silicon powder is too high, then the exposure of surface silica would be more difficult, which could hinder silicon nitridation and result in lower density for the SNBSC materials.

5.3.2.5 The Effects of Firing Temperature and Starting Composition

Densification of SNBSC materials relies on the formation of α-Si_3N_4 and β-Si_3N_4 as the bonding phases of SiC. Si_3N_4 is formed as a result of in-situ reaction between silicon and nitrogen at elevated temperatures and its formation is dependent on the nitriding temperature and the starting composition. The silicon nitridation process controls the density and strength of SNBSC materials. Figure 5.61 shows the relationship between density and nitriding temperature for SiC/Si mixtures containing different amounts of silicon. The general trend observed is that density of the materials increases with increasing reaction temperature, but the degree of the increase is different in different systems. After firing at 1350℃, the sample with a mass ratio of Si∶SiC = 15∶85 shows higher density than that with a mass ratio of Si∶SiC = 30∶70. At this low firing temperature, silicon powders in all samples only partially reacted with nitrogen to form Si_3N_4 (see Fig. 5.62). Because silicon has a lower density (2.3 g/cm^3) than both SiC and Si_3N_4 (3.1 – 3.2 g/cm^3), the partial nitridation will, in general, result in lower density of the samples that have higher silicon contents. The melting temperature of silicon is around 1400℃ above which silicon nitridation is accelerated. Thus density of the materials fired below 1400℃ is dependent on the SiC:Si ratio as well as the amount of Si_3N_4 bonding phases formed in the system. At 1450℃, silicon in all samples is fully nitrided and this has facilitated significant increases in density, particularly for the sample with a high silicon concentration.

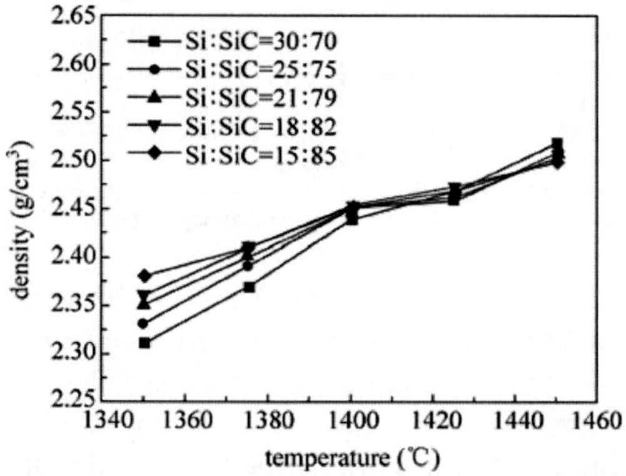

Figure 5.61 The relationship between density and nitriding temperature of SNBSC materials with various percentage of silicon in the starting compositions

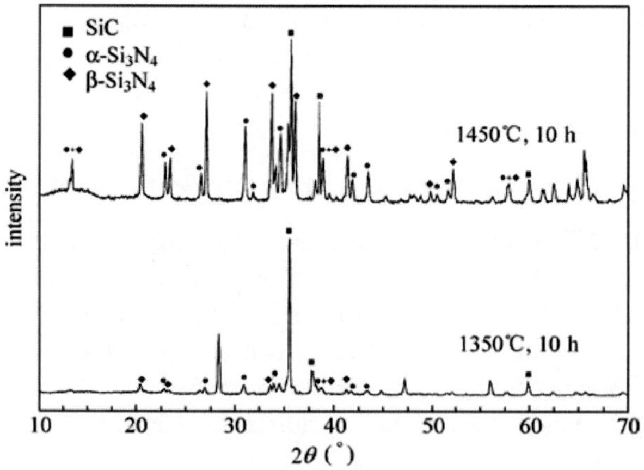

Figure 5.62 The comparison of XRD spectra of sintered samples at 1450℃, 10 h and 1350℃, 10 h

The ratio of β-Si_3N_4 and α-Si_3N_4 increased from 1350℃ to 1450℃ according to XRD analysis. The formation mechanisms of Si_3N_4 include direct reactions of nitrogen with all three phases of silicon, solid, liquid and gas (Boyer S. M., et al.,

1978; Barsoum M., et al., 1991). It is presumed that α-Si_3N_4 is mainly formed by a gas-phase reaction between gaseous silicon and nitrogen and that the major growth of β-Si_3N_4 occurs in the liquid phase and, to a minor extent, as the result of the reaction between solid silicon and nitrogen (Ziegler G., et al., 1987; Jennings H. M., et al., 1983; Jennings H. M., et al., 1988). The increase of β-Si_3N_4 is attributed to the melting of silicon above 1400℃. SEM images of specimens fired at 1350 and 1450℃ respectively have reflected their microstructure difference. As shown in Fig. 5.63(a), thin α-Si_3N_4 whiskers were produced through the gaseous reaction at 1350℃. At 1450℃ relatively coarse β-Si_3N_4 grains were developed (see Fig. 5.63(b)).

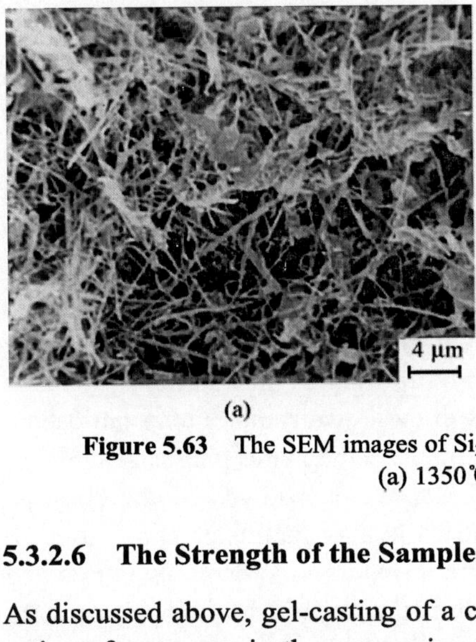

(a)

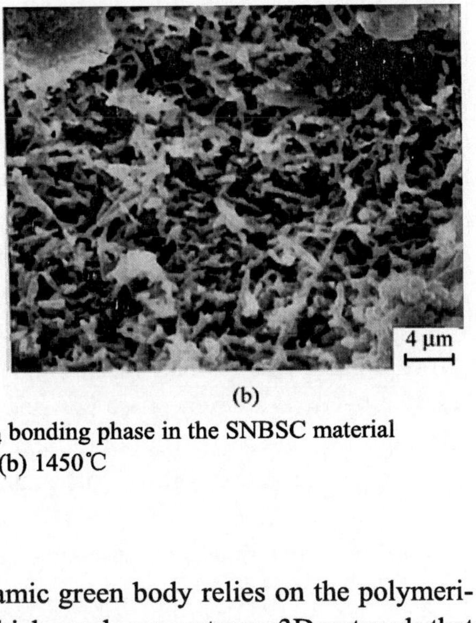

(b)

Figure 5.63 The SEM images of Si_3N_4 bonding phase in the SNBSC material (a) 1350℃; (b) 1450℃

5.3.2.6 The Strength of the Samples

As discussed above, gel-casting of a ceramic green body relies on the polymerization of monomer in the suspension, which produces a strong 3D network that holds ceramic particles together after drying. Thus the flexural strength of a gelcast green body would be proportional to the monomer content in the specimen. Table 5.17 shows the relationship between the green strength and the monomer content. The green strength increases from 1.32 to 2.11 MPa when monomer content varies from 0.38 to 2.95 wt% (based on the amount of ceramic powder). The bonding of the polymer network to SiC particles is clearly viewed from the fracture surface of the green sample containing 2.2 wt% monomer (see Fig. 5.57). Although a higher monomer content is helpful for achieving greater green strength, densities for both the green body after polymer burnout and the final material after firing at 1450℃ are inversely proportional to the initial monomer content (Table 5.17). As discussed earlier, removal of excessive polymer in the samples could produce more pores after polymer burnout. The results indicate that a high density for the green body after polymer burnout is essential for achieving high strength of SNBSC materials after nitridation.

Table 5.17 The effect of monomer content on fracture strength of green body and fired materials

polymer content in green body (wt%)	0.38	0.73	1.77	2.95
green strength before polymer burnout (MPa)	1.32	1.48	1.75	2.11
flexural strength sintered at 1450℃ for 10 h (MPa)	38.59 ± 1.4	34.3 ± 2.2	25.8 ± 2.2	13.5 ± 1.9
density after polymer burnout (g/cm^3)	2.570	2.554	2.473	2.325

A comparison was made between two SNBSC samples formed by gel-casting and slip casting respectively. The results are shown in Table 5.18. Both the samples had the same amounts of SiC and Si powders, but the gelcast sample contained additional 0.38 wt% of monomer whereas for the slip-cast sample about 4 – 5 wt% of clay was added to the SiC/Si slurry to improve its flowability for casting. It is seen that demolding of the gelcast sample required a much shorter time (<1 h) when compared with that for the slip-cast sample (>10 h). This would be a significant advantage with regard to the gel-casting technology for commercial production. After nitridation sintering at 1450℃ for 10 h, the sample formed by gel-casting shows higher density than that obtained by slip-casting. In addition, the XRD results unveiled different phase components in these two samples after nitridation. Although SiC and Si_3N_4 (in both α and β forms) are the major phases in samples, additional silicon oxynitride and glass phases were observed in the slip-cast product (see Fig.5.64). It is presumed that while clay facilitates slip-casting, it introduces oxides that are responsible for the formation of silicon oxynitride and glass after sintering. It would be interesting to compare the high-temperature performance of these two samples to further evaluate the two green body forming technologies. This remains for a future study of the subject.

Table 5.18 The comparison between gelcast and slip-cast SNBSC materials

	starting powders	time from cast to demold	mold materials	density after sintering(g/cm^3)	phases after sintering
gel-casting	SiC Si	20 – 60 min	nonporous materials	2.587	Si_3N_4 SiC
slip casting	SiC Si Clay	⩾10 h	plaster	2.576	Si_3N_4 SiC Si_2N_2O glass

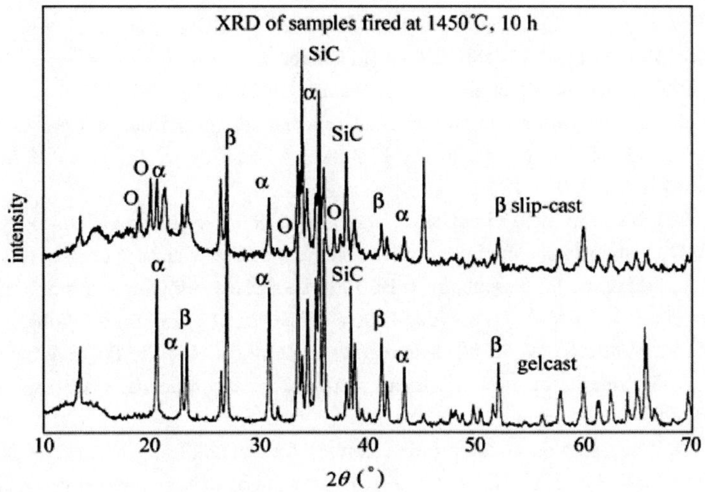

Figure 5.64 The comparison of XRD spectra between gelcast and slip-cast SNBSC

References

Baader F. H., Graule T. J., Gauckler L. J., 1996. Ind. Ceram., 16: 36.

Barsoum M., Kangutkar P., Koczak M. J., 1991. Nitridation kinetics and thermodynamics of silicon powder compacts. J. Am. Ceram. Soc., 74: 1248 – 1253.

Bergstrom L., Paugh R. J., 1989. Interfacial characterization of silicon nitride powders. J. Am. Ceram. Soc., 72(1): 103 – 109.

Bergstrom L., 1994. US Patent No. 5340532.

Boyer S. M., Moulson A. J., 1978. A mechanism for the nitridation of Fe-contaminated silicon. J. Mater. Sci., 13: 511 – 516.

Brow K., Pantano C. G., 1986. J. Am. Ceram. Soc., 69: 314.

Busca G., Lorenzelli V., Baratoon M. I., et al., 1986a. FTIR charaterization of silicon nitride Si_3N_4 and silicon oxynitride Si_2N_2O surfaces. J. Mol. Struct., 143: 525 – 528.

Busca G., Lorenzelli,V., Porcile G., et al., 1986b. FTIR study of the surface properties of silicon nitride. Mater. Chem. Phys., 14: 123 – 140.

Carisey T., Laugier W. A., Brandon D. G., 1995. J. Eur. Ceram. Soc., 15: 1.

Castanho M., Fierro J . L. G., Moreno R., 1997a. J. Euro. Ceram. Soc., 17: 383.

Castanho M., Moreno R., Fierro J . L. G., 1997b. J. Mater. Sci., 32: 157.

Cook R. F., Clarke D. R., 1988. Fracture stability, R-curves and strength variability. Acta Metall., 36(3): 555 – 562.

Crimp M. J., Johnson R. E., Halloran J. W., et al., 1986. Colloidal behavior of silicon carbide and silicon nitride. In: Science of Ceramic Chemical Process. Ed. by L. L. Hench. New York: John Wiley and Sons Inc., 539 – 549.

Dai J. Q., 2001. The effects of surface characterization of silicon nitride powders on preparation of concentrated suspensions and properties of sintered ceramics (Ph.D Thesis). Beijing: Tsinghua University.

Dai J. Q., Huang Y., Xie Z. P., et al., 2001. FTIR study of Si_3N_4 powder. Chinese Journal of Spectroscopy Laboratory, 18(1): 78 – 83 (in Chinese).
Dai J. Q., Xie Z. P., Huang Y., et al., 1999. J. Mater. Sci. Lett., 18: 1739.
Ezis A., 1974. The fabrication and properties of slip-cast silicon nitride. In: Ceramics for High Performance Applications. Ed. by J. H. Burke, A. E. Gorum, R. N. Katz. Brook Hill, Chestnut Hill, MA, 207 – 222.
Fagerholm H., Johansson L. S., Graeffe M., et al., 1996. Surface charge and viscosity of mixed Si_3N_4-Y_2O_3 suspensions containing lignosulphonate. J. Eur. Ceram. Soc., 16: 671 – 678.
Fagerholm H., Johansson L., Rosenholm J. B., 1994. A surface study on adsorption of lignosulphonate on mixed Si_3N_4-Y_2O_3 powder dispersions. J. Eur. Ceram. Soc., 14: 403 – 409.
Fang Y., Yu F., White K. W., 2000. Microstructural influence on the R-curve behavior of a 70% Si_3N_4-30% barium aluminum silicate self-reinforced composite. J. Mater. Sci., 35(11): 2695 – 2699.
Greil P., 1989. Processing of silicon nitride ceramics. Mater. Sci. Eng., A109: 27 – 35.
Greil P., Nitzsche R.,Friedrich H., et al., 1991. Evaluation of oxygen content on silicon nitride powder surface from the measurement of the isoelectric point. J. Eur. Ceram. Soc. 7: 353 – 359.
Giridhar R.V., Rose K., 1998. Conditions for thermal nitridation of Si in N_2-O_2 mixtures, J. Electrochem. Soc. Solid-State Sci. Technol., 135 (111): 2803 – 2807.
Garg A. K., De Jonghe L. C., 1990. Microencapsulation of silicon nitride particles with yttria and yttria-alumina precursors. J. Mater. Res., 5(1): 136 – 142.
Graule T. J., Gauckler L. J., Baader F. H., 1996. Ind. Ceram., 16: 31.
Hackley V. A., Malghan S. G., 1994. The surface chemistry of silicon nitride powder in the presence of dissolved ions. J. Mater. Sci., 29: 4420 – 4430.
Hackly V. A., Paik U., Kim B. H., et al., 1997. Aqueous processing of sintered reaction-bonded silicon carbide: 1, dispersion of silicon powder. J. Am. Ceram. Soc., 80(8): 1781 – 1788.
Hackley V. A., Wang P. S., Malghan S. G., 1993. Effects of soxhlet extraction on the surface oxide layer of silicon nitride powders. Mater. Chem. Phys., 36: 112 – 118.
Harame D. L., 1984. Integrated circuit chemical sensors (Ph.D. Thesis). Stanford: Stanford University.
Huang Y., Ma L. G., Yang J. L., et al., 2000. Surface oxidation to improve water-based gel-casting of silicon nirtride. J. Mater. Sci., 35: 3519 – 3524.
Huang Y., Zhou L. J., Tang Q., et al., 2001. Water-based gel-casting of surface coated silicon nitride powder. J. Am. Ceram. Soc., 83(4): 701 – 707.
James R. O., Parks G. A., 1980. In: Surface and Colloid Science (Vol. 2). Ed. by E. Matijevic. New York: John Wiley and Sons Inc.
Jennings H. M., 1983. Review on reaction between silicon and nitrogen. Part 1: Mechanisms. J. Mater. Sci., 18: 951 – 967.
Jennings H. M., Dalgleish B. J., Pratt P. L., 1988. Reaction between silicon and nitrogen. Part 2: Microstructure. J. Mater. Sci., 23: 2573 – 2583.
Joshi R. N., Ronald R. A., 1994. Metallorganic surfactants as sintering aids for silicon nitride in an aqueous medium. J. Am. Ceram. Soc., 77(11): 2926 – 2934.

Kim J. S., Schubert H., Petzow G., 1997. Sintering of Si_3N_4 with Y_2O_3 and Al_2O_3 added by coprecipitation. J. Eur. Ceram. Soc., 5: 311 – 319.

Kitahara A., Watanab A., 1992. Electric Phenomena of Interface (in Chinese). Beijing: Peking University Press, 49 – 62.

Kulig M., Greil P., 1991. Surface chemistry and suspension stability of oxide-nitride powder mixtures. J. Mater. Sci., 26: 216 – 224.

Laarz E., Lenninger G., Bergstrom L., 1997. Aqueous silicon nitride suspensions: effect of surface treatment on the rheological and electrokinetic properties. Key Eng. Mater., 132 – 136, 285 – 288.

Levine S., Neale G., Epstien J., 1996. J. Coll. Interface Sci., 57: 424.

Levine S., Smith A. L., 1971. Discuss. Faraday Soc., 52: 290.

Malghan S. G., 1992. Dispersion of Si_3N_4 powders: surface chemistry interactions in aqueous media. Colloids Surf., 62: 87 – 99.

Mezzasalma S., Baldovino D., 1996. Characterization of silicon nitride surface in water and acid environment: a general approach to the colloidal suspensions. J. Coll. Inter. Sci., 180: 413.

Moulson A. J., 1979. Review, reaction-bonded silicon-nitride: its formation and properties. J. Mater. Sci., 14: 1017 – 1051.

Natansohn S., Pasto A. E., 1991. Improved processing methods for silicon nitride ceramics. Am. Soc. Mech. Eng., Paper No. 91-GT-316.

Natansohn S., Pasto A. E., Rourke W. J., 1993. Effect of powder surface modifications on the properties of silicon nitride ceramics. J. Am. Ceram. Soc., 76(9): 2273 – 2284.

Nilsen K., Danforth S. C., Wautier H., 1987. Dispersion of laser-synthesized Si_3N_4 powder in nonaqueous systerms. In: Advances in Ceramics. Ceramic Powder Science, 21: 537 – 547.

Omatete O. O., Janney M. A., Nunn S. D., 1997. Gel-casting: from laboratory development toward industrial production. J. Eur. Ceram. Soc., 17: 407 – 413.

Omatete O. O., Janny M. A., Streklow R. A., 1991a. Amer. Ceram. Bull. 70: 1641.

Omatate O. O., Strehlow R. A., Armstrong B. L., 1991b. Forming of silicon nitride by gel-casting. In: Proceedings of the Annual Automotive Technology Development Contractors' Meeting. Society of Automotive Engineers, Warrendale, PA, 245 – 251.

Peuckert M., Greil P., 1987. Oxygen distribution in silicon nitride powders. J. Mater. Sci., 22: 3717.

Popper P., Ruddlesden S. N., 1960. The preparation, properties, and structure of silicon nitride. Trans. J. Br. Ceram. Soc., 60: 603 – 623.

Prabhakaran K., Pavithran C., 2000. Gel-casting of alumina from acidic aqueous medium using acrylic acid. J. Eur. Ceramic Soc., 20: 1115 – 1119.

Rahaman M. N., Boiteux Y., De Jonghe L. C., 1986. Surface characterization of silicon nitride and silicon carbide powders. Am. Ceram. Soc. Bull., 65(8): 1171.

Raider S. I., Flitsch R., Aboaf J. A., et al., 1976. Surface oxidation of silicon nitride films. J. Electrochem. Soc., 123: 560 – 565.

Ramis G., Busca G., Lorenzelli V., et al., 1989. FTIR characterization of high surface area silicon nitride and carbide. In: Surfaces and Interfaces of Ceramic Materials. Ed. by L. C. Dufour, C. Monty, G. Petot-Ervas. Dordrecht: Kluwer Academic, 173 – 184.

Rolf W., Gabriele S., 1997. Key Eng. Mater., 132 – 136: 374.

Sanchez R. T., Garcia A. B., Cesio A. M., 1996. IBID, 16: 1127.

Schwelm M., Kaiser G., Schulz W., et al., 1993. The effect of alcohol treatment on the rheology of Si_3N_4. J. Eur. Ceram. Soc., 11: 283 – 289.

Sheldon B.W., Haggerty S., 1988. The nitridation of high purity, lasersynthesised silicon powder to form reaction bonded silicon nitride. Ceram. Eng. Sci. Proc., 9(7 – 8): 1061 – 1072.

Shih W. H., Kisailus D., Shih W. Y., 1996. Rheology and consolidation of colloidal alumina-coated silicon nitride suspensions. J. Am. Ceram. Soc., 79(5): 1155 – 1162.

Stadelmann H., Petzow G., Greil P., 1989. Effects of surfaces purification on the properties of aqueous silicon nitride suspensions. J. Eur. Ceram. Soc., 5: 155 – 163.

Sun J., Gao L., Guo J. K., et al., 1996. Acta Metall. Sin., A9: 489.

Tang Q., Xie Z. P., Yang, J. L., et al., 1998. Coating of silicon nitride and its colloidal behavior. J. Mater. Sci. Lett., 17(14): 1239 – 1241.

Tuan W. H., Lai M. J., Lin M. C., et al., 1994. Mechanical performance of alumina as a function of grain size. Mater. Chem. Phys., 36(3 – 4): 246 – 251.

Yang J. F., Sekino T., Niihara K., 1999. Effect of grain growth and measurement on fracture toughness of silicon nitride ceramics. J. Mater. Sci., 349(22): 5543 – 5548.

Yang J., Oliveira F. J., Silva R. F., et al., 1999. Pressureless sinterability of slip cast silicon nitride bodies prepared from coprecipitation—coated powders. J. Eur. Ceram. Soc., 19: 433 – 439.

Young A. C., Omatete O. O., Janney,M. A., et al., 1991. Gel-casting of alumina. J. Am. Ceram. Soc., 74(3): 612 – 618.

Zhou L. J., Huang Y., Xie Z. P., et al., 2000. Gas-discharging reactions and their effect on the microstructures of green bodies in gel-casting of non-oxide materials. Mater. Lett., 45(8): 51 – 57.

Zhou Ninghuai, 1980. SI Chemical Data. Beijing: Higher Education Press, 123.

Ziegler G., Heinrich J., Wotting G., 1987. Review: relationships between processing, micro-structure and properties of dense and reaction-bonded silicon nitride. J. Mater. Sci., 22: 3041 – 3086.

Chapter 6 Application of New Colloidal Forming

Abstract The colloidal forming process based on gel-casting was applied for preparing microbeads of ceramics, improving the breakdown strength of rutile capacitor, developing the thin-wall rutile tube for ozone generator, and producing the refractory nozzle of zirconia (ZrO_2) and lead zirconate titanate (PZT). The microbeads of ceramics were used as grinding media, gel-pen-ball and healthcare product with far infrared function for human body. Owing to the low-cost associated with the production of the gel-beads as well as because of its superior properties, they have been widely used and marketed.

When compared with the extrusion forming, the breakdown strength of rutile capacitor prepared by gel-casting was improved greatly. The rutile ceramics with high dielectric constant and high breakdown strength have been used in thin-wall tube of ozone generator.

ZrO_2 refractory nozzle prepared by gel-casting has a high melting point above 2680℃ and a good thermal shock resistance to withstand the temperature rising from the room temperature to 2200℃ in several seconds. Hence, it is considered as a suitable material for making the nozzle for precision casting of Cu-Cr molten alloy.

When compared with die pressing, gel-casting can provide a more homogenous microstructure as well as better piezoelectric property of the PZT samples. By using the same sintering procedure, gelcast samples were observed to exhibit slightly stronger piezoelectric effects than the die pressed ones. This may be attributed to both density difference and pore size.

Keywords gel-casting, gel-bead forming, microbeads, gel-pen-ball, rutile capacitor, ozone generator, lead zirconate titanate, refractory nozzle

6.1 Ceramic Microbeads

6.1.1 The Forming Principle of Ceramic Microbeads Based on Gel-Casting

According to the interfacial tension principle, the gel-beads forming is a process that employs the rapidly solidifying slurry with water-soluble monomer. By adjusting

Y.Huang et al., *Novel Colloidal Forming of Ceramics*
© Tsinghua University Press, Beijing and Springer-Verlag Berlin Heidelberg 2010

the forming media and these parameters of slurry characteristics, the key processes of forming, dressing shape and shape maintenance can be accomplished. In this work, a new gel-beads forming of ceramics microbeads is proposed, and the preparation equipments are designed and manufactured.

The forming principle of ceramics microbeads:

The basic principle of this process is based on the interface tension between two different surfaces. When the slurry is dropped into the oil medium, the solidification occurs and then the beads are formed. The interface tension can be described by the surface tensions difference when two kinds of liquids have been mutually saturated, that is $\gamma_{12} = \gamma_1 - \gamma_2$. According to this formula, without the influence of the external force, the bead can always be formed when the γ_{12} is large and the time is long enough. The larger the γ_{12} is, the shorter the required forming time of the beads.

In preparation process, there are two kinds of mediums. One is water-based ceramic slurry ($\gamma_1 \approx 50$ mN/m), the other is oil-based medium ($\gamma_2 \approx 20$ mN/m). When the interfacial tension γ_{12} is about 30 mN/m, the slurry droplet can rapidly become a bead in the oil medium.

The design of the process parameters is as follows:

$$mg \approx f + f' \tag{6.1}$$

When falling into the medium, the bead is mainly influenced by three forces, namely, the downward gravitational force, *mg*, the upward buoyancy force f' and the viscous force f (as shown in Fig. 6.1).

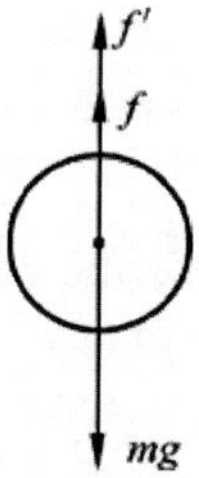

Figure 6.1 The applied forces on the beads

According to the Stoke's formula, when the bead moves in the liquid, the viscous force can be expressed as:

$$f = 6\pi r \eta v \tag{6.2}$$

where η is the liquid's viscous coefficient, v is the bead's velocity, r is the radium of the bead.

The buoyancy can be given as:

$$f' = \rho_0 V g \tag{6.3}$$

where V is the volume of the bead, ρ_0 is the density of the liquid.

It can be concluded from the above formula that once the size of the bead is determined, the factors that affect the falling time of the bead are the density and viscosity of the medium, i.e., the buoyancy and the viscosity of the liquid medium. So the medium chosen must be in line with the following two conditions (or at least one of them): ① the density of slurry is close to that of the medium (increasing the buoyancy); ② the medium viscosity (increasing the adhesion force).

The vertical downward gravitational force is larger than the sum of the buoyancy and viscous forces which are vertical upward when the beads fall into the liquid, so the beads exist in accelerated motion state. The viscous force is increased with the increase of the bead velocity. When the velocity of the bead reaches a certain value v_0, the sum of their values being zero, and then the bead will exist in a uniform speed state, which can be calculated using the following formula:

$$mg - \rho_0 V g - 6\pi r \eta v_0 = 0 \tag{6.4}$$

where v_0 is the end speed.

From the formula, it can be deduced that:

$$v_0 = \frac{2}{9} \cdot \frac{(\rho - \rho_0) g r^2}{\eta} \tag{6.5}$$

There exists a whereabouts distance S in which the bead moves from accelerated to uniform speed state. When the bead travels the distance S, it can be deduced from the two above formulas that,

$$S = 1.11 \times 10^4 \frac{r^4 \rho^2}{\eta^2} \left(1 - \frac{\rho_0}{\rho}\right) \tag{6.6}$$

Thus, the sphericity and diameter of the microbead can be determined by the balance status distance S.

The advantages of gel-beads forming:

The gel-beads forming is suitable for various ceramic materials. For the forming of ceramic microbeads of Al_2O_3, ZrO_2, Si_3N_4, SiC etc., this methods can be realized just by adjusting process parameters, and it don't need the replacement of equipments. When compared with other forming methods of ceramic microbeads, this method has shorter production cycle and higher forming efficiency. The microstructure of the products is uniform and dense. The decortications and fragmentations which occurrerd in traditional rolling method of beads are overcome. The hollow microbeads which was produced by melting method are solved. The gel-beads forming method has a simple process, short production cycle, and it is easy to adjust the process parameters (the diameter of microbeads is between 0.1 – 3 mm), and automatic and continuous production are easy to realize.

6.1.2 The Processing of Microbeads

The technical route of gel-beads forming for microbeads is presented in Fig. 6.2.

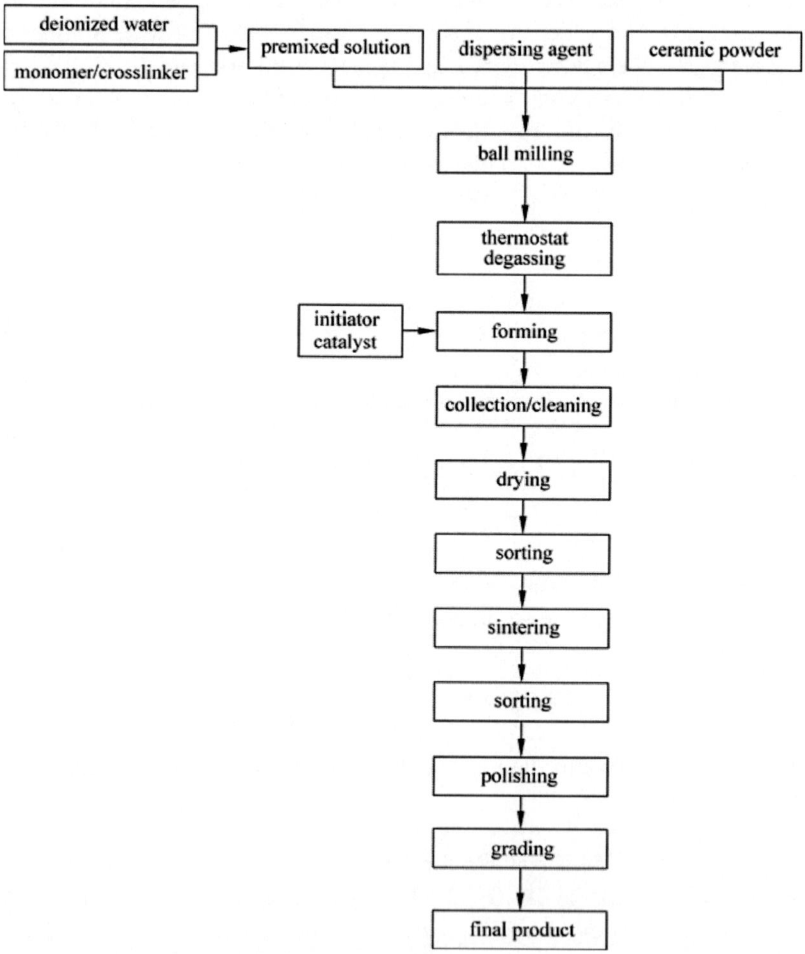

Figure 6.2 The technical route of gel-bead forming process

The process can be more specifically described as follows:

Step 1: the preparation of high concentrated ceramic suspensions with low viscosity and high solid loading. The ceramic powders, organic additives, dispersants are mixed in the aqueous solution uniformly. Generally speaking, the solid loading of ceramic powders in the aqueous suspension is greater than 50 vol%.

Step 2: by adding the curing agents, the suspension is dropped into the oil-based medium by dedicated equipment, and then it is solidified into a microbead through the synthetical effect of curing agents and temperature.

Step 3: after forming, cleaning and drying, then microbeads are sintered.

Step 4: by using the centrifugal rotation type polishing machine, the surface of sintered ceramic ball can be polished.

Step 5: after cleaning, the above mentioned balls are sieved into the final products with different diameter.

6.1.3 The Properties of Ceramic Microbeads

6.1.3.1 The Ceramic Microbeads Used for Grinding Media

(1) Ultrafine Grinding of Non-Metallic Minerals

Ultrafine grinding is a key technology in the deep processing of non-metallic minerals. With its rapid development, the ceramic microbeads, as an indispensable grinding media for ultrafine grinding process, show a more vast market potential and application fields. Many both overseas and domestic ultrafine materials, such as paint, printing ink, dye-based color, magnetic materials, electronic materials, coatings, optical materials, cosmetics, food, and pharmaceutical products, etc., need the ceramic microbeads to carry out ultrafine grinding, and homogenizing.

China's non-metallic minerals industry earns foreign exchange through export of 2.6 billion U.S. dollars in 2001 and 2.8 billion U.S. dollars in 2002. Kaolin and heavy calcium carbonate, as the paper coatings, play a vital role in the inorganic non-metallic minerals processing. The demand of ceramic microbeads that can be used as grinding media for ultrafine powders is growing. From the foreign markets, it can be seen that the annual output of heavy calcium carbonate and kaolin has reached 57 Mt, and that the annual amount of ceramic microbeads as the grinding media has approximately reached 8000 t. According to the statistics provided by China Paper-Making Association and China Inorganic Materials Association, at present, the market capacity of China's non-metallic mineral ceramic microbeads is about 8000 t/a. Only for the paper industry, it is 3000 – 5000 t/a and it is growing rapidly with the 15% growth rate.

No matter domestic or foreign, the market of paper coating is in a high-growth period, and the growth of its demand is fast. It also brings the high growth of the grinding media.

(2) Inks, Paints and Dope Industry

The grinding media in inks, paint and dope industry can be divided into two categories: ceramic microbeads and glass beads. The ceramic microbeads have some excellent characteristics such as high density, high grinding efficiency, and low wear-abrasion, pollution free and so on. The products can be divided into three categories, alumina, zirconium silicate and zirconia, which can be used for the grinding and disperse of high viscosity materials. At present, the annual output of inks, paints and dope in China has reached 3 Mt, yet the international market has reached 50 Mt.

(3) Machinery Equipment Industry

The ceramic materials have the excellent properties of resistance to heat, acid, alkali and wear, which are not observed in the metal products. Ceramic bearing ball, which has a wide application in the field of aerospace, military, electronics, metallurgy, energy and so on, is the most advanced alternative to the traditional steel bearing ball. With the development of the machine tool technology, the spindle speed becomes more and more high. For example, the maximum spindle speed had been improved from 5000 r/min in the early 1980's to the current 30000 r/min. Because the ordinary steel bearings can not meet the requirements, it is necessary to develop new ultra-high-speed bearing, which are used under the condition of high temperature, vacuum, non-magnetic or strong corrosive medium.

(4) Deburring and Polishing of Workpiece

Deburring is an indispensable aspect in the workpiece processing. In this step, the workpiece is put into the polishing equipment, and then in the course of operation, the cutting force which generated during vibration and rotation of workpiece, is used to achieve such effects, such as deburring, derusting, and high brightness and smoothness of its surface. The workpieces are machined into different products such as: micro-motor shafts, metal parts, die castings and copper, etc., using the different polishing equipments. The main equipment has a centrifugal polishing machine, vibration polishing machine, water swirl polishing machine and sandblasting equipment. According to statistics, the demand for cylindrical grinding media in the domestic market is 300 – 500 t/a and the international market sales are greater.

In addition, the ceramic microbeads have been widely used as a part of advanced ball-point pen and other wear-resistant pieces.

According to the diameter distinction of high-performance ceramic beads, the application fields of them are divided as follows in Table 6.1.

Table 6.1 The application fields of ceramic beads with high-performance

No.	diameters (mm)	applications
1	0.4 – 0.8	sandblasting, derusting, surface treatment
2	0.4 – 3.0	ultrafine grinding of non-metallic minerals, paint, dope, printing ink, food, pharmacy
3	0.3 – 1.2	ball-pen beads
4	>3.0	bearing (hybrid and complete), decorations, ball screw, bike ball
5	>10	grinding media, ball valve with acid and alkali resistant and high temperature resistance

The following are the performance requirements of the grinding media:

The widespread applications of ceramic microbeads are based on its excellent performance. With the expending of application, there is a need for the ceramic microbeads with higher quality.

The main application of ceramic microbeads in industrial production is the grinding media in the ultrafine grinding device. The average diameter of these microbeads is less than 6 mm in the mixing mill, and less than 1 mm when used for ultrafine grinding. The diameter of the grinding media has a direct impact on the grinding efficiency and the particle size. The greater the size of media, the larger the particle size of powder and the higher the production yield. Generally, the size of the grinding media is determined in accordance with the raw material and particle size of final powder. To improve the grinding efficiency, the size of the grinding media must be 10 times larger than the average feeding size.

The density of grinding media also plays an important role in grinding efficiency. The larger the density of the media, the shorter the grinding time. So, it is necessary to consider the density when selecting the material system. To increase the grinding effect, the hardness of grinding media must be higher than the grinding materials. Usually, it is better for the Mohs hardness larger than 3 for the grinded materials. The density and diameter of the grinding medias which are used commonly are shown in Table 6.2.

Table 6.2 The density and diameter of the grinding media

grinding media	glass (leaded)	glass (unleaded)	alumina series	$ZrSiO_4$ series	ZrO_2	steel ball	chrome ball
density(g/cm^3)	2.5	2.9	3.6 ± 0.3	3.7 ± 0.1	5.9 ± 0.1	7.6	7.8
diameter(mm)	0.3 – 3.5	0.3 – 3.5	0.3 – 3.5	0.3 – 1.5	0.3 – 3.5	0.2 – 1.5	1.0 – 2.0

The grinding media is one of the key factors which decide the grinding efficiency. Figure 6.3 shows the performance requirements of ceramic microbeads as grinding media. The main factors that impact the performance of microbeads include the following:

(1) Wear resistance. It is directly related to the property of the microbead. In the grinding process, the wearing of the microbeads should be decreased, because the grinded material will be polluted by the wear, and at the same time, the grinding equipment may be damaged by the broken microbeads.

(2) Sphericity. If the sphericity of the microbeads is better, the distribution of roundness is narrower and the particle size distribution of grinded materials is more uniform.

(3) Density. Under the same grinding conditions, the microbeads with higher density have larger impact force as well as produce the excellent grinding effect.

(4) Breaking strength. It is one of the most important performance indicators. It shows the greatest impact force and can be used to judge the occasion which the microbeads are employed.

(5) Color. The main application areas of the ceramic microbeads are paper and paint industry, hence, many white microbeads are manufactured.

(6) Different applications. For different situations, microbeads with different diameters are required, as shown in Table 6.1.

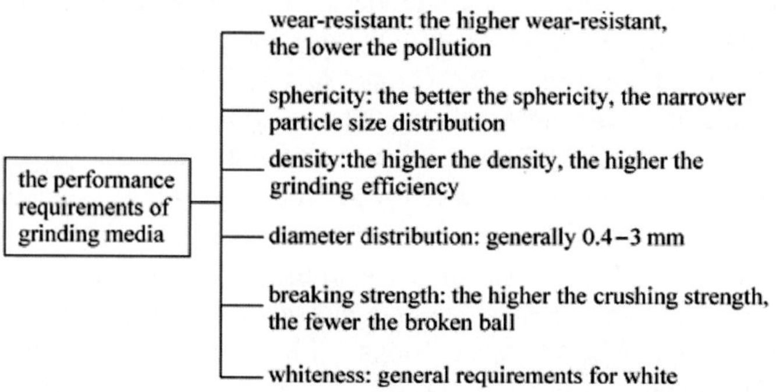

Figure 6.3 The performance requirements of the grinding media

6.1.3.2 The Ceramic Microbeads Used for Ball-Pen

Over the past two decades, the development of ball-pen manufacturing industry in China has attracted worldwide attention. The production scale and export capacity has been ranked first in the world. While developing pen products, the evolution and improvement of the related parts and components are also desired by pen-making industry. The performance of the ball-pen depends mainly on the quality of the pen head and ink. The quality of the pen head includes its inherent structure and its coordination with the bead holder, the bead and the ink. The beads, as an important component of the ball-point pen, influence the writing performance based on its material and machining accuracy.

In the 1970s, the materials selection of the beads was examined by the technical staff of our domestic pens industry. Owing to its characteristics of high hardness, wear-resistance and corrosion-resistance, ceramics were considered to be the best material. But at that time, because of the poor preparation process and machining technology of high-performance ceramics, the obtained Al_2O_3 ceramic beads, with poor surface finish and low yield, were unable to satisfy the demand of commercialization. Currently, ceramic beads have been equipped in the top quality ball-point pen in Japan, Switzerland and Germany, whose pens industries are well developed. However, the forming technology and processing of ceramic microbeads have always been kept as a secret. Nowadays, the ceramic body of the beads molded by the new process of gel-beads forming has obtained a breakthrough in the process aspect of molding and tooling. The obtained ceramic micro-beads sized by 1.2, 1.0, 0.9, 0.8,0.7, 0.6, 0.5, 0.38 mm have passed the test which carried out by the related technology department of China, satisfing the technical requirement of the pens industry.

The tungsten carbide alloy is now widely used as material for the ball-pen beads. These beads are produced with powder metallurgy technology. The metallurgical structure consists of tungsten carbide phase and the binder phase. The ZrO_2, with three kinds of crystalline, does not have much practical engineering value because of the volume change during the crystalline transformation at different temperature regions. Tetragonal zirconia polycrystals (TZP) is a kind of material with single-phase tetragonal zirconia. The 3Y-TZP is a kind of TZP with 3 mol% yttria as a stabilizer. The contents of yttria range from 1.75 to 3.5 mol%. Its strength and fracture toughness, respectively, are up to 1.5 GPa and 15 $MPa \cdot m^{-1/2}$. 3Y-TZP has higher room temperature mechanical properties in the fileds of ceramic materials at present, and hence, it is known as ceramic steel. This material has been widely used in harsh loading conditions and harsh industrial environments, and is therefore considered to be the best replacement materials for tungsten carbide pen beads.

Figure 6.4 shows the micrographs of cross-section and surface of 3Y-TZP ceramic beads produced by Hebei Yonglong Bangda New Materials Company.

Figure 6.5 presents the micrograph of cross-section of the tungsten carbide (WC) alloy beads produced by a company in Shanghai.

It can be seen from the Fig. 6.4 that the microstructure of 3Y-TZP is uniformly dense; all the grains are all small equiaxed-shaped and all the grain size at 0.3 – 0.5 μm. It can be seen from Fig. 6.5 that tungsten carbide grains are bonded and surrounded by the binder phase, with uneven grain size and the larger grain flakes reaching 3.0 – 5.0 μm.

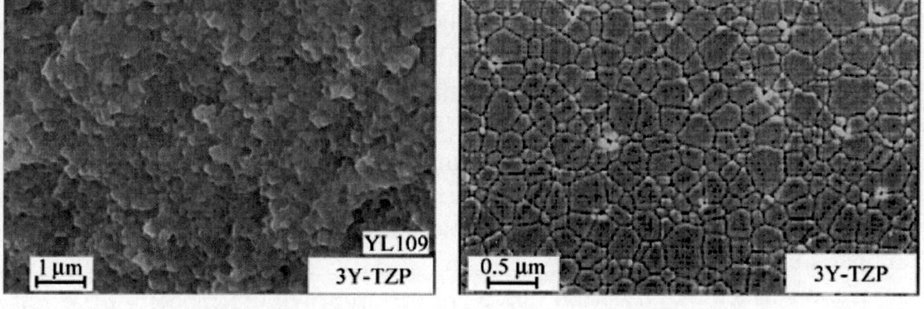

(a) (b)

Figure 6.4 3Y-TZP ceramic beads
(a) cross-section (fracture surface); (b) surface (polished and etched)

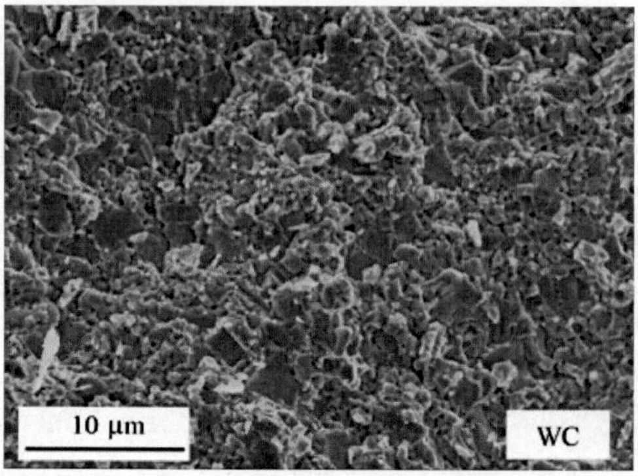

Figure 6.5 The cross-section of WC alloy beads

As the zirconia-ceramic grain size is smaller and more homogeneous than the tungsten carbide, and as the surface of the zirconia beads can reach a very high surface finish after processing, the abrasion of the bead holder will be very small. At the same time, zirconia-ceramic also has good insulation properties and non-magnetic characteristics, so it can avoid the electrochemical corrosion engendered when the bead roll in the bead holder and effectively reduce the damage to the bead holder. Therefore, zirconia-ceramic beads can easily maintain the stability gap between the bead and the bead holder. The writing life is long and the lineation is uniform.

When writing with a ball-point pen, because of the tangential friction force, the bead rolls in the bead holder, adheres and carries ink to the surface of the paper and finishes the writing process. The good coordination between the bead and the parts which contact with it, is the premise of continuous writing and results in uniform ink out. According to the description about the transport condition in fluid mechanics, a good affinity between the solid and liquid is needed to make them achieve a balance.

At the interface of the bead, air and the ink, a tangent plane is made along the contact surface of solid wall and liquid. The angle between the tangent plane and the solid wall in the liquid is the contact angle of the two phases, as shown in Fig. 6.6. The liquid infiltrates the solid when θ is an acute angle, otherwise the opposite occurs. The occurrence of the infiltration depends on the characteristics of the liquid and the solid. The better infiltration, the better affinity, and the liquid can easily adhere to the surface of the solid. The contact angles of different inks and bead materials have been measured at room temperature by the CS400 contact angle apparatus. Two other kinds of ceramic materials that have good mechanical properties and chemical stability have been measured at the same time, as a reference for comparison.

Use of the surface modification treatment can improve the capacity of affinity with ink.

The ink, bead and bead holder are three major factors which decide the writing performance of ball-point pen. A ball-point pen with excellent writing performance

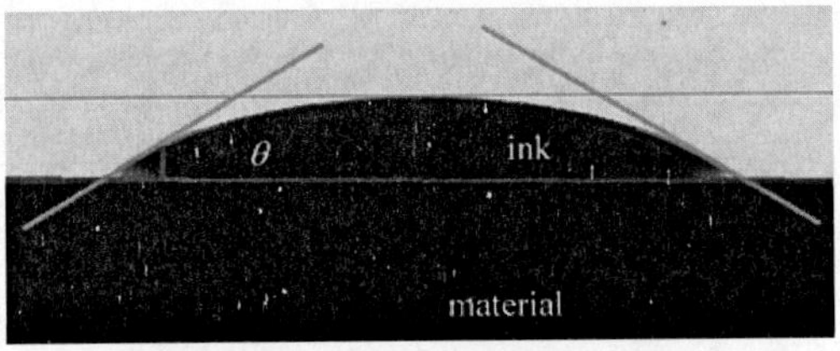

Figure 6.6 The ink contact angle

is a perfect outcome of the combination of the three aspects. When using it, because of the tangential friction force, the bead rolls in the bead holder, adheres and carries ink to the surface of the paper and finishes the writing process. So it is essential that the bead and ink have perfect affinity for the ink to adherent to the beads. An earlier test shows that, the contact angle between the ink and 3Y-TZP ceramic beads is smaller than that between the ink and the traditional tungsten carbide beads. Therefore ceramic beads satisfy the basic requirements as a ball-point pen. The zirconia ceramic has many merits such as wear-resistant, corrosion-resistant, non-conductive, non-magnetic, dense microstructure. The zirconia ceramic beads have little abrasion with the bead holder, and the gap between the two is capable of maintaining long-term stability. The writing length of the line is 2 – 4 times greater than that of the traditional beads. Moreover, zirconia ceramic bead with small and uniform grains, can be machined with high precision. In addition, the comfortable feeling is another outstanding feature.

However, the zirconia-ceramic beads prepared only by grinding and polishing process have some drawbacks such as poor ink capacity, hydroplaning and disconnections when writing. In this study, a surface modification technology for ceramic is analyzed, intending to improve the affinity of the beads with the ink, increase the adaptive capacity of the beads to different inks and bead holders, expand its application scope. The result shows that, the ceramic beads after surface modification treatments have better affinity with ink, and have managed maintain a long-term writing characteristic and comfortable writing feeling.

The 3Y-TZP beads produced by Heibei Yonglong Bangda New Materials Company are treated by surface modification process. Before the treatment, the diameter of the bead is 0.7 + 0.001 mm, the roundness tolerance is less than or equal to 0.20 μm, and the surface roughness (R_a) is less than or equal to 7 nm.

The process of milling and surface treatment is shown as Fig. 6.7.

To improve the comparability of the test results, all the measured beads are assembled with the bead holder that of the same size and materials in the same assembly machine. A kind of commercial neutral ink is used. The line length and the volume of ink refer to the national standard: GB/T2625 – 2003 Neutral pen and ink cartridge. Tianjin RH-01C zoned circular writing instrument and analytical balance are employed. The S-450 SEM is used to observe the surface morphology changes of the samples before and after treatment. AJ-Ⅲ atomic force microscope (AFM) is used to test the sample surface morphology and roughness.

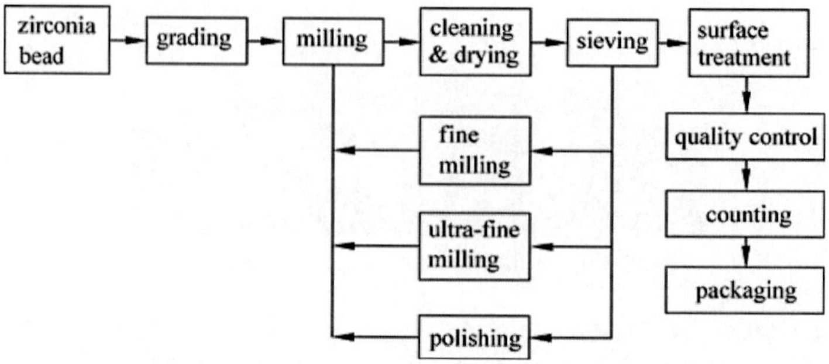

Figure 6.7 The process of milling and surface treatment

The contact angle of different bead material and ink was measured by the CS400 contact angle apparatus at room temperature. The results are shown in Table 6.3. It can be seen that the contact angle of the ink and the ceramic materials are all smaller than those of the ink and tungsten carbide. It is decided by the inherent characteristic of the material. The contact angle of zirconia ceramics and ink is the smallest, so the affinity between the ceramic bead and the ink is the best, and it is very much easier to bring out the ink for smooth and even writing.

Table 6.3 The test result of contact angle

materials of the bead		WC	Al_2O_3	Si_3N_4	ZrO_2(3Y-TZP)
contact angle θ(°)	neutral ink	28.6	27.2	24.9	22.5
	oil-based ink	33.7	31.0	30.1	27.0
affinity		increasing →			

The writing performance testing of zirconia ceramic beads:

Lead brass, alpaka and stainless steel are the most commonly used bead holder materials. In this study, the 0.5 mm ceramic beads produced by Hebei Yonglong Bangda New Materials Company were assembled with the bead holder made by different materials, and the beads were tested at National Pens Testing Center. Table 6.4 shows the actual length of the lines wrote by using different pen point.

Table 6.4 The test results of the length of the lines wrote using zirconia ceramic beads pen

beads holder	ink	writing length (m)										
		1	2	3	4	5	6	7	8	9	10	average
lead brass	neutral	564	593	575	569	671	627	577	568	589	666	600
alpaka	neutral	2211	2183	2175	2166	2219	2194	2237	2208	2216	2235	2204
	water-based	1313	1274	1181	1243	1195	1301	1186	1207	1215	1217	1233
stainless steel	neutral	2031	2069	2087	2117	2102	2048	2093	2101	2027	2111	2079

It can be seen from the Table 6.4 that the length of writing gets significantly improved after using zirconia ceramic microbeads assembled with the bead holder made from different materials. Lead brass has a good machining performance and low price, but it is susceptible to be eroded. The average length of the lines wrote using the ceramic microbead pen is 600 m that fully meets the international standard. The length of the lines wrote using the ceramics microbeads which is assembled with alpaka is 2200 m, which makes alpaka as the best alternative to much more expensive stainless steel. The zirconia ceramic microbeads pen presents universality because it is applicable to bead holders made using different materials as well as to a variety of ink.

The influence of surface treatment on writing performance:

Table 6.5 shows the average of the ten test results of the 0.7 mm tungsten carbide and zirconia ceramic microbeads pen assembled with a bead holder made up of lead brass.

Table 6.5 The average of the ten test results

0.7 mm bead		R_a (nm)	ink amount (mg/m) average of 10 times	line length (m) average of 10 times	poor writing	remark
WC		12 – 15	1.53	208	not	bead holder is made by lead brass
ZrO_2	before surface treatment	<7	1.45	605	exist	
	after surface treatment	8 – 10	1.66	517	not	

It can be seen from Table 6.5 that the average ink amount of the ceramic bead before modification treatment is only 1.45 mg/m which is less than that of the tungsten carbide bead (1.53 mg/m). Although the surface roughness of the ceramic bead pen after surface treatment is still smaller than that of the tungsten carbide one, the ink amount increases to 1.66 mg/m. The tungsten carbide bead takes a lot of wear and tear with the bead holder, which causes the line to become uneven when the length reaches to about 150 m. When the length reaches 200 m, the pen can not write normally because the bead nearly breaks away from the bead holder. With regard to the ceramic bead without surface treatment, slipping and breaking occurs at the beginning, but the line length achieved is about 3 times of that of the tungsten carbide bead. Besides, the ink streams evenly, especially after the length reaches 200 m. The surface treatment can prevent slipping and breaking at the beginning, and the line length of the ceramic bead achieved is about 2.5 times of that of tungsten carbide bead. From this it can be seen that the ceramic bead pen after modification treatment has smaller surface roughness, higher ink amount, and is better than the tungsten carbide bead of the same specifications because of the more even line and prevention of slipping and breaking.

The influence of surface treatment on surface microstructure:

For the ceramic bead surface without surface treatment, the R_a is no more than 7 nm. Figure 6.8 is the surface morphology of the ceramic bead after surface treatment, (a) is the SEM photo, (b) is the AFM one, and R_a of the surface is about 8 – 10 nm. Figure 6.9 is the surface morphology of the WC bead through AFM of which the surface R_a is about 15 nm. Figure 6.10 is the surface morphology of the ceramic bead after mechanical grinding of which the surface R_a is no more than 7 nm.

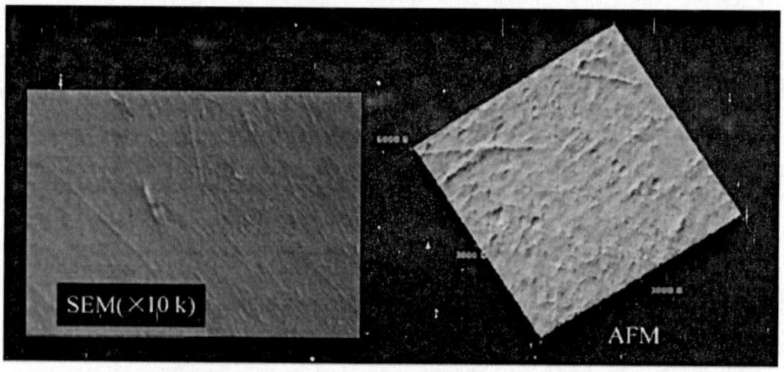

(a) (b)

Figure 6.8 The ceramic beads after surface treatment

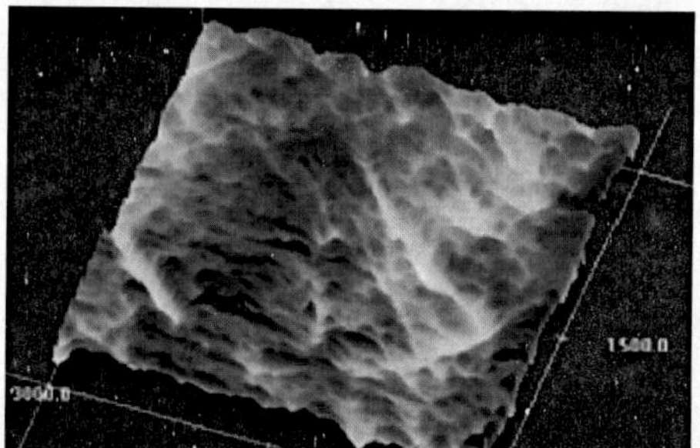

Figure 6.9 The surface morphology of WC beads

It can be seen that zirconia ceramic beads have dense microstructure and smooth surface. From Fig. 6.8 it can be seen that after surface treatment the ceramic bead could forms micro-grooves and stripes whose width is less than 0.2 μm on the smooth surface. Figure 6.9 shows the surface morphology of tungsten carbide beads. Because the crystal grains of tungsten carbide are about 10 times larger than those of zirconia and because the hardness of its alloy coupling phase among crystal grains is smaller than that of tungsten carbide crystal grains, the crystal grains easily form large pits when abrasive machining occurs during the

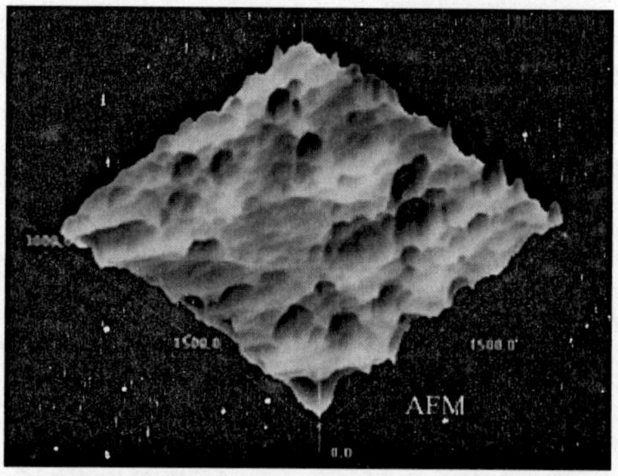

Figure 6.10 The ceramic beads after mechanical grinding

coupling phase. Although these pits can improve the affinity of the ink toward the tungsten carbide beads, the microstructure like peaks and valleys, takes a lot of wear and tear with the bead holder and doesn't fit well with the bead holder whose hardness has large difference from that of the bead. Figure 6.10 shows the ceramic beads after mechanical grinding. Its surface endures mechanical damage, and fracture occurs among the crystal grains, so the smooth base plane gets damaged and the crystal grains bump down significantly which is easy to take a lot of wear and tear with the bead holder.

The surface of ceramic beads after surface treatment has both smooth base plane and micro-grooves which can improve its affinity toward the ink. It's neither different from the pits abraded between large tungsten carbide crystal grains and coupling phase nor the inequalities of the crystal grains after mechanical grinding. The microstructures just like plains and rivers can ensure less wear between the smooth base plane and the bead holder while the affinity toward the ink is improved through the rivers, so it's an ideal method of surface treatment.

6.1.3.3 The Ceramic Microbeads with Far Infrared Function for Health Care

Many international research institutes have made the following conclusions regarding far infrared radiation: the far infrared radiation that has the almost same wavelength as human body's radiation can be totally absorbed by human body. The far infrared radiation can penetrate through the skin to stimulate the inner organs of the body. It can accelerate blood circulation, promote metabolism, remove humidity and algidity, and increase blood and nutrition supply. When the feet are in the far infrared energy field, the powerful energy can reach everywhere in the human body through the meridians in the feet. Therefore it can bring out the healthcare effects after the above description.

The far infrared ceramic beads are made of high purity inorganic ceramic raw material by using gel-beads forming. They have many specific electrical, magnetic and thermal functions. And they can radiate far infrared rays with the specific wavelength and high emissivity. The far infrared radiation can be totally absorbed by human body. So, the ceramic beads are very peculiar far infrared radiation materials. After heating, the ceramic microbeads with far infrared function will bring out the health-care effects just like accelerating blood circulation, promoting metabolism, dredging meridians, removing humidity and algidity, and increasing blood and nutrition supply.

6.2 Improving the Breakdown Strength of Rutile Capacitor

Gel-casting is a near-net-shape method combining physical chemistry for polymer, colloidal chemistry and ceramic technology (Omatete O. O., et al., 1991; Young A. C., et al., 1991b). The main steps (Janney M. A., et al., 1991; Janney M. A., et al., 1992; Janney M. A., 1990) of gel-casting are as follows: Firstly, the ceramic powder (e.g. alumina or Si_3N_4) along with small quantities of gel initiators, catalyst, monomers, crosslinker, and sintering aids is thoroughly mixed to form the homogeneous suspension with high solid loading and low viscosity. Secondly, gelation is initiated after the suspension is casted into a non-pore mold and the temperature is elevated. Thus the 3D network structure resulting from this *in-situ* polymerization holds the ceramic particles together. Finally, after unmolding and drying the green body possesses high homogeneity and high strength, and after sintering such bodies have good mechanical property (Bossel C., et al., 1995; Huang Y., et al., 2000). As a promising colloidal *in-situ* forming technique, it has been highly concerned and utilized in the forming of many sorts of ceramic material systems (Nunn S. D., et al., 1996; Sun L., et al., 1996; Waesche B., et al., 1997).

The rutile capacitor exhibits an excellent mechanical and electrical performance, such as high dielectric constant, low loss angle and high breakdown strength. It is an important material especially applied as a high power capacitor. It has been widely used in high frequency appliances, such as broadcast emitters, radar, high frequency welding machines, smelting furnaces etc. For these applications the properties such as dielectric constant, loss angle and breakdown strength are considered to be very important (Li B. R., et al., 1995; Kingery W. D., et al., 1976). Usually, the dielectric constant of rutile ceramic is 60 – 80 ranging from 0.5 Hz to 5 MHz at room temperature. For rutile ceramic the tangent of loss angle under 1 MHz is about $(4-5)\times 10^{-4}$ (testing temperature is 20 ± 5℃), such a low loss angle is beneficial to decrease the energy loss of the rutile ceramic under high frequency. Another important property of dielectric materials is the ability to withstand large field strengths without electrical breakdown. According to the national standard of the People's Republic of China, the breakdown strength of the rutile ceramic applied under DC field should be above 10 kV/mm.

In the past decades, rutile capacitors with complex shapes and large sizes were mainly produced through extrusion forming and subsequent machining (Yang J. L., et al., 1999). However, the whole production cycle is too long (about 2.5 months), and the machining of dried green bodies will result in a great deal of pollution and wastage of raw materials. It is difficult to produce rutile capacitors with a large size, complex shape or containing numerous additives in the raw materials by using the traditional process. To overcome these disadvantages, the present authors developed a new route based on gel-casting (Ma L. G., et al., 2001) by which the production cycle can be reduced to three weeks, and the dust caused by machining can be eliminated. Therefore, both production costs and the pollution can be reduced greatly, while the functional property can be improved greatly, especially the breakdown strength.

The composition of the rutile mixture applied to extrusion forming is shown in Table 6.6. The commercial rutile powder and additives (Coke-oven Plant of Shanghai, China) are of industrial purity. The particle sizes of additives are at the micrometer level. The reagents that are used in this study are as follows.

Table 6.6 The composition of the rutile mixture

main starting materials	additives (wt%)				
TiO_2	CaF_2	H_2WO_4	$BaCO_3$	ZnO	bentonite
100	2	2	1	1	5

The rutile mixture according to Table 6.6 was calcined for 2 h at 1170℃ and then ball-milled for 50 h, obtaining the calcined rutile mixture with $d_{50} = 2$ μm. The gel-casting process of the rutile mixture is described in Fig. 6.11. First, the calcined rutile mixture was dispersed into a premix solution, which has been prepared by dissolving AM and MBAM in deionized water. With the aid of a dispersant, the concentrated slurry with low viscosity was obtained after ball-milling for 24 h. The slurry was then degassed in the vacuum for 7 min after the addition of the initiator and catalyst. All the above steps were operated at room temperature. Afterwards, the slurry was casted into a steel mold which has been heated in an oven at 70℃. After coagulation, the demolded green bodies were dried, debindered and sintered.

Disk-shape green bodies of $\phi 30$ mm × 2 mm were prepared for electrical properties measurements. Silver slurry was pasted on the two main sides of the sintered bodies to form electrodes. The dielectric properties of these samples were measured at 1 MHz under room temperature, and the breakdown strength was measured in silicon oil by applying a DC field between the two opposite electrodes.

6.2.1 The Influence of Sintering Additives on the Flow Behavior

Generally, many kinds of additives are added to the rutile powder, such as $BaCO_3$, ZnO, CaF_2, H_2WO_4, and bentonite, to improve the sintering behaviors and the electro-properties. Bentonite, usually employed as an extrusion aid, can also lower

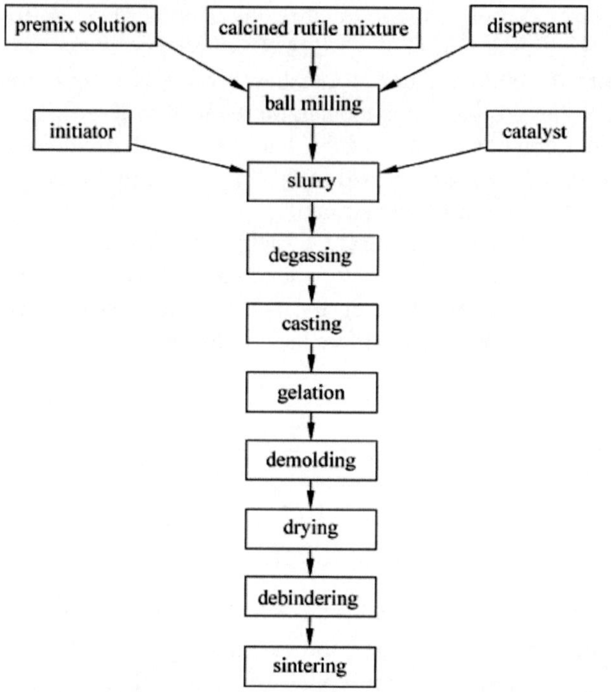

Figure 6.11 The flow chart of gel-casting of calcined rutile mixture

the sintering temperature, so it was considered as a sintering aid in our experiment. At relatively lower temperature, fusible phase resulted from $BaCO_3$, ZnO and CaF_2 maybe come into being, causing the decrease of sintering temperature. H_2WO_4 was found to lead to fine crystal grain of rutile, hence it will play a role in increasing the dielectric properties of rutile.

Since zeta potentials of all these components are greatly different, it is difficult to disperse all these powders suitable for gel-casting. Rutile powder containing different aids in Table 6.6 was dispersed into deionized water respectively. To maintain definite solid loading, the added quantity of the deionized water was decided by means of content of different sintered aids and their density. H_2WO_4 particle is difficult to disperse into basic medium owing to their acidity, while the concentrated rutile suspension must be prepared in basic condition. According to Fig. 6.12, $BaCO_3$, ZnO and CaF_2 have a slight influence on the viscosity of the rutile suspension, while bentonite has a relatively large influence. From the viscosity measurement, concentrated suspension of pure rutile shows a obvious shear-thinning behavior (Mikulasek P., et al., 1998; Mikulasek P., et al., 1997), which is similar to that of the pure alumina gel-casting slip. However, according to Fig. 6.13, the rheological characteristics of rutile mixture slurries prove to be shear thickening owing to the existence of additives, and the more bentonite is contained, the more obviously the viscosity changes. When absorbing water, bentonite particles dilate up to 30 times (Du H. Q., et al., 1986) changing the volume fraction greatly, which results in the increase of slurry viscosity and the shear thickening phenomenon. Since the viscosity of the slurry with 5 wt% of bentonite is too large (about 6 Pa·s), the slurry is not suitable for gel-casting.

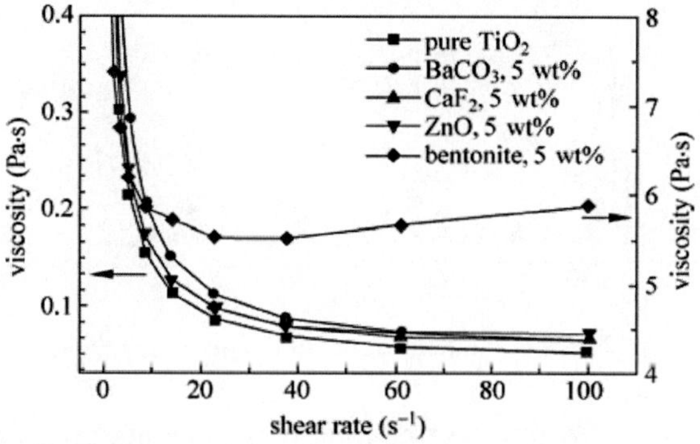

Figure 6.12 The influence of the sintering additives on rheological behavior of rutile suspension

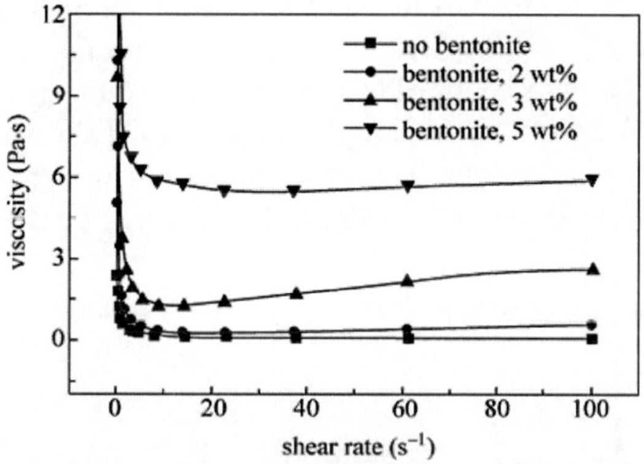

Figure 6.13 The influence of the addition of bentonite on rutile suspension flow behavior

6.2.2 The Calcining of the Rutile Mixture

In general, calcining can remove volatile impurities, chemicals bonded water, physical-absorbed water and improve the purity of the raw materials (Liu K. S., 1990). Furthermore, it can also help to form stable crystal phases and densify particles, which result in improved properties of the ceramic parts.

In our experiment, the rutile mixture was calcined at high temperature and subsequently ball-milled for 50 h. The XRD pattern (as shown in Fig. 6.14),

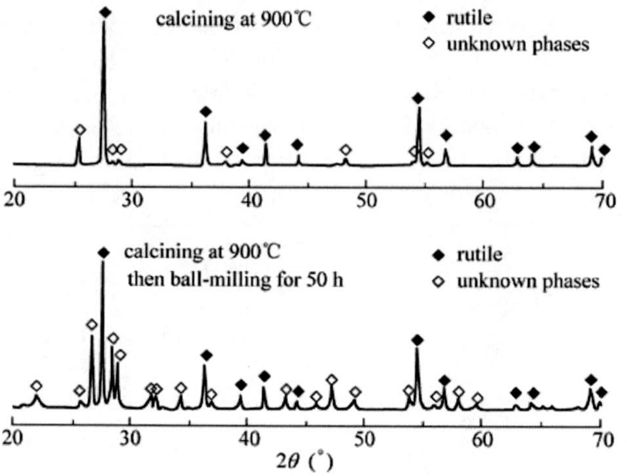

Figure 6.14 The XRD patterns of the rutile mixture calcined at 900℃

revealed some other peaks besides rutile peaks for the mixture calcined at 900℃ (In order to observe the diffraction peak obviously, the ratio of rutile to sum of additives is changed to 1:1.). When compared with the XRD pattern for the un-milled mixture, more complex peaks, which do not represent rutile, are found in the XRD pattern of the ball-milled mixture. This means that new phases appear during the ball-milling procedure. However, almost no other peaks besides rutile are found in the XRD pattern for the rutile mixture calcined at 1170℃ (as shown in Fig. 6.15); even after ball-milling, other peaks fail to appear. Therefore, a conclusion can be drawn that when calcined at 1170℃, phases in the mixture

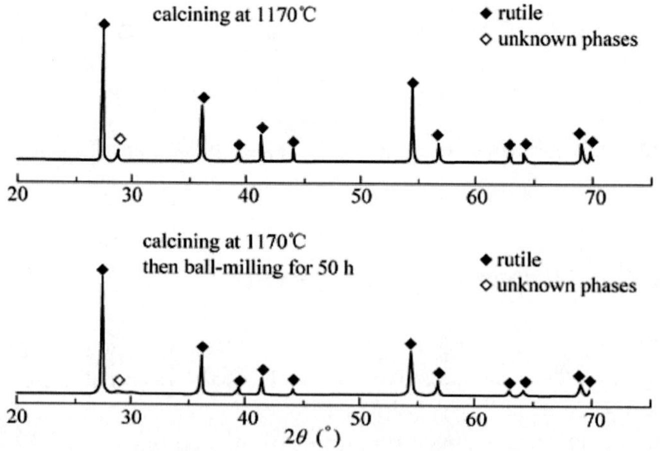

Figure 6.15 The XRD patterns of the rutile mixture calcined at 1170℃

will be unified even after ball-milling. In summary, when calcined at 1170℃, the additives can form fusible phase unobservable by X-rays; hence the harmful influence of H_2WO_4 and bentonite mentioned earlier can be removed, the interaction between the rutile mixture and water can also be avoided. Therefore, the suitable suspension for gel-casting can be obtained easily.

6.2.3 The Rheological Behavior of the Calcined Rutile Mixture

Figure 6.16 shows relationship between zeta potential and pH of rutile mixture calcined at 1170℃. When a dispersant (a PMAA-NH_4 solution) is added, zeta potential of rutile mixture is greatly improved. The dispersant is found to result in the change of the particle surface charge as well as electrochemical double layer at the solid/liquid phase because of its adsorption at the particle interface. The negative groups on the polymer chain make the zeta potential more negative, imparting stability to the particles against flocculation. The adsorption obviously decreases the elecrtrophoretic mobility, shifts IEP of system to low pH and improves the absolute value of zeta potential (Strauss H., et al., 1993; Chen X. B., et al., 1998).

Figure 6.17 includes curves for rheological behavior of the rutile mixture calcined at 1170℃ at different solid loading. The figure indicates that the viscosity of the slurry increases gently with increase in solid loading, and that the slurries are of shear-thinned and subsequently shear-thickened. When the solid loading of the slurry is more than 52 vol%, the slurry changes from shear-thinning to shear-thickening very fast, and its lowest viscosity is higher than 1 Pa·s, hence the slurry does not fit gel-casting. While the solid loading is less than 52 vol%, it is applicable for gel-casting owing to its relatively low viscosity (shear rate <80 s^{-1}).

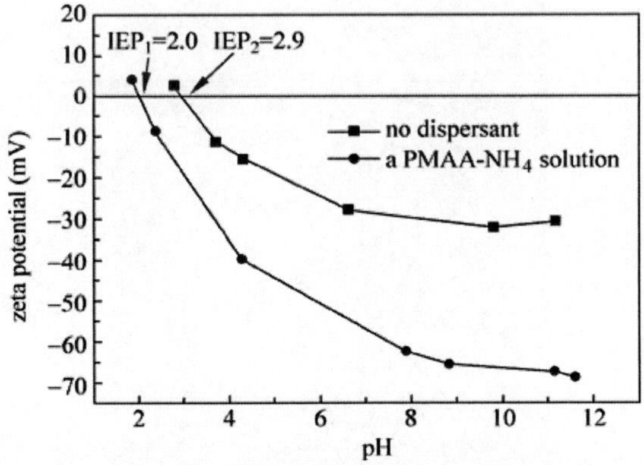

Figure 6.16 The relationship between zeta potential and pH

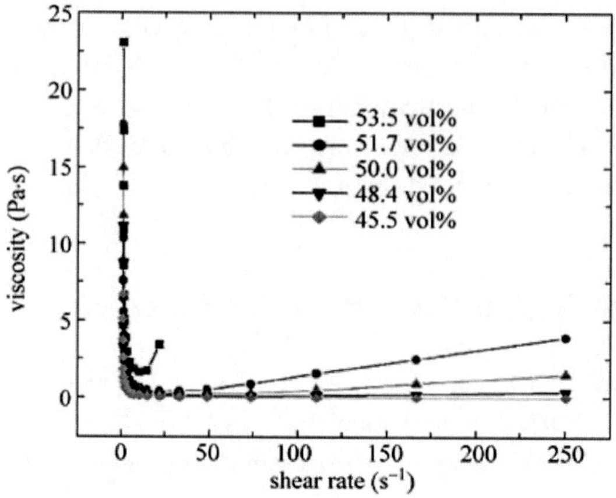

Figure 6.17 The rheological behavior of the rutile suspension at different solid loadings

6.2.4 The Gel-Casting of the Calcined Rutile Mixture

Figure 6.18 is a micrography of the green bodies gelcasted from the slurry of the rutile mixture calcined at 1170℃ with 51.7 vol% solid loading. It can be seen that the monomer has been polymerized efficiently, and that the homogeneous microstructure can be achieved. Figure 6.19 indicates that all close pores in the rutile parts are removed, which results in a good microstructure after sintering. The functional properties were measured respectively, which is listed in Table 6.7. Among these parameters, the breakdown strength is increased largely from original 12.6 kV/mm to more than 23.6 kV/mm.

Figure 6.18 The micrography of the green bodies gelcasted from the slurry of the rutile mixture

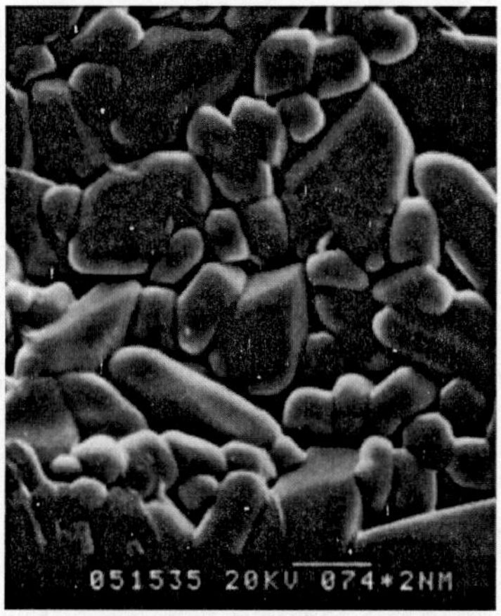

Figure 6.19 The microstructure of the parts produced by gel-casting

The flaw minimization and processing optimization of ceramic materials can be successfully accomplished by colloidal processing (Lange F. F., 1989; Pugh R. J., et al., 1994). During the gel-casting processing, the aggregates of powders could be removed in a well dispersed suspension and homogeneous green bodies can be obtained from the suspension through adequate degassing (Zhou L. J., et al., 2000), hence the microstructure of gelcast parts after sintering is homogeneous. It is well known that the dielectric break down phenomenon of dielectrics is structure-sensitive, and that the breakdown strength values vary widely according to the characteristics of the microstructure of dielectric materials (Yamashita K., et al., 1980). By gel-casting, since the microstructure of the sintered bodies of rutile mixture can be improved, and the breakdown strength can be significantly enhanced.

Table 6.7 The properties of the rutile sintered body formed by gel-casting

tested sample	ε_1	$\tan\delta_1$			E_B (kV/mm)	E_B (kV/mm)[a]
		(20 ± 2)℃	(85 ± 2)℃	H_2O		
1	81.9	0.5	0.3	0.3	25.0	13.2
2	80.9	0.4	0.2	0.2	22.0	12.7
3	81.1	0.4	0.2	0.2	22.5	13.2
4	80.3	0.3	0.1	0.3	22.6	12.9
5	80.2	0.4	0.2	0.4	22.7	11.3
6	80.9	0.6	0.4	0.5	25.0	12.2
7	81.0	0.4	0.2	0.5	24.6	14.1
8	80.8	0.4	0.2	0.4	23.6	11.7
9	81.7	0.4	0.2	0.5	22.7	12.3
10	80.6	0.5	0.3	0.5	25.2	12.2
average value	80.9	0.4	0.2	0.4	23.6	12.6

[a] The breakdown strength E_B of the rutile sintered body by extrusion forming process.

6.3 The Thin-Wall Rutile Tube for Ozone Generator with High Dielectric Constant

Ozone is one of the strongest oxidants, which are extensively applied in water treatment, chemical oxidation, food protection and medical treatment etc. In recent years the requirement for high capacity ozone generators is rapidly increasing. Previous studies indicate that the yield of ozone is directly proportional to the dielectric constant and wall thickness of the dielectric body in the ozone generator (Chu Jinyu, et al., 2002; Wang Jian, 1990). The dielectric constant of rutile ceramic is about 80 – 100 which is 15 – 30 times greater than that of glass and enamel. Furthermore, the rutile ceramic possesses high breakdown voltage. These two characteristics indicate that the rutile ceramic is the optimal material for dielectric body of ozone generators. But the forming process of large size thin-wall ceramic tube is very complicated. Technical difficulty and high cost are the main reasons that restrict application of the rutile ceramic in ozone generators.

A concentrated 52 vol% TiO_2 suspension with low viscosity was prepared by adding the TiO_2 powder and dispersant to the premix solution, followed by ball milling for 24 h.

A stainless steel mold was used to form the thin-wall ceramic tubes. An antisticking agent was pre-coated on the inwall of the mold, which was preheated at about 80 – 100℃. The large size thin-wall rutile tubes are prepared by the machine of colloidal injection molding of ceramics. The processing parameters: injection speed is about 15 – 20 mm/s, injection pressure 0.3 MPa, pressure holding time 1 – 2 min.

6.3.1 The Experiment Results

Appropriate sintering schedule is important for obtaining high performance ceramics. Thermal analysis revealed that there is a strong exothermic effect at 357℃, the corresponding decomposition of organic substances in the green body requires a slow temperature rising speed from 300℃ to 400℃. Within the temperature ranges from 950℃ to 1050℃ and from 1100℃ to 1200℃, the shrinkage rates of the TiO_2 body are more rapid than ever, and hence, the heating-up process should be mild at these two stages to avoid formation of closed gas porosity. At temperature from 1200℃ to 1300℃, liquid phase will appear in the body and ceramic forming reaction occurs, so control of holding time is very important at this stage. Figure 6.20 shows the sintering schedule curve of the thin-wall rutile tube. From Fig. 6.21 that presents the SEM fractograph of the sintered body, it can be seen that the tube possesses fine crystal grain and high degree of densification.

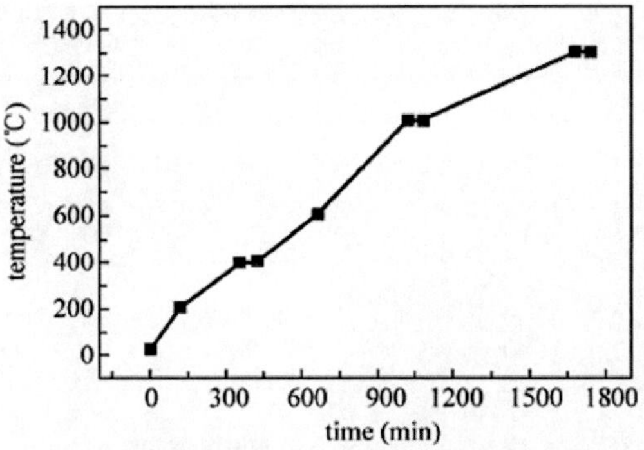

Figure 6.20 The sintering schedule curve of the thin-wall rutile tube

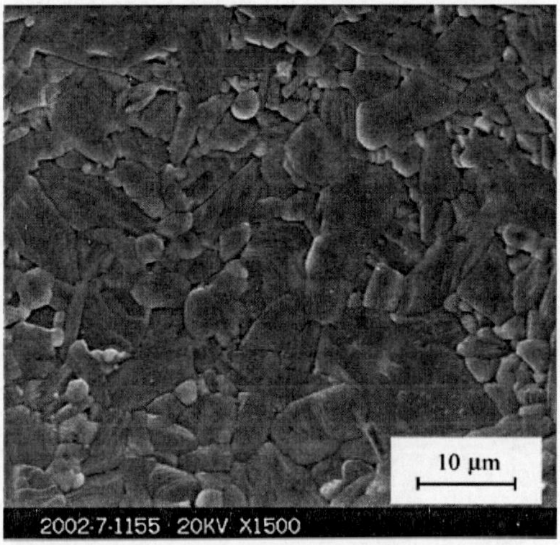

Figure 6.21 The fractograph of the sintered rutile body

The ozone generator requires a dielectric body with high dimensional precision. The thin-wall rutile tubes prepared through CIMC can meet the demand. The wall thickness of the sintered tube is 1.5 – 2 mm and its inner diameter is about 26 mm, its length can reach about 300 mm. The largest eccentricity of the tubes is 0.15. Figure 6.22 shows the configuration of the large size thin-wall rutile tubes prepared by CIMC for the ozone generator.

The dielectric constant of the dielectric body is the important factor that determines its electric performance. In this research, a frequency spectrometer was used to measure the dielectric constant of the obtained rutile tube. Figure 6.23 shows the variation of the tubes' dielectric constants with frequencies at different temperatures. This figure indicates that the tubes' dielectric constants are about 85 at low frequency and the values rise constantly with the increasing of frequency. Figure 6.24 shows the variation of the tubes' dielectric constants with temperatures

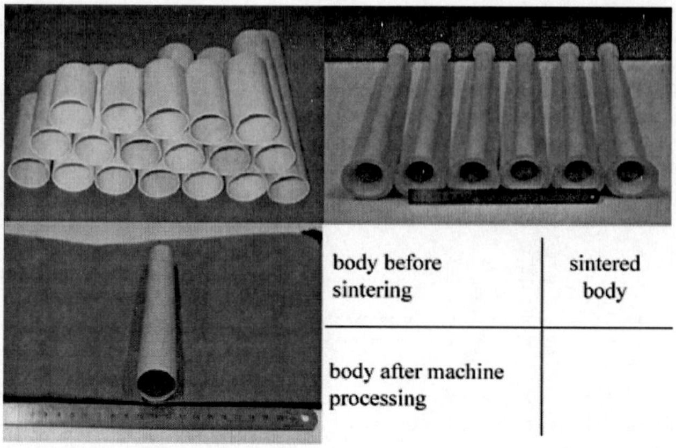

Figure 6.22 The configuration of the rutile tubes for the ozone generator

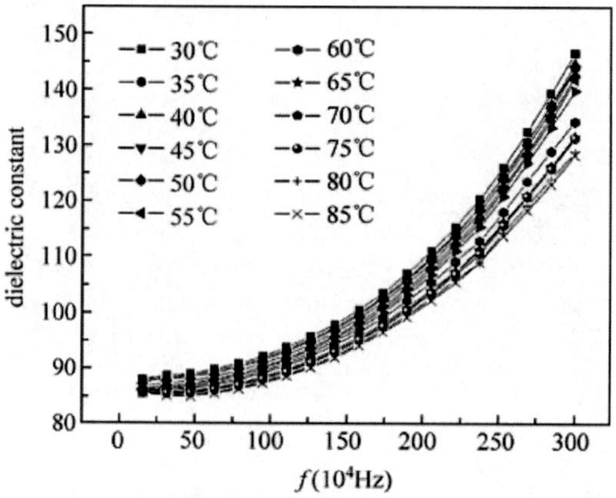

Figure 6.23 The variation of the dielectric constants with temperatures at different frequencies

at different frequencies. It tells that the variation of tubes' dielectric constant is inconspicuous when the temperature increases. Therefore, the tubes' dielectric performance will not be influenced when the temperature in discharge chamber varies owing to corona discharge. This characteristic can make the ozone generator work stably and safely.

The breakdown voltage of the tubes was also measured in this study. The average spot breakdown voltage of 20 samples is 16.8 kV/mm, and the value of surface is about 10.1 kV/mm. This performance can perfectly meet the demand of the ozone generator.

The obtained tubes were applied in the ozone generators. Under the following experimental conditions: power supply of 300 W; frequency of 7000 Hz; input voltage of 220 V and current of 0.3 A; the yield of ozone was about 50 mg/L. At the

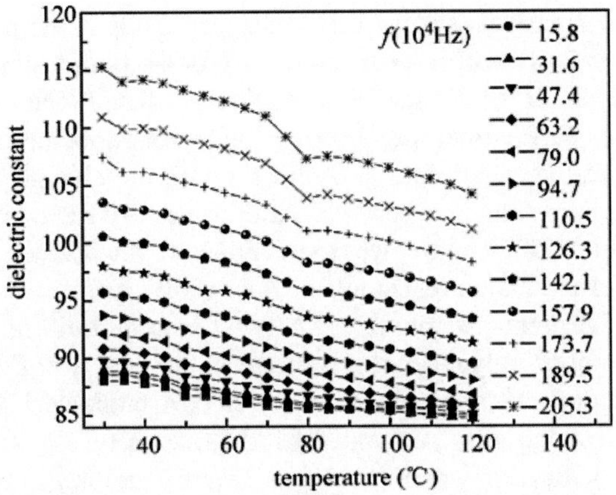

Figure 6.24 The variation of the dielectric constants with frequencies at different temperatures

same time, the generator's Corona Voltage sharply decreased from 4000 – 10000 V to 800 – 1000 V.

6.4 The Refractory Nozzle of Zirconia

The refractory nozzle for precision casting of a 50% Cu – 50% Cr molten alloy is required with a high melting point above 2200℃ and a good thermal shock resistance to withstand the temperature rising from the room temperature to 2200℃ in several seconds. To control the molten alloy flux accurately, the nozzle should also possess a precise dimension that remains unchanged during casting.

Ziconia, melting at 2680℃, is usually used as a refractory material at an extremely high temperature. Owing to its microcracks induced by martensitic phase transformation, ZrO_2 materials have high toughness and good resistance to thermal shock and flux scour. So, it's a suitable material to make the nozzle for precision casting of Cu-Cr molten alloy. To avoid the deformation caused by the shrinkage of green body during sintering, some coarse particles in the ingredient is useful.

The gel-casting is a near-net-shaping technique used for the manufacturing of homogeneous ceramic green bodies, owing to its characteristic of *in-situ* consolidation of suspension. However, few researches on the gel-casting of suspensions with coarse particles have been reported. For these suspensions with coarse particles, settlement stability as well as aggregation stability should be considered. In this study, the effects of various kinds of dispersants on the rheological behavior and the settlement stability of zirconia suspension with coarse particles were studied, and such a suspension can be employed to produce a refractory nozzle for precision casting of Cu-Cr molten alloy by gel-casting.

Three kinds of zirconia used in this study are ZrO_2 A with a size range of 0.2 – 0.074 mm, ZrO_2 B with a size range of 0.044 – 0 mm, and ZrO_2 C with a size of $d_{50} = 1$ μm. ZrO_2 A and B are electric fused CaO-stabilized cubic structures, and C is TZP containing 3 mol% Y_2O_3. By choosing cubic ZrO_2 as the major raw material, we were able to avoid the volume change caused by transformation during sintering and service. And some Y-TZP containing complex phase was used as one of the ingredient to manufacture the nozzle, which is believed to improve the thermal shock resistance of the ceramic parts.

Three kinds of dispersants, namely TAC (tri-ammonium citrate), arabic gum and APA (ammonium-polyacrylate), were used to prepare ZrO_2 suspensions. They separate the particles in suspensions by electrostatic, steric and electrosteric repulsive forces respectively. AM and MBAM, in the mass ratio of 50 : 1, were dissolved in the deionized water to obtain 15 wt% premix solution. ZrO_2 was added into the premix solution in the proportion of A : B : C = 4 : 8 : 1. The mixture was ball-milled for 48 h to obtain a suspension.

To measure the sediment stability, the suspension was cast into a ϕ20 mm glass tube with the height of 30 cm, and the height of the supernatant relative to the limpid liquid layer was measured at a certain interval to indicate the sediment stability of the suspension.

6.4.1 The Rehological Behaviors of ZrO_2 Suspensions with Different Dispersants

In ZrO_2 suspensions, the trend of particle flocculation owing to van der Waals attraction is inhibited because of the repulsion caused by the dispersants. However, owing to the Brownian motion of solvent molecules, some particles may overcome the repulsion to move closer to each other and form stable or metastable particle clusters. Solvent enclosed in the clusters loses its fluidity to some extent and becomes non-free solvent. The viscosity of suspensions indicates the volume of the free solvent, and thus, the more clusters in the suspension result in the higher viscosity of the suspension.

For the suspension stabilized by TAC, particles are separated by electrostatic repulsion, which is governed by zeta potential on the particle surface. Appropriate TAC ratio (0.2 wt% to the ZrO_2 mass) yields a higher zeta potential absolute value and few particle clusters, and then a low viscosity of the suspension. Exceeding TAC amount may suppress the double-layer thickness near the particle surface and reduce zeta potential absolute value because of the high ion strength, and increase the viscosity of the suspension. The ZrO_2 suspension stabilized with TAC shows a shear-thinning behavior. When sheared, the particle clusters are destroyed by the velocity gradient in the suspension, and solvent enclosed in these clusters is released and becomes free solvent, so the viscosity of the suspension decreases. As the shear rate rises, more and more clusters are destroyed. When the shear rate reaches more than 300 s^{-1}, almost all the clusters are destroyed, and the viscosity of the suspension remains, by and large, unchanged and independent of TAC content (see Fig. 6.25(a)).

The viscosity of the suspension stabilized with arabic gum changes with the shear rate in another way (see Fig. 6.25(b)). If the amount of arabic gum is not sufficient to cover the surface of particles in the suspension, in the case of 1 wt% to ZrO_2 mass, there may not be adequate repulsion between particles to separate them, and clusters may be formed by bridging effect. These clusters will be broken when sheared, and the suspension shows a shear-thinning behavior. If the amount of arabic gum is enough to cover the surface of particles (2 wt % and 4 wt% to ZrO_2 mass), there are few clusters in the suspension, and then the suspension shows an inconspicuous shear-thinning behavior. In this case, at a higher shear rate, osmotic pressure in the zone among the particles, which is caused by the concentration difference of macromolecular chains and which avoids the particles to move closer in static state, will prevent the particles from moving apart. As a result, the viscosity of the suspension increases with the rise of the shear rate, and the behavior of shear-thickening is observed.

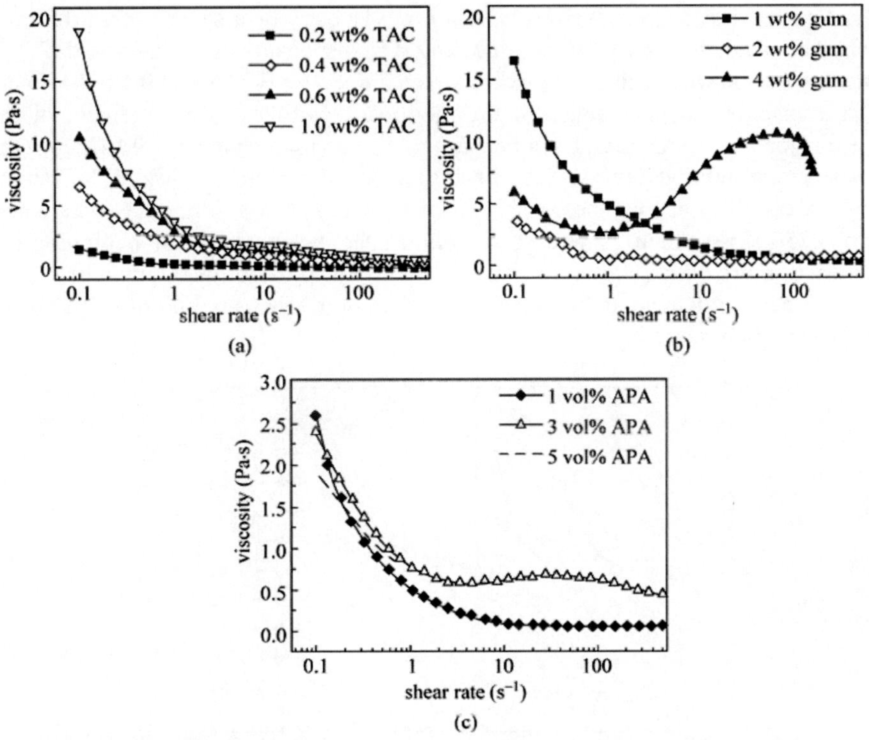

Figure 6.25 The rheological curves of 55 vol% ZrO_2 suspensions with different dispersants

(a) dispersant TAC; (b) dispersant arabic gum; (c) dispersant APA

The ZrO_2 suspension stabilized with APA (see Fig. 6.25(c)) shows an average rheological behavior of those stabilized with arabic gum and TAC. The shear thinning behavior of this suspension is not as severe as that of those stabilized with TAC at low shear rates because of the steric repulsion, and the shear thickening behavior is also not as severe as that of those stabilized with arabic gum at high shear rates because of the existence of electrostatic repulsion between APA chains.

6.4.2 The Sediment Stability of ZrO_2 Suspension with Different Dispersants

The sediment volumes of 55 vol% ZrO_2 suspensions stabilized with different dispersants were compared in Fig. 6.26. During the incipient stage of settlement, ZrO_2 particles undergo an accelerated course, and the increasing velocity of the particles improves the velocity gradient near them. Hence, in the shear-thinning suspensions, the settlement of the particles decreases the viscosity of the suspensions; on the other hand, the decreased viscosity accelerate the particles according to Stokes' equation. As a result, settlement equilibrium is achieved in a short period for the suspensions with a low viscosity or a severe shear-thinning behavior. And this can be illustrated by ZrO_2 suspensions stabilized with TAC, which have the worst settlement stability (see Fig. 6.26). Because of the impediment of the macromolecular chains to the motion of particles, the settlement of the suspensions with APA and arabic gum is lesser than that with TAC. For the suspensions stabilized with APA, the suspension containing 1 vol% APA has the highest viscosity at low shear rate but most severe shears thinning (as shown in Fig. 6.25(c)), and then its settlement height changes slowly at the beginning but increases more quickly after 180 min. The suspension stabilized with arabic gum has the best settlement stability because of its high viscosity and obvious shear-thickening behavior.

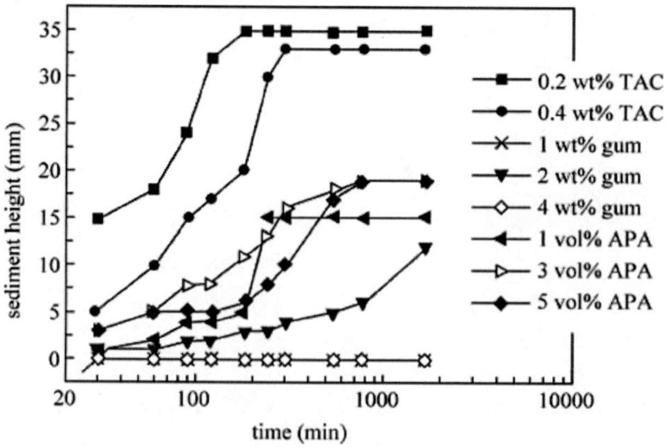

Figure 6.26 The sediment height of ZrO_2 suspensions with various dispersants

6.4.3 The Preparation of ZrO_2 Refractory Nozzles

Although ZrO_2 suspension with arabic gum has the best settlement stability, it is not the best choice to prepare the refractory nozzle, since the solid loading of ZrO_2 suspension, which is important to avoid the deformation during the sintering of green bodies, couldn't be increased to more than 60 vol% only by the steric stabilization of arabic gum. As a compromise of fluidity and settlement stability, ZrO_2 suspension with 69 vol% solid and 1 vol% APA was prepared to shape the refractory nozzle for precision casting of Cu-Cr alloy. 2 μL initiator (8 wt% $NH_4S_2O_8$ solution) and catalyst (tetramethylethylenediamine) were added into 1 mL suspension, and then stirred the suspension for 3 min to make it disperse uniformly. The stirred suspension was cast into a mold, and kept at about 40℃ for about 30 min to ensure solidification of the suspension. The demoulded green body was dried at 110℃ for 24 h, and then was sintered at 1580℃ for 2 h. The prepared nozzle can satisfy the requirement of precision casting of Cu-Cr molten alloy, and its microstructure, phase composition and some final properties are shown in Fig. 6.27, Fig. 6.28 and Table 6.8 respectively. As shown in Fig. 6.27, there are some pores in the nozzle, which may be responsible for the good thermal shock resistance of the nozzle. On the other hand, the underfired nozzle may shrink to counteract the dilation caused by the elevated temperature when it is used, which also improves the thermal stability. XRD pattern (see Fig. 6.28) shows that besides some $CaZr_4O_9$, which is a metastable phase synthesized during the production of electric fused CaO-stabilized ZrO_2, the nozzle mainly consists of cubic and tetragonal ZrO_2, and the complex phases also contribute to the thermal stability. The good resistance to flux scouring may result from the coarse fused Ca-ZrO_2 particles.

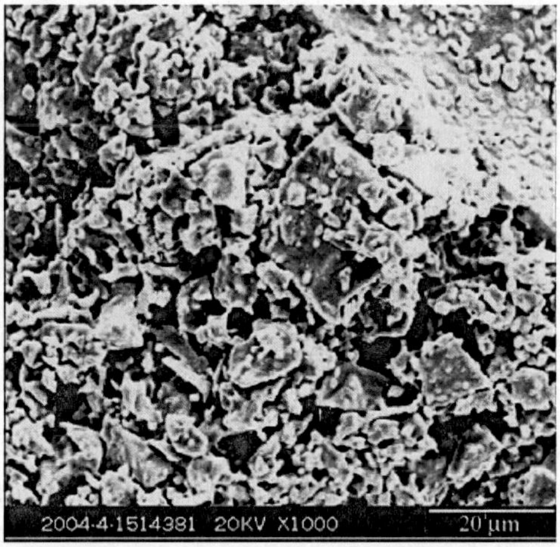

Figure 6.27 The SEM micrograph of ZrO_2 nozzle

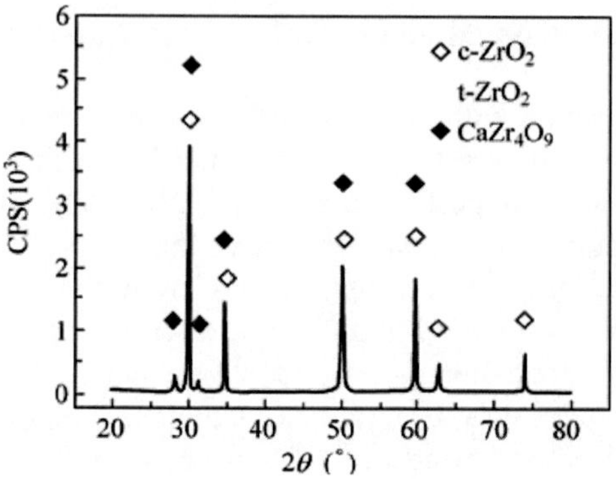

Figure 6.28 The XRD pattern of ZrO_2 nozzles

Table 6.8 Some properties of ZrO_2 refractory nozzles

density (g/cm^3)		three-point flexural strength (MPa)			sintering linear shrinkage (1580℃, 2h)
top of the nozzle	bottom of the nozzle	green body	sintered	sintered, 1100℃ water cooled	
4.85	4.93	20.6	35.9	23.6	4.17%

6.5 Water Based Gel-Casting of PZT

In the last several decades, piezoelectric materials such as PZT have found an increased demand in a wide field of applications including actuators, sensors, and transducers. The hardness and fragility of the material makes it difficult for mechanical processing and fabrication is often a time-consuming task. In some methods used for producing the material, such as tape casting or injection molding, the ceramic phase is dispersed in a nonaqueous solvent and a significant volume fraction of organic binders are required to support the structure. Because organic substances bring about a series of material problems, such as debinding issues during sintering and in homogeneity of the sintered parts, an aqueous solvent with low organic content is highly desirable in the processing of the materials.

Gelcast green bodies also have high homogeneity and high mechanical strength, which is of great advantage for handling of the parts before firing and for producing large castings. Since its invention in 1991 a wealth of literature describing the application of gel-casting to the forming of structural ceramics, such as alumina (Prabhakaran K., et al., 2000; Cai K., et al., 2002), zirconia (Bengisu M., et al., 2002), silicon carbide (Vlajic M. D., et al., 2002; Yi Z. Z., et al., 2002), Si_3N_4 (Huang Y., et al., 2000), and mullite (Liu Y. F., et al., 2001), etc., has been published. However, little has been reported about its application in the production of piezoelectric ceramics.

In this study the colloidal chemistry of PZT powders in an aqueous solvent and the rheological behavior of the PZT suspension are reported. Microstructures and piezoelectric properties of gelcast samples from suspensions with different solid loading were also investigated and compared with those of the die pressed samples. It was found that the gelcast samples exhibited stronger piezoelectric effects than die pressed ones using the same sintering method. This may be owing to their microstructural difference.

Near the morphotropic phase boundary (MPB) composition, PZT ($Pb(Zr_{0.52}Ti_{0.48})O_3$) prepared by the mixed-oxide processing technique was chosen for this study. Stoichiometric amounts of Pb_3O_4, ZrO_2, and TiO_2 were mixed together in ethanol and thoroughly ball-milled for 24 h. After drying the mixture was calcined at 850℃ for 2 h, then ground and passed through 200 mesh sieves. The ultimate powder was characterized by XRD before being used. Particle size distribution of the powder was measured by X-ray sedimentation method using a BI-XDC type analyzer (see Fig. 6.29).

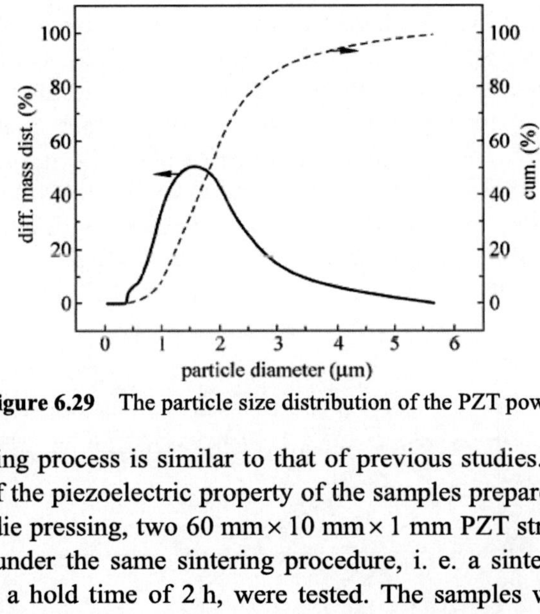

Figure 6.29 The particle size distribution of the PZT powder

The gel-casting process is similar to that of previous studies. To compare the homogeneity of the piezoelectric property of the samples prepared by gel-casting with those by die pressing, two 60 mm × 10 mm × 1 mm PZT strips derived from both methods under the same sintering procedure, i. e. a sintering temperature of 1250℃ and a hold time of 2 h, were tested. The samples were divided into 6 identical 10 mm×10 mm × 1 mm square segments, respectively, then polarized in a silicon oil bath by applying the same electric field of 3 kV/mm at 120℃ for 30 min. Piezoelectric coefficient d_{33} of the samples was determined by using a ZJ-3A piczomcter produced by Institute of Acoastics, Academia Sinica. The pore size distribution of the green ceramics was measured using the mercury porosimetry (Autopore 9220).

6.5.1 The Colloidal Chemistry and Rheological Behavior

The solid loading required for gel-casting is higher than that required either for spray drying or slip casting. It is preferred that gel-casting slurries be at least 50 vol% solids. Low solid loading and hard agglomerates in ceramic suspensions can result in deleterious surface and bulk defects in the finished ceramics. Therefore, a key factor for the successful production of ceramics by gel-casting technique is to produce a flowable and stable slurry with possible high solid loading. Homogenous dispersion of the powder in the premix solution and stability of the suspension are mainly determined by attractive and repulsive forces between the particles in the system. The former generally arises from the van der Waals forces while the latter can arise from electrostatic repulsion or steric repulsion of polymeric or surfactant material present on the particle surfaces (Israelachvili J., 1992). The magnitude of the van der Waals forces is mainly determined by the nature of the particle surface and the solvent, so it is essentially a constant. On the other hand, the repulsive forces can be modified over a wide range by the addition of dispersants. Electrostatic repulsion is directly dependent on the zeta potential of the powders. The higher this value with the same polarity, the more important the electrostatic repulsion between the particles. On the other hand, when this value is close to the IEP, the particles tend to flocculate.

The zeta potential of different PZT aqueous suspensions at different pH are shown in Fig. 6.30. The zeta potential value of pure PZT suspension changes from 33.4 mV at pH = 1.7 to –35.1 mV at pH = 11.9 with an IEP at about pH = 7.2, suggesting that neutral environment may be disadvantageous in order to acquire good dispersion. Addition of 0.5 wt% TAC results in much more negative potential values. A value of 2.2 mV for pure PZT suspension at pH = 7

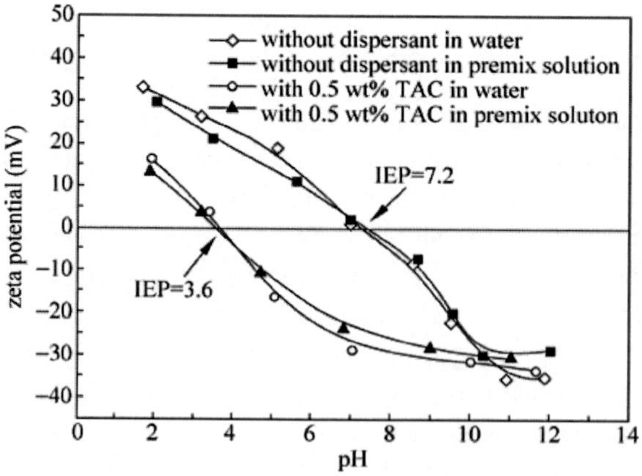

Figure 6.30 The influence of dispersant and monomer on the zeta potential of the PZT suspension

is decreased to about –28.7 mV, and the IEP is also shifted to about pH = 3.6. Addition of monomer shows little effect on the IEP position of the two systems, but it slightly decreases the absolute value of the zeta potential. This might be ascribed to the screening of the charged PZT particles caused by the uncharged AM molecules (Barnes H. A., et al., 1989). With regard to the two systems containing TAC, the zeta potentials decrease slightly with the increase of pH from the range of neutral environment to about pH = 11.7, suggesting that a basic environment might have little effect on the decrease in the viscosities of the two systems.

The influence of solid loading on the apparent viscosity for several PZT slurries is shown in Fig. 6.31. The viscosity data are obtained at the shear rate of 0.1 s^{-1}. Generally, there exists a complex nonlinear relationship between viscosity and solid loading, which is closely related to many factors, such as the continuous phase viscosity, particle-size distribution, and particle shape, etc. The curves in Fig. 6.31 have the typical shape of a plot of solid loading versus viscosity: low viscosity at low solid loading and sharp increase at high solid loading. At one particular time point, more and more particles would be added to suspensions to "jam up", causing continuous 3D contact throughout the suspension, thus, making flow impossible, i.e. the viscosity tends to infinity. The particular phase volume at which this happens is called the maximum packing fraction ϕ_{m} (Barnes H. A., et al., 1989). Before adding dispersant the system has a measurable viscosity at about 31 vol% solid loading. Adding a little more PZT to approximately 32 vol% will lead to a "solidified" slurry, thus, the viscosity will increase sharply and make it practically impossible to measure the viscosity using a rheometer with a concentric-cylinder setting. The ϕ_{m} value is thus determined to be 32 vol%. Adding 0.8 and 1.5 wt% of TAC increase ϕ_{m} to about 47 and 53 vol%, respectively, suggesting that TAC is very effective in increasing PZT solid loading. While increasing further the dispersant amount to about 2.2 wt% will decrease the maximum packing fraction again, indicating that 2 wt% of TAC in the suspension has achieved the maximum coverage of the PZT powder and that excess dispersant may be harmful in decreasing viscosity. The relationship between the viscosity and the solid loading for monodispersed suspension can be modeled by the Krieger-Dougherty (K-D) model as follows:

$$\eta = \eta_0 \left(1 - \frac{\phi}{\phi_{\mathrm{m}}}\right)^{-[\eta]\phi_{\mathrm{m}}} \tag{6.7}$$

where η is the viscosity of the suspension and η_0 is the viscosity of the solvent. The true solid loading of the powder dispersed in the suspension is represented by the variablc ϕ. Thc intrinsic viscosity $[\eta]$ of the suspension is a function of particle geometry; a value of 2.5 is appropriate for spherical particles. A maximum packing fraction of 0.63 ± 0.002 is appropriate for random close packing at low shear rate. The K-D model curve with $[\eta] = 2.5$ and $\phi_{\mathrm{m}} = 0.63$ is also plotted in the Fig. 6.31. Although there are obvious discrepancies between the experimental curves and the K-D model curves in Fig. 6.31, the experimental curves have the typical shape of a plot of solid loading versus viscosity as that of K-D model: low viscosity at low solid loading and sharp increase at high solid loading. The discrepancy may partly be ascribed to the inhomogeneity of the particle size. On the other hand, particle flocculation will lead to a lower value of ϕ_m because the flocs themselves are not close-packed (Barnes H. A., et al., 1989; Starov V., et al., 2002) and they can trap part of the continuous phase, thus, leading to a bigger "effective phase volume" and a greater viscosity than that of the primary particles. Therefore, the discrepancy also suggests that the PZT suspension without TAC is composed of a large amount of agglomerates. The results also reveal that TAC can effectively break up the agglomerates in the suspension.

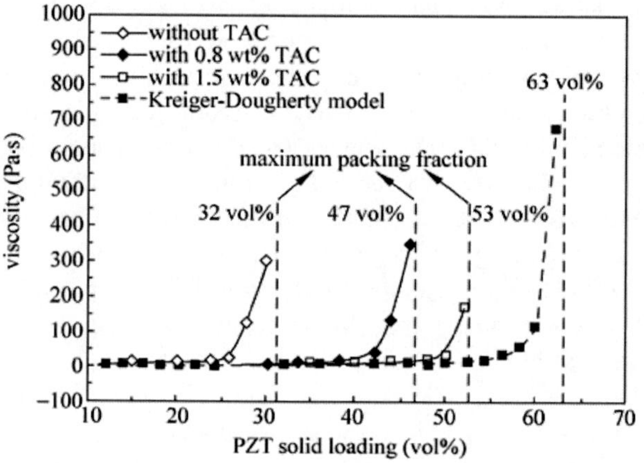

Figure 6.31 The influence of solid loading on the viscosities of the PZT suspensions with and without dispersant at a steady shear rate of 0.1 s^{-1}

Figure 6.32 shows the apparent viscosity as a function of shear rate for different PZT slurries. Here, 30 vol% solid loading was used for systems having no dispersant in case of "jam up". Adjustment of the pH to basic condition has little effect on the decrease of viscosity, while addition of TAC can greatly decrease the viscosity. This indicates that electrostatic stabilization mechanism alone is not sufficient for good dispersion of the PZT powders. Effect of the dispersant can be attributed to both electrostatic and steric stabilization mechanisms. Because of the relative short chain of TAC macromolecule than that of the polymeric dispersant, the latter stabilization mechanism can be more precisely denoted as a semi-steric mechanism. Generally, a viscosity below 100 Pa·s is indicative of a flowable slurry, thus, a TAC content of 1.5 wt% is suitable for casting and excess of TAC has little effect on the viscosity. Consistent with the zeta potential results, increasing pH of the system containing TAC to about 12.4 shows little effect on the viscosity. All the systems show shear-thinning behaviors at low shear rates.

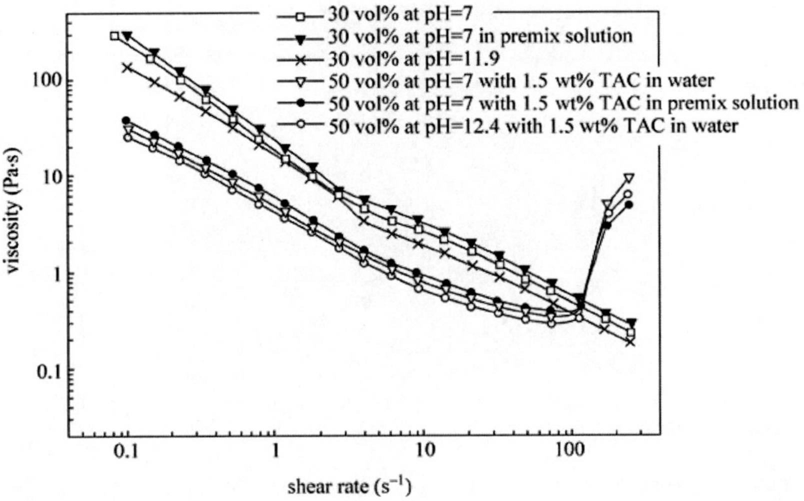

Figure 6.32 The influence of dispersant and monomer on the viscosity of the PZT suspension

Shear-thinning behavior indicates that the flow can bring about a more favorable particle spatial rearrangement. For the concentrated suspensions, it was generally accepted and has been verified that the flow-induced structure is a layered structure, which can result in a low resistance of the particle movement between different layers, and thus, a low viscosity is obtained (Barnes H. A., 1989). Due to the damage of the layered structure at very high shear rate, critical shear rate (γ_c), the viscosity increases again and a shear-thickening behavior appears in the systems containing TAC. This shear-thickening behavior can also be ascribed to the increase of the solid loading. Furthermore, addition of AM shows little (a η of 30.5 Pa·s increases to 38.6 Pa·s at the shear rate of 0.1 s^{-1}) effect on the viscosity. These results are consistent with the zeta potential results, indicating that the dispersants work by increasing the stability and resisting particle agglomerates by both electrostatic and semi-steric stabilization mechanisms. After developing concentrated PZT suspension with low viscosity, complex shaped PZT objects are successfully fabricated (illustrated in Fig. 6.33).

6.5.2 The Microstructure and Properties

The SEM micrograph of the fracture surfaces of gelcast PZT green body and those of sintered PZT samples are shown in Fig. 6.34. From Fig. 6.34(a) we can see there exists a little linear shape organic substance between the PZT particles, which is responsible for the high strength of the green body. Figure 6.34(b) and (c) are derived at a solid loading of 30 vol% (without dispersant) and 50 vol%, respectively. Because the former is formed from a low solid loading suspension

Figure 6.33 The illustration of the ultimate PZT ceramic parts fabricated by the gel-casting approach

with many agglomerates, a porous and cracked morphology is obtained. On the other hand, Fig. 6.34(c) indicates that after increasing the solid loading to 50 vol% and breaking up the agglomerates by adding TAC, a homogeneous microstructure

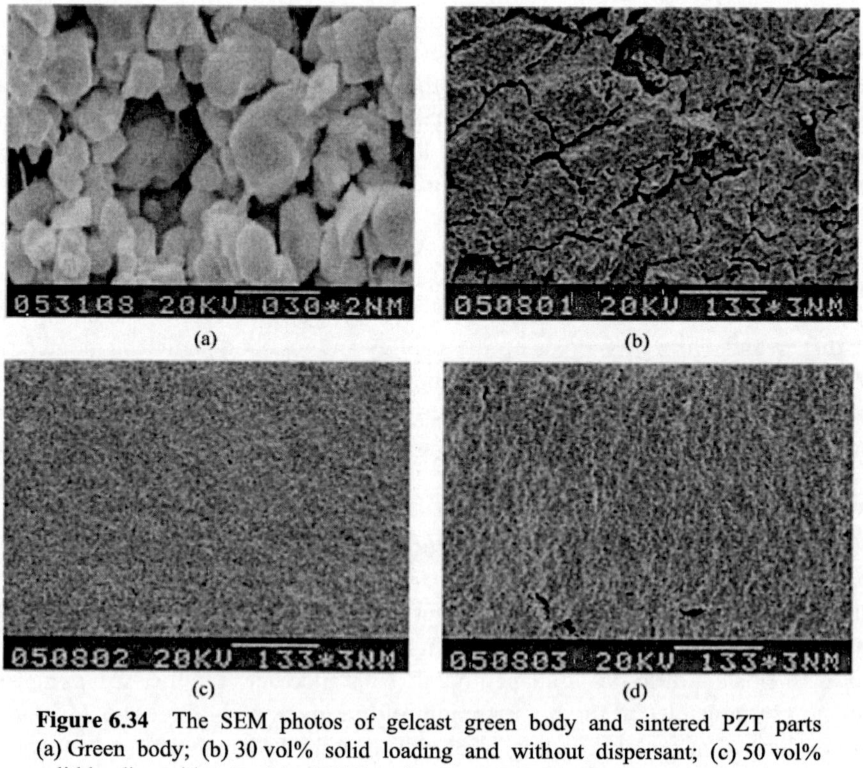

(a) (b)

(c) (d)

Figure 6.34 The SEM photos of gelcast green body and sintered PZT parts (a) Green body; (b) 30 vol% solid loading and without dispersant; (c) 50 vol% solid loading with 1.5 wt% of TAC as the dispersant; (d) die pressed samples can be obtained. Moreover, Fig. 6.34(c) also shows a more homogeneous microstructure than the dry pressed sample shown in Fig. 6.34(d). The density and shrinkage value of the PZT samples prepared under different conditions are consistent with this finding (see Table 6.9). As expected, increasing solid loading results in less shrinkage, higher density and green strength. Although gel-casting results in a slightly greater shrinkage than die pressing, a higher density was also obtained, thus, a more uniform structure is observed in Fig. 6.34(c) than that in Fig. 6.34(d). In addition, because the 50 vol% slurry is relatively concentrated and it can solidify instantly after casting at 80℃, no sedimentation was observed. Consequently, no obvious density difference between the different sections of the ceramic parts has been observed. Moreover, based on our observation, no cracking has occurred for the gelcast body after drying, even though no special drying conditions, such as temperature and humidity, were maintained.

Table 6.9 The comparison of some parameters of PZT samples prepared under different conditions

samples	green strength (MPa)	sintered density (g/cm^3)	total shrinkage after sintering (%)
gel-casting at 30 vol% solid loading	17	7.228	18.6 (with cracks)
gel-casting at 50 vol% solid loading	20	7.762	15.9 (without cracks)
die pressing[a]	3.6	7.661	14.8 (without cracks)

[a] Containing 1.5 wt% of PVA as the binder and pressed at 80 MPa.

The d_{33} coefficient results of the samples prepared by gel-casting (50 vol% solid loading) and by die pressing are illustrated in Fig. 6.35. A more homogenous piezoelectric property of gelcast samples than that of die pressed ones was observed by contrasting the S. D. of the two series of data (listed in Table 6.10). This also suggests that a higher homogeneity can be obtained by gel-casting than by die pressing. We also found that gelcast samples have a slightly higher average d_{33} value than die pressed ones. This is consistent with a previous study and may be attributed to the density difference (Kiggans J. O., et al., 1999). Furthermore, study of their microstructures by examining the pore size distributions of two kinds of green bodies (see Fig. 6.36) revealed that the gelcast sample has a smaller median pore size than the die pressed sample, which is indicative of a more homogeneous microstructure and may partly account for the piezoelectricity difference. The results in Fig. 6.34 and Table 6.10 are in good agreement with those obtained from SEM examination and density measurement, suggesting that the homogeneously dispersed PZT particles in the suspension obtained by the addition of dispersant and thoroughly milling were effectively conserved in the green body during the *in-situ* polymerization and subsequently in the sintered samples.

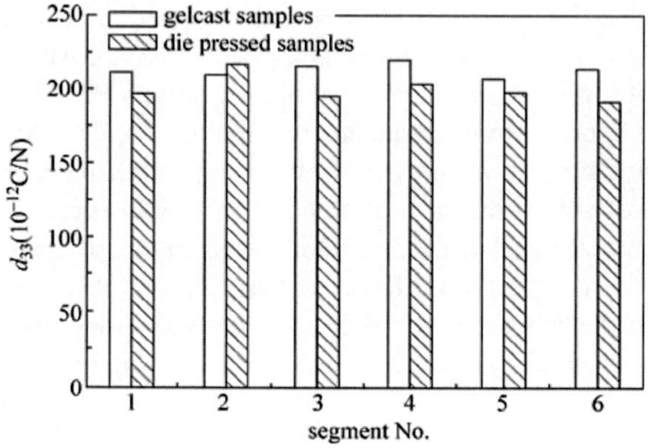

Figure 6.35 The comparison of piezoelectric properties of PZT samples derived from different methods

Table 6.10 The comparison of piezoelectric property of PZT samples prepared by gel-casting with a 50 vol% solid loading and those prepared by die pressing with a pressure of 100 MPa

samples	average d_{33} $(10^{-12}C/N)$	S. D.* of d_{33} $(10^{-12}C/N)$
gel-casting	213.2	3.933
die pressing	200.2	8.513

* S. D.: standard deviation.

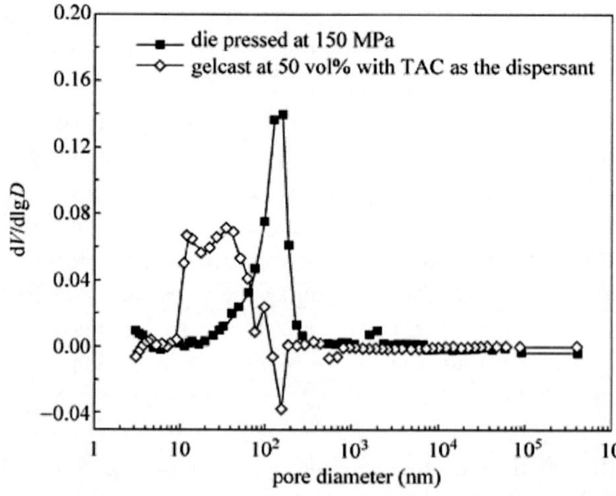

Figure 6.36 The comparison of pore size distributions of PZT green bodies obtained from different methods

References

Barnes H. A., 1989. J. Rheol., 33: 329.

Barnes H. A., Hutton J. F., Walers K., 1989. An Introduction to Rheology. Oxford: Elsevier.

Bengisu M., Yilmaz E., 2002. Ceram. Int., 28: 431.

Bossel C., Dutta J, Houiet R., et al., 1995. Processing of nano-scaled silicon powders to prepare slip cast structural ceramics. Mater. Sci. Eng., A204 (1 – 2): 107 – 112.

Cai K., Huang Y., Yang J. L., 2002. Key Eng. Mat., 224: 643.

Chen X. B., Cheng H. M., Ma J. M., 1998. A study on the stability and rheological behavior of concentrated TiO_2 dispersions. Powder Technol., 99(2): 171 – 176.

Chu Jinyu, Wu Chundu, 2002. Ozone Technology and Its Application. Beijing: Chemistry Industry Publishing House.

Du H. Q., Tang S. Q., 1986. Ceramic Material and Recipe. Beijing: Light Industry Press.

Huang Y., Ma L. G., Tang Q., et al., 2000. Surface oxidation to improve water-based gel-casting of silicon nitride. J. Mater. Sci., 35(14): 3519 – 3524.

Israelachvili J., 1992. Intermolecular & Surface Forces (2nd ed.). London: Academic Press.

Janney M. A., 1990. Method for forming ceramic powders into complex shapes. US Patent 4894194.

Janney M. A., Omatete O. O., 1991. Method for molding ceramic powders using a water-based gel-casting. US Patent 5028362 – 98.

Janney M. A., Omatete O. O., 1992. Method for molding ceramic powders using a water-based gelcatsing. US Patent 5145908 – 97.

Kiggans J. O., Tiegs T. N., Montgomery F. C., et al., 1999. Ceramic Engineering and Science Proceedings. American Ceramic Soc., 20: 19 – 26.

Kingery W. D., Bowen H. K., Uhlmann D. R., 1976. Introduction to Ceramics. John Wiley & Sons Inc., 913 – 931.

Lange F. F., 1989. Powder processing science and technology for increased reliability. J. Am. Ceram. Soc., 72(1): 3 – 15.

Li B. R., Wang X. Z., Zhang X. L., 1995. Inorganic Dielectric (in Chinese). Wuhan: Science and Technology University of Huazhong Press, 50 – 56.

Liu K. S., 1990. Principle of Ceramic Technology. Guangzhou: Science and Technology University of Hua'nan Press.

Liu Y. F., Liu X. Q., Wei H., et al., 2001. Ceram. Int., 27: 1.

Ma L. G., Huang Y., Yang J. L. et al., 2001. Improving the break down strength of rutile capacitor by gel-casting. J. Mater. Sci. Lett., 20(14): 1285 – 1288.

Mikulasek P., Wakeman R. J., Marchant J. Q., 1997. The influence of pH and temperature on the rheology and stability of aqueous titanium dioxide dispersions. Chem. Eng. J., 67(2): 97 – 102.

Mikulasek P., Wakeman R. J., Marchant J. Q., 1998. Crossflow microfiltration of shear-thinning aqueous titanium dioxide dispersions. Chem. Eng. J., 69(1): 53 – 61.

Nunn S. D., Kirby G. H., 1996. Green machining of gelcast ceramic materials. Ceram. Eng. Sci. Proc., 17(3): 209 – 213.

Omatete O. O., Janney M. A., Sterklow R. A., 1991. Gel-casting—a new ceramic forming process. Ceram. Bull., 70(10): 1641 – 1649.

Prabhakaran K., Pavithran C., 2000. J. Eur. Ceram. Soc., 20: 1115.

Pugh R. J., Bergströ m L., 1994. Surface and Colloid Chemistry in Advanced Ceramics Processing. New York: Marcel Dekker.

Starov V., Zhdanov V., Meireles M., et al., 2002. Adv. Colloid Interface, 96: 279.

Strauss H., Heegh H., Sreienitz I., 1993. Effect of PAA adsorption on stability and rheology of TiO_2 dispersion. Chem. Eng. Sci., 48(2): 323 – 332.

Sun L., Gao L., Guo J. K., et al., 1996. Gel-casting of nano-size Y-TZP. Acta. Mater., 9(6): 489 – 492.

Waesche B., Steinborn G., 1997. Influence of slip viscosity on the mechanical properties of high purity alumina made by gel-casting. Key. Eng. Mater., 132 – 136 (1): 374 – 377.

Wang Jian, 1990. Measurement of Ozone Concentration, Yield and Power Consumption for Ozone Generator. Trade standard of town building in PRC: CJ/T3028, Beijing.

Yamashita K., Koumoto K., Takata M., et al., 1980. Jpn. J. Appl. Phys., 19: 867.

Yang J. L., Xie Z. P., Ma J. C., et al., 1999. Study on gel-casting processing of high power rutile capacitor. Key Eng. Mater., 161 – 163: 517 – 520.

Young A. C., Omatete O. O., Janney M. A., 1991a. J. Am. Ceram. Soc., 74: 612.

Young A. C., Omatete O. O., Janney M. A., et al., 1991b. Gel-casting of alumina. J. Am. Ceram. Soc., 74(31): 612 – 618.

Vlajic M. D., Krstic V. D., 2002. J. Mater. Sci., 37: 2943.

Yi Z. Z., Xie Z. P., Huang Y., et al., 2002. Ceram. Int., 28: 369.

Zhou L. J., Huang Y., Xie Z. P., 2000. Gel-casting of concentrated silicon carbide suspension. J. Eur. Ceram. Soc., 20: 85 – 90.

Chapter 7 The New Methods and Techniques Based on Gel-Casting

Abstract The reliability and the cost of the products, which are closely related to the preparation process, are the two major factors in the industrial production of high performance ceramics. The preparing process of ceramic powders, the forming and sintering process are thus becoming increasing important in the recent decades in the ceramic research field. As a precondition of materials design and realization of the high performance ceramic products of desired shapes, the ceramic forming process usually plays a crucial role in determining the reliability and the cost of the products. The traditional methods of producing the ceramic products have been widely used in the field of fine ceramics, but they do not meet the requirements of high performance ceramics. Therefore, enormous research efforts have been directed to devise new methods for the ceramic forming process. In this chapter, some new methods and techniques based on gel-casting are introduced, such as solid freeform fabrication (SFF) of alumina ceramic parts; freeform fabrication of aqueous alumina-acrylamide gel-casting suspensions; ceramics with special porous structures fabricated by freeze-gel-casting; solidifying concentrated Si_3N_4 suspensions for gel-casting by ultrasonic effects; laser machining technique of Al_2O_3 green ceramic, etc.. The forming mechanism is analyzed and the characteristics of the green and sintered bodies of ceramic parts using the new forming methods or new techniques are introduced.

Keywords SFF; freeze-gel-casting; ultrasonic effects; laser machining technique

In this chapter, the investigation status and the new achievements about the novel forming techniques based on gel-casting such as SFF, freeze-gel-casting, as well as gel-casting by ultrasonic effect and the novel laser drilling technique for green ceramics are reviewed and summarized in detail.

SFF is the near-net-shape approach to produce parts or prototypes directly from a computer aided design (CAD) file without hard tooling, dies or molds. It has attracted increasing interest because of the advantages of SFF over conventional fabrication methods. To date, several SFF techniques have been developed. Laminated object manufacturing (LOM) process uses tape-cast ceramic sheets to

Y.Huang et al., *Novel Colloidal Forming of Ceramics*
© Tsinghua University Press, Beijing and Springer-Verlag Berlin Heidelberg 2010

fabricate ceramic parts (Griffin G., et al., 1994). Ceramic-binder extruded filaments are being used as feedstock in fused deposition of ceramics (FDC) processing (Agarwals M. K., et al., 1996; Danforth S. C., 1995). The stereo lithographic (SLA) techniques are being developed to produce ceramic parts by ultraviolet curing of a highly concentrated suspension of ceramic particles in a photo-polymerizable liquid (Griffin M. L., et al., 1996; Hinczewski C., et al., 1998). However, many of these approaches directly rely on the ceramic feedstocks that contain a large number of organic species which results in debinding issues. Selective laser sintering (SLS) is a form of SFF. In this process, a thin layer of powder is spread across the build platform by a roller mechanism. The first slice or cross section of the object is selectively drawn on the layer of powder using a CO_2 laser. The laser energy heats the powder to a temperature just below the melting point of the material and bonds the powder into a solid mass. The laser power is modulated so that powder only in the areas described by the object's geometry is fused. The build platform is lowered and the roller deposits another layer of powder across the build platform. The steps are repeated until the complete object is formed. The primary advantage of SLS process is the flexibility in selecting material systems when compared with other SFF techniques (Pham D. T., et al., 1998). Freeze casting technique is a good processing technique for fabrication of porous ceramics. It can produce porous materials with controllable porosity and open interconnected pores. Particularly, the morphology and the size of the pore can be tailored in a certain range by easily altering the parameters in the processing, such as slurry concentration, freezing temperature and cooling rate. Moreover, freeze casting is an environment friendly method. In the preparation process, ceramic slurries with different solid loading are first frozen, and then the frozen bodies are dried in a lyophilizer to sublimate the frozen medium, finally, the porous samples are obtained by sintering the green bodies at different temperatures. Till date, aqueous and non-aqueous slurries have been successfully used to fabricate porous materials by freeze casting. Gel-casting is a novel near-net-shape forming method for fabricating complex-shaped ceramic bodies. It was invented by Albert C. Young et al. (Young A. C., et al., 1991; Omateta O. O., et al., 1991) and is gaining increasing attention. In this process, monomer, crosslinker, initiator, catalyst and ceramic powder are thoroughly mixed in water to form a homogeneous suspension with high solid loading and low viscosity, then by means of *in-situ* polymerization a macromolecular network is created to bind the ceramic particles together. When compared with other casting methods, gel-casting results in much more homogeneous materials with little density difference over the parts and produces green body with higher flexural strength.

With these new methods and techniques, high performance ceramic parts with high reliability and complex shapes can be mass prepared by automation. This may pave the way for the industrial production of advanced ceramics in the near future.

7.1 Development Overview and Application of SFF

7.1.1 Development Overview of SFF

The SFF, which appears in the 1980s, has triggered a huge change in the way of thinking about manufacturing technology and production efficiency. Many countries in the world have given great importance to the new forming technology and it was also considered to be a major breakthrough in the filed of manufacturing.

In today's society, with the continuing diversification demand for high performance materials and low-cost manufacturing technology, the product competition become more and more intensive and the update cycle become shorter. Thus, it needs the designer to design the new products rapidly and to create a product sample in the shortest time. Using the traditional method to prepare samples, many kinds of machine tool, mold, mechanical appliances as well as high-level skilled workers are needed. It is time-consuming, high cost, long cycle, and can't adapt the rapid change about the ceramic industrialization. As an advanced preparation technology, solid freeform fabrication began to develop based on the thinking of discrete analysis and accumulation.

The SFF is a comprehensive embodiment of multi-discipline and technology, the emergence of which accompanied the high-speed development of computer, laser, digital control and material technology as well as the combination of various disciplines. The concept of SFF appeared in the late 1970s, but the thinking prototype about using the layered manufacturing principle to deposit 3D objects could be traced back to 19th century. In 1892, the idea of using layered manufacturing to form topographic map was proposed by J. E. Blanther in his patent. In 1902, the principle of using photosensitive polymer to manufacture plastic was mentioned by Carol Bease in his patent, and it was also the preliminary envisage of the first SFF—stereolithography. Now, with the development of material science, SFF has been developed rapidly. It has been used in the fields of electrical engineering, toys, light industry, aviation, space, medicine, film, modeling of artificial organs, archaeology, and industrial design and so on. It not only provides convenient for manufacturing, but also brings new ways and techniques for materials preparation. It is an advanced manufacturing technique that generates accurate geometrical objects directly from a three dimensional computer image without part-specific tooling or human intervention. This means that designers have the freedom to produce physical models of their drawings more frequently, allowing them to check the assembly and function of the design as well as to discuss downstream manufacturing issues with an easy-to-interpret, unambiguous prototype. Consequently, errors are minimized and product development costs and lead time are substantially reduced. With the emergence of SFF technology, the industrialization of advanced ceramics is predicted to have a bright future.

However, since green bodies made by the technique will contain a large amount of organic binder, long and complicated burnout heat treatments are required to eliminate it. Gel-casting is also a near-net-shaped fabrication technique by employing polymerization of monomer. The gel-casting system contains only a small amount of binder. In this work, a new freeform fabrication approach combining SFF and gel-casting is described, which uses selective gelation of concentrated aqueous gel-casting suspensions. The forming mechanism of this method is similar to AM gel-casting system which is studied widely. The process of the SFF of alumina ceramic parts is also introduced.

7.1.2 Application of SFF

7.1.2.1 The SFF of Alumina Ceramic Parts through a Novel Lost Mold Method

(1) The Combination of SLS and Gel-casting

In order to put forward a more efficient SFF approach, a novel lost mold method is successfully designed by combining SLS and gel-casting technique. Figure 7.1 shows a graphic depiction of the process. Firstly, a composite polymer powder is developed for SLS to fabricate sacrificial molds having a negative of the desired structure for gel-casting. Then, homogeneous alumina slurry with high solid loading and low viscosity is poured into the molds. During the sintering of the ceramic parts, the polymer mold is entirely removed. Because the solidified green bodies have high strength even at elevated temperature, the desired geometry of the ceramic parts is successfully retained after sintering. In this way, many complex alumina parts have been produced successfully.

In this study, AM and MBAM are dissolved into the deionized water to obtain a premix solution. TAC is selected as the dispersant to produce well dispersed homogeneous ceramic suspension. The slurry containing Al_2O_3 powder and the dispersant in the premix solution are thoroughly ball-milled for 24 h. After being de-gassing for 15 min, the polymerization is initiated by the addition of the initiator and catalyst. The slurry is cast into the sacrificial plastic molds, which is fabricated by an AFS SLS machine. The zeta potential of the ceramic particles is determined by a zeta plus analyzer (Brookheaven Instrument, USA) and the viscosity of the alumina slurries is determined by an advanced MCR300 rheometer (Physica, German). The details about the sintered mold structure and pyrolysis of the polymers are determined by TGA in air using a Dupont Thermal Analyzer 2000. The microstructure of the laser sintered polymer mold is observed by a HITACHI S-450 SEM.

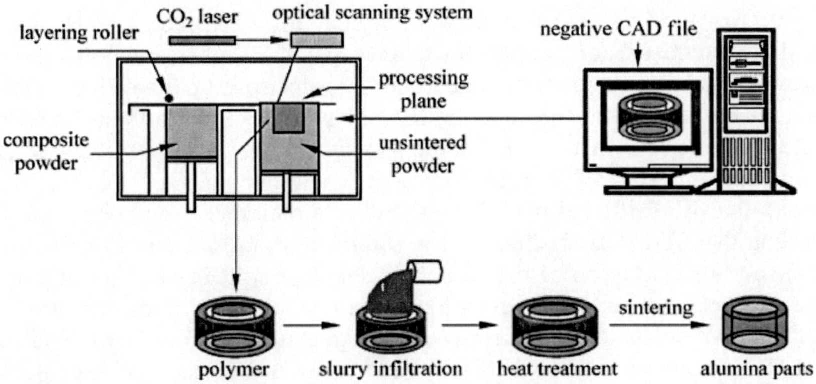

Figure 7.1 The schematic chart of the steps followed in the lost mold approach for fabricating alumina ceramics

(2) The Rheological Property of the Al_2O_3 Suspension

It is of prime importance to be able to obtain a stable and homogeneously dispersed ceramic suspension to acquire high quality ceramic parts by gel-casting. The rheological behavior or viscosity is related to the suspension property, which is determined by attractive and repulsive forces of the particles in the system. Attractive van der Waals forces can be counteracted by repulsive forces resulting from either overlapping of the electrical double layers (electrostatic stabilization) and/or layering of materials adsorbed on the surface (steric stabilization). The magnitude of the van der Waals forces is mainly determined by the nature of the particles and the solvent while the repulsive forces can be modified over a wide range by the addition of dispersants (Ferreira J. M. F., et al., 1992). The repulsive force can be characterized by the zeta potential and the potential of the particles as a function of pH is shown in Fig. 7.2.

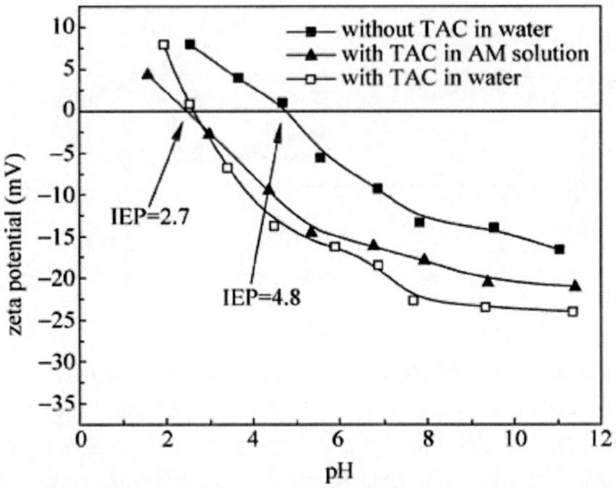

Figure 7.2 The zeta potential versus pH for different alumina suspensions

The 0.1 vol% Al_2O_3 suspensions were regulated by HCl and NaOH to fixed pH and ultrasonicated for 2 min prior to analysis. The higher this potential with the same polarity is, the more important the electrostatic repulsion between particles. Contrarily, when close to the IEP, the particles tend to flocculate. From Fig. 7.2 it can be seen that the IEP of the powder is at about pH = 4.8 and it has a zeta potential of –10 mV at pH = 7. With the addition of TAC (1 wt% of the alumina), the IEP shifted at pH = 2.7 and the zeta potential is doubled at pH = 7, suggesting that TAC has an electrostatic stabilization mechanism to bring about good dispersion. Addition of the monomer does not shift the IEP but it slightly decreases the relative value of the zeta potential. This indicates that the uncharged AM molecule either screens the charge that is developed by the powders in solution or preferentially gets absorbed onto the surface of the alumina particles. The viscosity of the alumina slurries as a function of the shear rate is illustrated in Fig. 7.3. It can be seen that TAC can greatly decrease the viscosity, which is in good agreement with the zeta potential results. The shear thickening behavior of the slurries with TAC at higher shear rate is owing to the increase of solid loading (Barnes H. A., et al., 1989). For a dispersant, there exists an optimum concentration at which just the adequate dispersant exists to provide maximized coverage of the ceramic powder and any excess dispersant may be harmful in decreasing viscosity (Mcnulty T. F., et al., 1999). The optimum concentration of TAC is about 0.3% of Al_2O_3 powder.

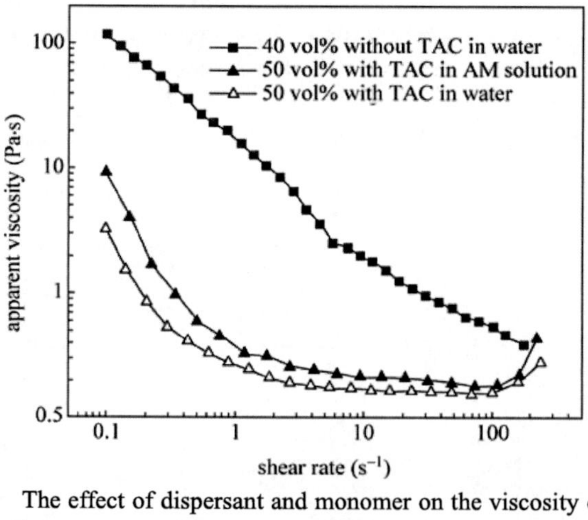

Figure 7.3 The effect of dispersant and monomer on the viscosity of different ceramic slurries

(3) The Selection of Mold Materials and Heating Procedure

A large range of materials can be used in SLS, such as nylon, ABS, sand, wax, metals and polycarbonate, etc. However, some problems must be considered when selecting a material. During heat treatment the sacrificial mold should be burned out clearly without leaving any residue that may be detrimental to the properties of the ceramic parts. Therefore, all organic constituent materials system should be adopted. Here a polystyrene based composite powder is developed to make the sacrificial mold. In fact, polystyrene is frequently used in lost mold casting owing to its thermal degradation property (Chang A. S., et al., 1999). Because a general polystyrene product may result in large geometry distortion upon heat treatment, the composite powder is specially processed by reactive blending of PS, rubber, antioxidant, impact modifier and other additives through twin crew extruder. It is also necessary to cool the part when it becomes too hot in order to prevent distortions in the final piece. Therefore wax is not a suitable material because it needs a long cooling cycle on the machine although it can also be removed easily. From the SEM photograph of the cross section of a sintered mold shown in Fig. 7.4, a porous network structure can be observed, which is helpful in preventing distortion.

Figure 7.4 The SEM photograph of the microstructure of the sintered polymeric mold materials

The details of the TGA curves (4℃/min in air) of the sintered molds and that of a green cast Al_2O_3 ceramic sample (50 vol% solid loading) are shown in Fig. 7.5.

The mold material starts evaporating at about 85℃ and it losses 94% of its weight at 85 – 300℃. At a higher temperature, a slow burnout occurs which ends at about 510℃ without leaving any residue. The gel-casting sample starts evaporating at about 120℃ and before reaching 500℃ only a 6% mass loss occurs owing to the trapped water and the burnout of the cross linked polymerized AM network. Because of the lower mass loss temperature and the porous structure of the polymeric mold, green ceramic structures have sufficient strength to hold the structure during evaporation. As a result, the geometry of the ceramic part is well preserved. The mold material should evaporate first before the evaporation of the binder to prevent collapse of the ceramics caused by the deformation or flow of the plastic mold may happen during heating. In fact, collapsing occurs when other polymer materials such as ABS are used as the mold materials (see Fig. 7.5). The binder removal and detailed sintering procedure

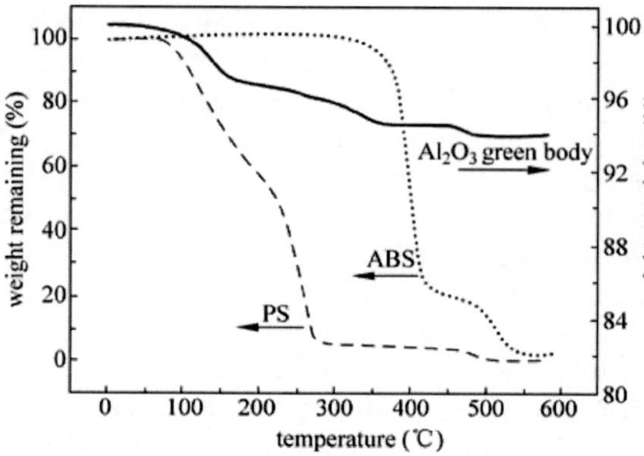

Figure 7.5 The TGA curves of the polymeric mold, ABS resin and a cast alumina sample

of dry ceramic powder infiltrated molds were determined based on the TGA results. The heating procedure consists of three steps. During the first step of the procedure (from room temperature to 600℃), mold material and crosslinked polymer binder evaporate. In the second step, 1 h of constant temperature at 600℃ is maintained to remove the remnants of the decomposed polymers. At higher temperature, densification of Al_2O_3 occurs. A final sintering temperature of 1550℃ and a hold time of 2 h are used for the alumina samples. The ultimate sintered ceramic parts are shown in Fig. 7.6.

Figure 7.6 The illustration of the ultimate alumina ceramic parts fabricated by the lost mold approach

Generally, the mass ratio of polymer mold to Al_2O_3 slurry is about 5:100 before sintering. The polymeric mold material can be fired with a relatively low-powered laser in a short time. A polymer mold with a dimension of $2\times 2\times 2$ cm^3 needs just 30 min to be fabricated. Using adaptive slicing that generates different slice thicknesses between 0.08 and 0.3 mm based on the local slope of the part, the build rate can also be increased. Therefore the cost of the mold is acceptable and this is an economical SFF method for fabricating ceramic parts.

7.1.2.2 The Freeform Fabrication of Aqueous Alumina-Acrylamide Gel-Casting Suspensions

(1) The Processing of Fabricating Ceramic Parts

The ceramic powder used in this work was also α-Al_2O_3. AM, MBAM, $(NH_4)_2S_2O_8$, TEMED and dibastic ammonium citrate were applied respectively as monomer, crosslinker, initiator, catalyst and dispersant. All these reagents were chemically pure.

In the process, the suspension was prepared firstly. The dispersant with 0.5 wt% dry powder weight was completely dissolved in the premix solution (5.20 wt%), which had been prepared by dissolving AM and MBAM in deionized water, and then Al_2O_3 powder was added. The suspension was degassed for about 10 min under vacuum after milling for 24 h. The solid loading of the final suspension was 50 vol%.

Secondly, similar to the tape casting process, a blade was used to spread the viscous ceramic suspensions to produce a layer with the desired thickness before polymerization over a build platform. A heating system was used to adjust the gelation rate. A nitrogen system was also employed to avoid oxygen that inhibits the polymerization process. According to computer information, localized gelation of selected areas of the suspension layer was then induced by a jet of droplets of the deionized water containing initiator and catalyst. When a layer of the part was formed, the build platform was lowered and a new layer of suspensions was deposited onto the already polymerized part to continue the manufacturing process. This process was repeated until the part was built. Figure 7.7 shows the schematic representation of selective gelation process. After formation, the green parts were washed in water, and then dried at room temperature for 24 h. According to the TGA result, the debinding was performed with a heating rate of 1℃/min up to 550℃ soaking for 2 h, and then at 3℃/min up to 1580℃ holding for 2 h.

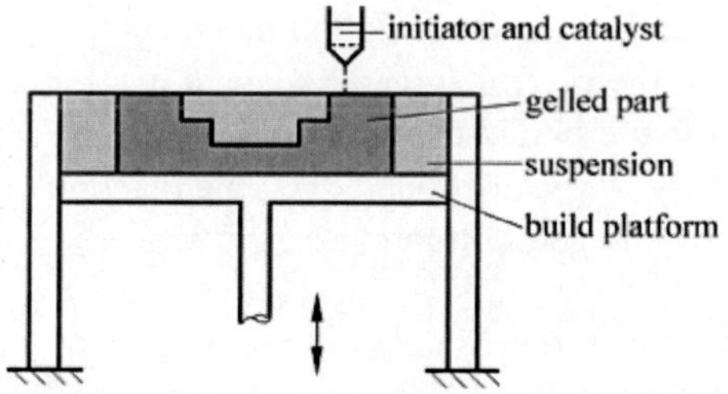

Figure 7.7 The schematic representation of selective gelation process

(2) The Rheological Properties of the Alumina Suspensions

The homogeneous and high loaded suspension are important for this method. To obtain ceramic parts with superior properties, the solid loading of ceramic powder must be at least 50 vol%. Furthermore, the viscosity of the suspension must be as low as possible to allow a good recoat of the suspension on the gelled layers. Usually to obtain good homogeneity and low viscosity of the suspension, which are in contradiction with a high solid loading, the use of a dispersant is necessary. Here dibastic ammonium citrate was used as the dispersant in alumina suspension system. To determine the optimal concentration of dispersant, the dispersion of the alumina powder in the remix solution was studied. Suspensions containing 50 vol% of alumina was prepared with various amounts of dispersant varying from 0.3 to 0.7 wt% with respect to the powder. Figure 7.8 shows the rheological behavior of the Al_2O_3 suspension at different concentration of the dispersant.

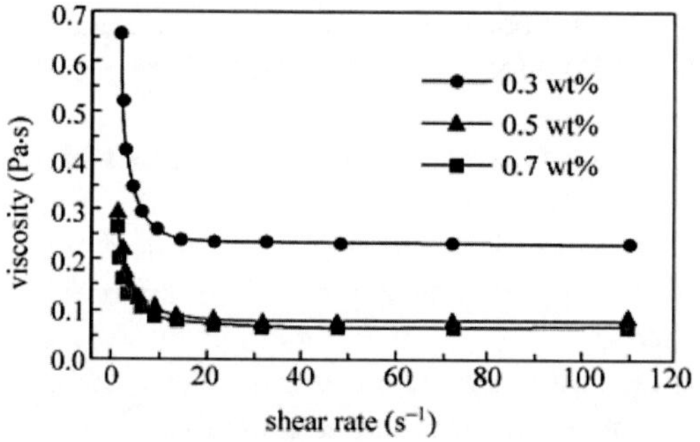

Figure 7.8 The rheological behavior of the Al_2O_3 suspension at different concentration of the dispersant

From the Fig. 7.8, it can be seen that the minimum of viscosity, corresponding to the best state of dispersion of Al_2O_3 particles, was obtained with an addition of 0.5 wt% of dispersant. In the range of shear rate tested, the suspensions exhibit a shear thinning behavior that is favorable for tape casting of the green layers. It could be observed that the viscosity of the suspensions was low enough to be utilized in the gelation process.

(3) The Properties of Alumina Ceramic Parts

For the fabrication of the ceramic green bodies, a heating system was important to this selective gelation method. At room temperature, the reactive degree of this gel-casting system is very low. But at 60 – 70℃, the reaction is strong. By adjusting the system temperature, the gelation rate, i.e., the process time, could be adjusted at a wide range.

Several nozzles with different diameters were used to fabricate the ceramic parts. Earlier experiments showed that the nozzle diameter had a significant effect on the gelation layer thickness. When the diameter was large (e.g., 0.2 mm), the build layers up to 1 mm thickness could gel completely. But when the diameter of the nozzle was 0.08 mm, the edge of the gelled object was 0.02 mm. When the layers were thicker (0.5 mm), the interlayer defects and bad interlayer adhesion were observed by using a nozzle with a smaller diameter. In fact a decrease of the nozzle diameter was observed to cause a reduction in flow rate, leading to a decrease in the diffusion distance of the initiator and catalyst in the suspensions, thus, the gelation layer thickness would decrease at the same time. Therefore, the build layer thickness should correspond to the diameter of the nozzle.

Various ceramic green parts were prepared by this forming process (as shown in Fig. 7.9). The thickness of individual green layers was varied from 200 μm to 1 mm. Some rectangular alumina bars approximately $6 \times 5 \times 50\ mm^3$ were also prepared, using 200 μm thick layer suspension by this method. The density of the green rectangular bars fabricated by this forming method was 53.2%. The anisotropic shrinkage behavior was observed in bars where the shrinkage was greater in the build direction than in the build plane. The shrinkage for the rectangular alumina bars was 14.3% in the x-y plane and 16.7% in the z direction.

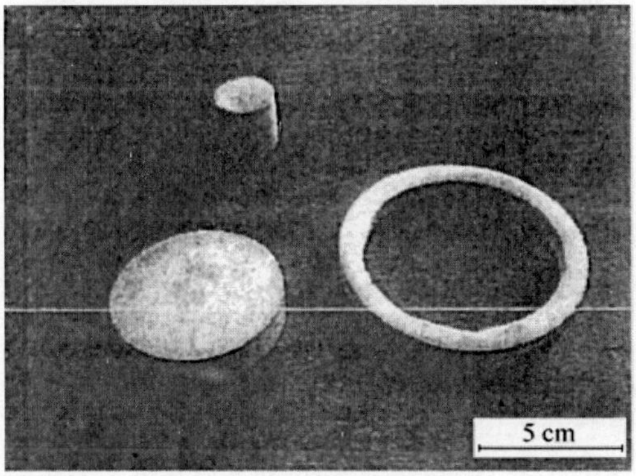

Figure 7.9 The green parts fabricated by selective gelation

The density of sintered bars was 91.2% of the theoretical density which was still insufficient for making ceramic parts. But the microstructure of the sintered bars was rather homogeneous, as shown in Fig. 7.10.

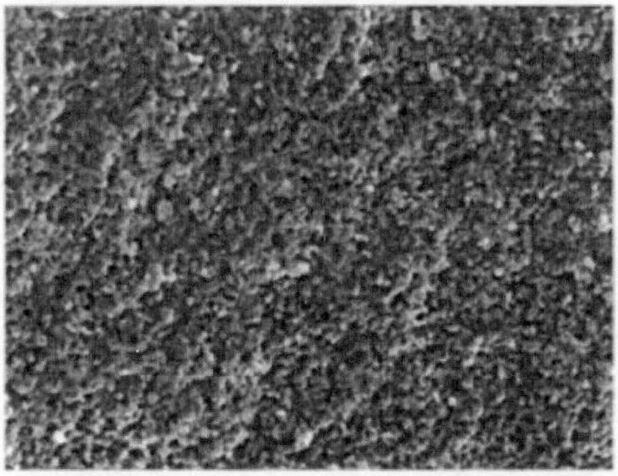

Figure 7.10 The SEM micrograph of fracture surface of sintered samples

The final density may be increased by improving the process parameters in the future. The sintered bars were observed by SEM to evaluate the microstructures for build-related defects or to determine the evidence of layerwise build strategies. There were no discernible signs of layerwise fabrication. It was difficult to distinguish the individual laminated layers inside the sample (see Fig. 7.11). The microstructure perpendicular and parallel to build direction were similar.

Figure 7.11 The SEM micrograph of a polished sintered sample surface (the arrow indicates the build direction)

The rectangular alumina bars were machined into test bars of $4 \times 3 \times 36\ mm^3$ and polished with a diamond paste, and then the flexure strength tests were carried out either perpendicular or parallel to the build direction. The strength of the specimen bars was 284 MPa in the perpendicular direction and 281 MPa in the parallel direction. The mechanical properties were similar regardless of the test direction.

A new technique was developed for freeform fabrication of ceramics that contained low amount of organic binders. Ceramic parts could be formed by this kind of localized suspension gelation method. This study proved that it was possible to obtain rather dense ceramic parts with a homogeneous microstructure.

7.2 Development Overview and Application of Freeze-Gel-Casting

7.2.1 The Combination of Gel-Casting and Freeze-Casting Technique

An increasing number of applications that require porous ceramics have appeared in the last decades, for example, the filtration of molten metals and gases, high-temperature thermal insulation, support for catalytic reactions, etc. The advantages of using porous ceramics are usually controllable permeability, large surface area, low density and high specific strength, which can be tailored for each specific application by controlling the microstructure of the pores. According to the microstructure, porous ceramics can be classified into several general categories based on pore distribution: disordered, long-range ordered and gradient, among which the last two categories have recently begun to attract much attention because they are expected to have both higher fluid permeability and higher mechanical strength. Processing methods used for the production of porous ceramics were recently reviewed by Gauckler et al., including replica, sacrificial template and direct foaming methods, while many novel methods are being developed.

As a wet shaping technique, gel-casting is widely used in advanced ceramics manufacturing, because high strength green bodies can be obtained and complex component geometries can be achieved. Besides the application of fabricating dense ceramics, the gel-casting technique has also been used for fabricating porous ceramics, among which direct foaming and pore-forming agent additives (Tari G., 2003; Xie Z. P., et al., 2001) are the two main methods. In the former method, pores are introduced by incorporating air into a suspension, that subsequently solidifies and retains the structure of air bubbles. Pores in ceramics acquired by this method usually have relatively large sizes in the order of several hundreds of micrometers and are spherical shaped. In the later method, a suspension containing ceramic powders and natural or synthetic pore-forming agents is prepared and subsequently processed by gel-casting. The pore forming agent is then burned out at temperatures between 200℃ and 600℃, leaving pores with the same shape. Strong green bodies can be acquired by this gel-casting method, while complex pore shape or special structures are difficult to be achieved.

Another special wet-shaping technique to fabricate porous ceramics is freeze-casting. Particular features of this technique are the steps of freezing a slip in a mold, demolding in the frozen state, followed by freeze drying and sintering. The frozen suspension liquid not only acts as a binder between the powder particles but also acts as the template of the pore channels. By controlling the freezing direction and temperature gradient, aligned pore channels and a porous gradient can be achieved (Studart A. R., et al., 2006; Araki K., et al., 2005; Koh Y. H., et al., 2006). Besides water (Fukasawa T., et al., 2001; Zhang Fazhi, et al., 2006), camphene (Meng G. Y., et al., 2000; Menon M., et al., 2006) and naphthalene-camphor eutectic (Guha S., et al., 2001), which can be frozen and sublimated at relatively high temperatures, were developed as alternative systems. A major problem of freeze-casting is the low strength of green bodies; when the frozen suspension is volatilized, the green bodies become very frangible and difficult to handle, and further efforts are underway to improve the green bodies' strength (Michael Scheffler, et al., 2005).

Both the gel-casting and freeze-casting techniques have their own merits and drawbacks. In this work, the advantages of these two methods are combined. To accomplish this purpose, tert-butyl alcohol (TBA) was chosen as a candidate for the vehicle of the freezing process capable of near-room-temperature manufacturing. At the following step, polymerization of AM was realized in TBA slurry, which greatly improved the strength of ceramic green bodies. We also investigated the effect of temperature on the microstructure and mechanical properties of green bodies and final sintered ceramics, and obtained porous ceramics with unique long-range ordered and gradient porous structures. The features and the mechanisms of this freeze-gel-casting method were analyzed, and are presented in the following section.

7.2.2 Fabrication of Ceramics with Special Porous Structures

7.2.2.1 Procedures of Freeze-Gel-Casting Technique

Alumina with a mean particle size of 0.2 μm was used as the ceramic powder. The TBA (chemical purity, China) was used as the vehicle; and the main properties are shown in Table 7.1. A premix solution of monomers in TBA was prepared with a concentration of 14.5 wt% of AM and 0.5 wt% of MBAM. Since the liquid phase in the slurry is a multicomponent system, i.e., a mixture of TBA with dispersant and AM monomer, the freezing temperature of this solution is lower than pure TBA, decreasing to about 8℃. The 40 wt% water solution containing initiator ($(NH_4)_2S_2O_8$) was prepared for the process of gelation. Citric acid and acetic acid were added to produce stable Al_2O_3 suspensions in liquid TBA (Menon M., et al., 2006).

Table 7.1 Some physical properties of TBA (compared with water)

	liquid density (g/mL)	boiling temperature (℃)	saturated vapor pressure at 40℃(kPa)	freezing temperature (℃)	volume increase during freezing (%)	morphological characteristics after freezing
water	1	100	3.4	0	10	dendritic crystal
TAB	0.79	82.5	6.4	25.3	2	straight prisms

Figure 7.12 shows the schematic illustrations of the TBA-based freeze-gel-casting method for the fabrication of porous ceramics with unidirectional pore channels.

Firstly, slurries with 25 vol% solid loading were prepared by ball-milling the alumina powder in warm premixed TBA/AM solution at 25℃ for 24 h, containing 1 wt% (to 100 wt% Al_2O_3 powder) of citric acid as dispersant, and with pH = 3 adjusted by acetic acid (see Fig. 7.12(a)). The prepared slurries were found to be stable and possessed good flowability for casting. Initiator solution (40 wt% in water) was then added at appropriate proportions to alumina suspension while stirring, and the temperature was maintained at 15℃ so that polymerization of AM could not be promoted (see Fig. 7.12(b)). The slurries prepared were then poured into polyethylene molds with a diameter of 12 mm packed by a heat barrier layer at bottom and surrounding side. The whole specimen was then placed in a freezing atmosphere at around 0℃ for over 1 h. The TBA froze from top to bottom, forming complete unidirectional crystalline prisms penetrating the whole body as the template of pore channels (Fig. 7.12(c)). The frozen bulks were easily removed from the molds at a low temperature (3℃, see Fig. 7.12(d)), and then heat-treated at 85℃ (see Fig. 7.12(e)). In this thermal treatment step, melting of TBA, gelation of AM and evaporation of TBA rapidly occurred continuously. Frozen TBA melted firstly, while AM rapidly polymerized at the same time under the influence of initiator, then TBA evaporated and formed the pore channels. This procedure occurred with a rapid speed from the exterior to the interior of the samples. Porous green bodies with high strength could be obtained by these freeze and gelation steps, and the final ceramic parts were fabricated by the processes of drying, deorganizing and sintering (see Fig. 7.12(f)).

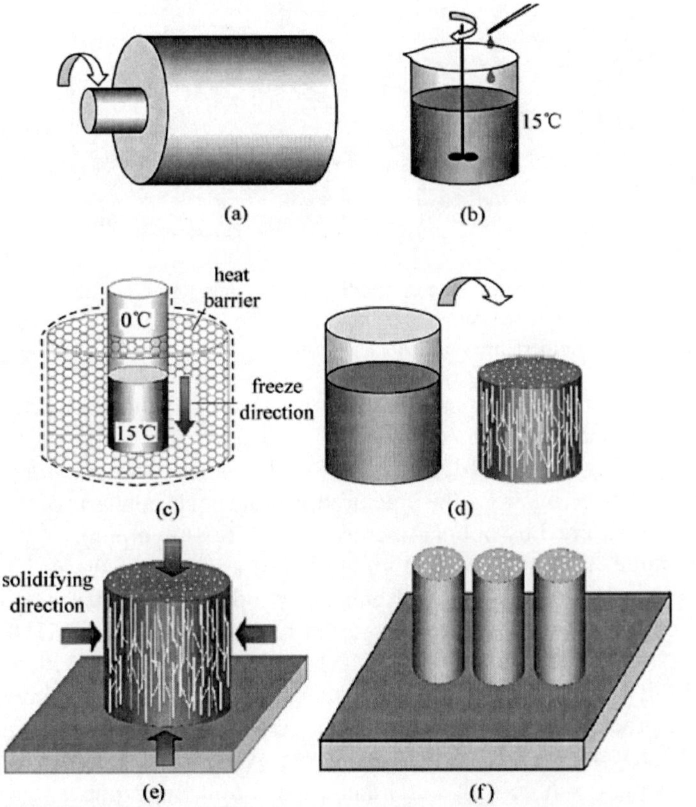

Figure 7.12 The complete process for the fabrication of ceramics with long-range ordered porous structure by freeze-gel-casting technique
(a) slurry preparison (ball milling, 30℃, 24 h); (b) adding initiator at 15℃; (c) orientated freezing; (d) demolding at 3℃; (e) heat treating for solidification and volatilization (85℃); (f) drying (40℃), deorganizing (500℃, 2 h), sintering (1500℃, 2 h)

The process of fabrication gradient porous ceramics is similar to that described earlier, with the main difference in the freezing step. Molds containing slurry and packed by heat barrier layers were placed under conditions with a temperature gradient. One end was set below the freezing point of TBA and the other end was exposed to air at a temperature in its melting range. Therefore, the slurry froze from the cold bottom with crystals of TBA. On the other end, solidification and volatilization occurred near the hot surface. These two procedures proceeded simultaneously in opposite directions and formed a novel pore gradient structure. The step to fabricate gradient frozen parts is illustrated in Fig. 7.13, in which two different temperature conditions, about 12℃/mm and 5℃/mm in (a) and (b), respectively, were put into effect.

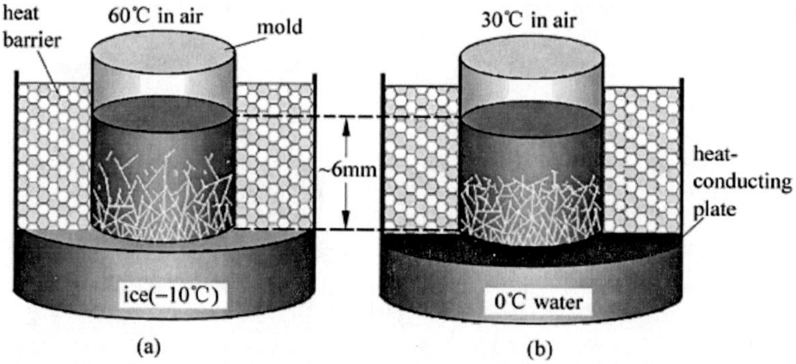

Figure 7.13 The freezing step under different temperature conditions to achieve novel gradient porous structures, with higher temperature gradient of 12℃/mm (a), and with lower temperature gradient of 5℃/mm (b)

The porosities of sintered samples were measured using the water displacement method based on the Archimedes' principle. The microstructures were observed by using the SEM (JEOL, JSM-6400, Japan) and the pore structures were directly observed from the fractured samples. The pattern of frozen TBA solution was examined by optical microscopy (BX50, Olympus).

To determine the unidirectional compression strength of the ordered porous alumina (green bodies and sintered bulks), cylinders with a diameter of 12 mm and height of 25 – 30 mm were fabricated and tested on a CSS110 materials testing machine (Changchun Testing Machine Institute, China) with a crosshead speed of 1 mm/min. By setting the dense side under tension, the maximum bending strength of gradient porous specimens was determined by three-point bending tests (test bars of $4 \times 3 \times 36\ mm^3$) using an AG-2000 G Shimadzu universal materials testing machine with a crosshead speed of 0.5 mm/min.

7.2.2.2 The Characteristics of Freeze-Gel-Casting Process

To date, water and camphene have been successfully used as freezing vehicles. Each vehicle has its own benefits and drawbacks. In this study, TBA was used as the freezing template to achieve pore channels, because of its advantages over other freeze-casting vehicles. First, TBA can be removed via rapid volatilization (80℃, 30 min), without any collapse or obvious shrinkage of the green body. However, in the case of water- or camphene-based freeze-casting, they have to be removed by the sublimation drying method, which takes a long time (over 48 h). Second, unlike the dendrite structure of frozen water or camphene, TBA normally presents a form of long straight ice prisms without any branches at its freezing temperature, which is important for specific applications such as fabricating unidirectional porous ceramics used as infiltrators. Third, the freezing process can be carried out at near room temperature (below 8℃); thus complex treatment can be handled.

To understand the mechanisms of pore formation using frozen TBA as a template, its freezing behavior was observed by optical microscopy. TBA containing 15 wt% AM was dropped onto a pre-cooled (about 0℃) slide glass, and then frozen immediately. Figure 7.14(a) shows the typical microstructure on the front edge of the frozen AM/TBA solution. It can be seen that TBA prisms grew straight in one direction, and did not intersect each other. It should be noted that AM had no obvious influence on the pattern of frozen TBA, and precipitation of AM under low temperature was not observed.

Using TBA as a template in its frozen state, green bodies with pore channels were fabricated. Unlike the circular cross-sections in aqueous or camphene based freeze casting, these pore channels had straight polygonal cross-sections, which can be clearly seen in Fig. 7.14(b). This channeled structure existed throughout the whole green body and was inherited in the subsequent processes.

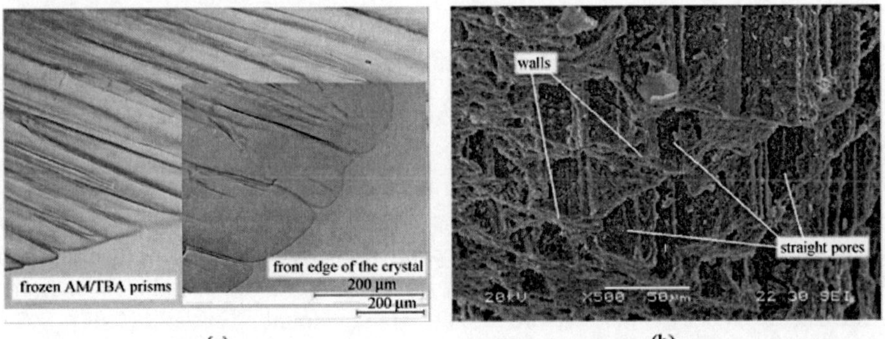

(a) (b)

Figure 7.14 The mechanism of pore formation using TBA as the template (a) growth of TBA ice prisms; (b) pore channels in dry green body (obtained from slurry with solid loading of 25 vol%)

As a critical step in this freeze-gel-casting technique, the following process of gelation was induced by heat treatment at a high temperature (85℃). Two processes occurred at this higher temperature: first, the liquid TBA crystals melted, and the

AM polymerizes simultaneously, converting the liquid medium into an alcogel. Subsequently, the liquid TBA from the alcogel evaporated, rather than subliming as in the case of typical freeze-casting. An important aspect is the speed of polymerization of AM in TBA, for which the heating temperature is a critical factor. Figure 7.15 shows the gelation process in TBA/AM/Al_2O_3 slurry, where the sharp rise of temperature indicates that the polymerization of AM occurs in slurry, owing to the exothermic nature of the reaction. The slurries containing 4 wt% $(NH_4)_2S_2O_8$ initiator could be gelated at temperatures above 30℃. The more initiator added, the shorter the polymerization time was. When the temperature increased to 60℃, the polymerization occurred with a very rapid speed and no obvious induction time was required.

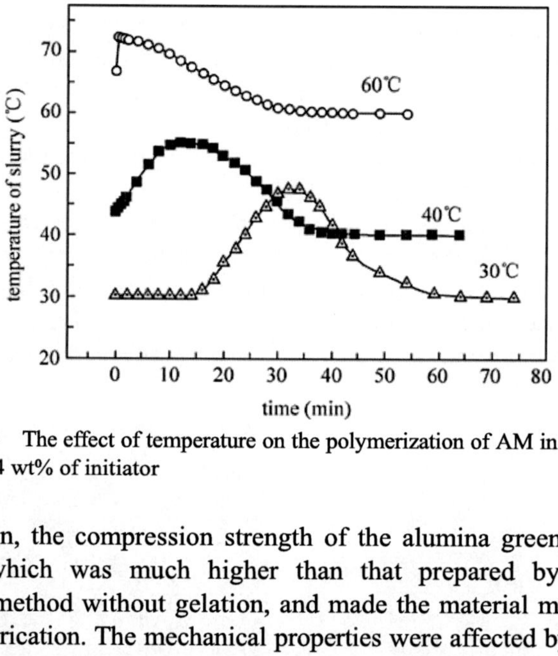

Figure 7.15 The effect of temperature on the polymerization of AM in TBA-based slurry with 4 wt% of initiator

After gelation, the compression strength of the alumina green body reached 5 – 10 MPa, which was much higher than that prepared by the common freeze-casting method without gelation, and made the material more convenient to handle in fabrication. The mechanical properties were affected by the initiator's proportion. Figure 7.16 illustrates the dependence of the compression strength of green bodies with unidirectional pores on the amount of initiator in ceramic suspensions. Under the temperature of 60℃, if the amount of initiator ($(NH_4)_2S_2O_8$) in slurry was less than 1 wt%, then there was no significant solidification. When 3 wt% of initiator was added, the green bodies solidified rapidly by heat treatment, and reached the maximum compression strength of 10 MPa. On the other hand, the strength decreased when excessive initiator was added to the slurry, because the modular chain length of polyacrylamide was restricted when too much chain growth occurred in the meantime.

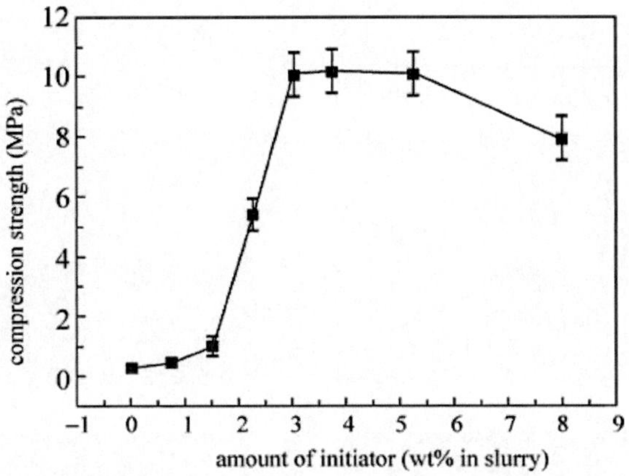

Figure 7.16 The influence of the amount of initiator ($(NH_4)_2S_2O_8$) on the strength of the porous green bodies

7.2.3 The Microstructure and Properties of Porous Alumina Ceramics

7.2.3.1 Long-Range Unidirectional Porous Structure Alumina

Since it is feasible to form pore channels in ceramic bodies as replicas of TBA ice prisms, unidirectional freezing was employed to control the growth of the solvent. The whole process is shown in Fig. 7.12, in which the steps of freezing and heat treatment can be completed rapidly (for about 5 min in this case).

Figure 7.17 shows typical microstructures of the unidirectional pore channels in an alumina body after sintering. Penetrating channels were aligned unidirectionally over a long range, and were observed at the vertical, horizontal and 45° angle orientation to the TBA freezing direction, which are shown in Figs. 7.17(a), (b) and (c) respectively. From the figure, it can be noted that the walls surrounding the channels were almost fully dense (see Fig. 7.17(d)) after sintering.

By comparing the microstructure shown in Fig. 7.14(b) with that shown in Fig. 7.17(c), the pore channels in the green bodies did not shrink obviously after sintering at 1500℃. The unidirectional channeled pore structure existed throughout the sintered body, except for a thin dense outer layer of about 0.5 mm thickness on the side wall that touched the mold. The channel sizes remained almost the same from top to bottom. The porosity and mechanical properties of these sintered samples acquired from the slurry of 25 vol% solid loading are summarized in Table 7.2.

These physical properties can be explained by the formation of open pore channels and fully dense alumina walls. It can be concluded from the measurements of total porosity and open porosity that most (>90%) of the pores in the

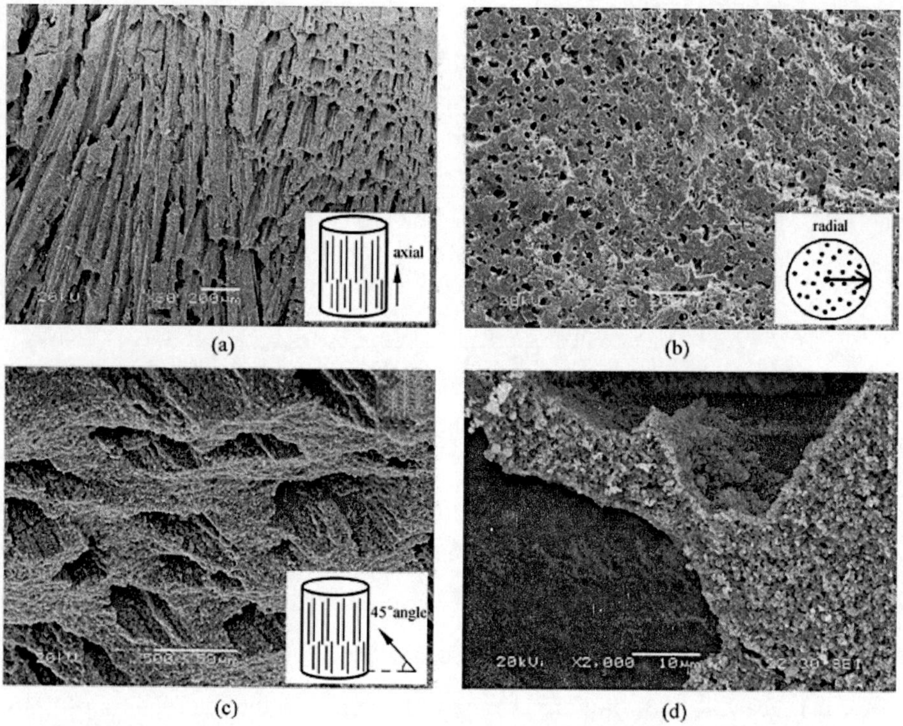

Figure 7.17 The microstructures of porous ceramics with long-range unidirectional pore channels fabricated by the freeze-gel-casting method observed at different orientations (a) vertical to freezing direction; (b) parallel to freezing direction; (c) at 45 ° angle direction; (d) the dense walls formed between pores

Table 7.2 The properties of sintered alumina with long-range ordered pore channels acquired by freeze-gel-casting

total porosity (%)	open porosity %)	closed porosity (%)	density of walls (% TD)	axial compression strength (MPa)	radial compression strength (MPa)
58 ± 3	52.3 ± 2	5.7 ± 2	94 ± 2	153 ± 30	54 ± 14

sintered bodies were open pores, indicating that the pore channels were almost fully open to the surface. It also can be seen that the ceramic walls had about 94% density, which is in good agreement with the SEM images. These dense walls were formed because during the freezing of TBA, most of the ceramic particles in the slurry were discarded by the growing ice prisms and became concentrated between the crystals (Shanti N., et al., 2006). Because of the high density of these walls, the ceramics' compression strength in an axial direction reached 150 MPa, after being sintered at 1500℃ for 2 hours.

7.2.3.2 The Gradient Porous Structure of Alumina

The freeze-gel-casting of alumina/TBA/AM slurry has also been used to fabricate gradient porous ceramics. In the freezing step shown in Fig. 7.13, the TBA near the top surface at the melting temperature range can easily be volatilized with an increasing solid loading. When the temperature is sufficiently high (>50℃), polymerization of AM will occur at the same time. On the other hand, the slurry freezes in the inner region from the cold bottom side. In this way, a pronounced pore gradient can be fabricated, with more and larger pores at the bottom, and lesser and smaller pores at the top.

As illustrated in Fig. 7.13(a), gradient porous alumina with a thickness of over 5 mm was successfully fabricated with a freezing temperature gradient of 12℃/mm and sintered at 1500℃. SEM observations revealed that the samples retained their gradient porous structures with a long range (⩾ 3 mm, see Fig. 7.18). Figures 7.18(b) and 7.18(c) show the dense and porous regions at a higher magnification. It is obvious that the pores become smaller and fewer with larger distance from the bottom plate. Sintered parts showed almost complete densification without any noticeable defects, either in the dense region near the top surface or in the alumina walls in the porous region near the bottom (see Fig. 7.18(d)).

As the low solid loading of 25 vol% was used to prepare the alumina/TBA/AM slurry, in a normal sense, the sintered sample without any special treatment should have a high degree of porosity. However, the top surface was exposed to air to evaporate the molten TBA, where the solid loading was higher. After sintering, this layer showed a density of over 90%, with some small pores, as shown in Fig. 7.18(b). At the freezing side, a high porosity of over 65% was achieved, and the concentrated alumina walls were sintered without destroying the porous structure. In the middle position, the evaporation of the molten TBA and the freezing of the slurry were synchronized, and hence, the porosity was on a medium scale.

To understand the function of the temperature on the formation of the gradient porous structure, a freezing process with a smaller temperature gradient (5℃/mm) was applied (see Fig. 7.13(b)). On comparing the porous structures achieved by the two different temperature conditions, obvious differences in the pore gradients were observed. The microstructure of sample acquired under 5℃/mm was also revealed (see Fig. 7.19).

By setting the bottom of the sample as the starting position, the porosity in each relative position was measured, and the results are shown in Fig. 7.20. It can be seen that in contrast with the method (a) (see Table 7.3), which had a larger temperature gradient, the pore gradient acquired by the method (b) was less uniform over the thickness of the sample, and there is a great divide in the middle of the height. These results suggest that the gradient of the pore structure can be

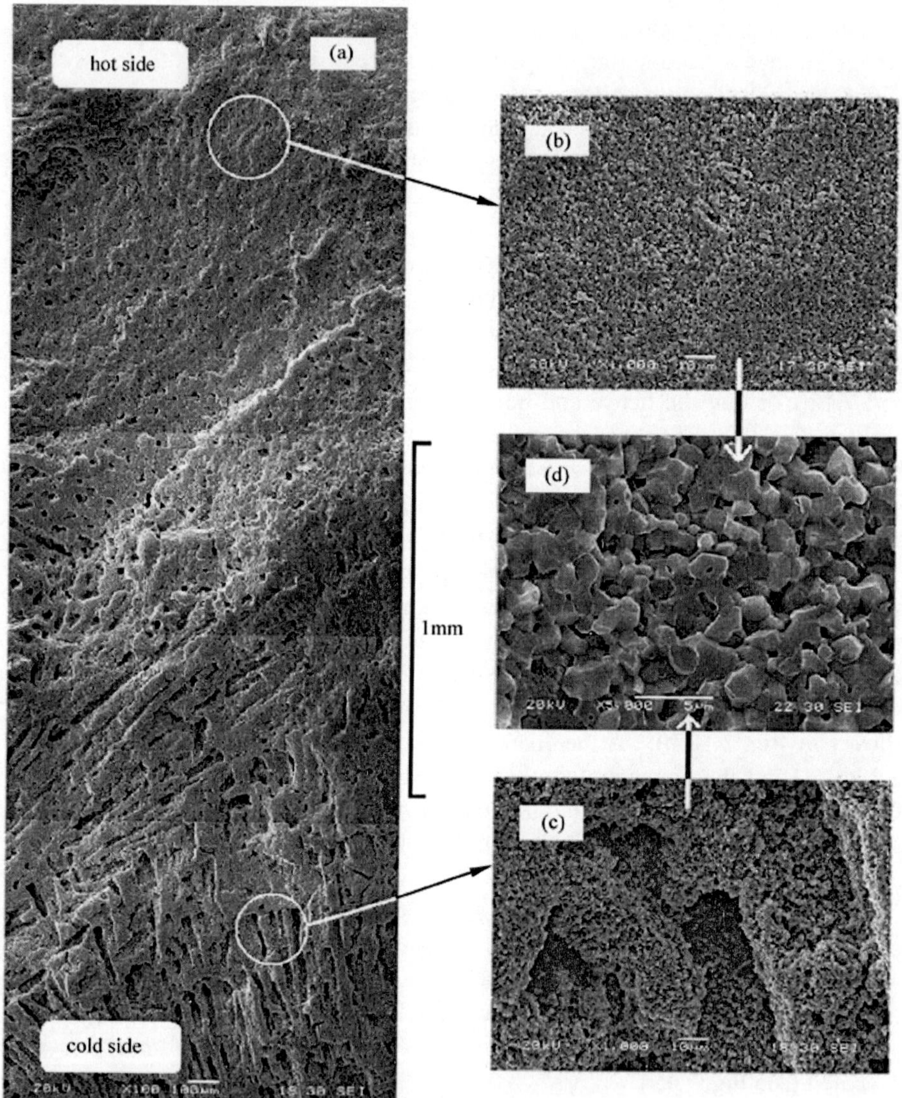

Figure 7.18 The gradient porous alumina ceramics fabricated by the freeze-gel-casting technique under the freezing temperature gradient of 12℃/mm

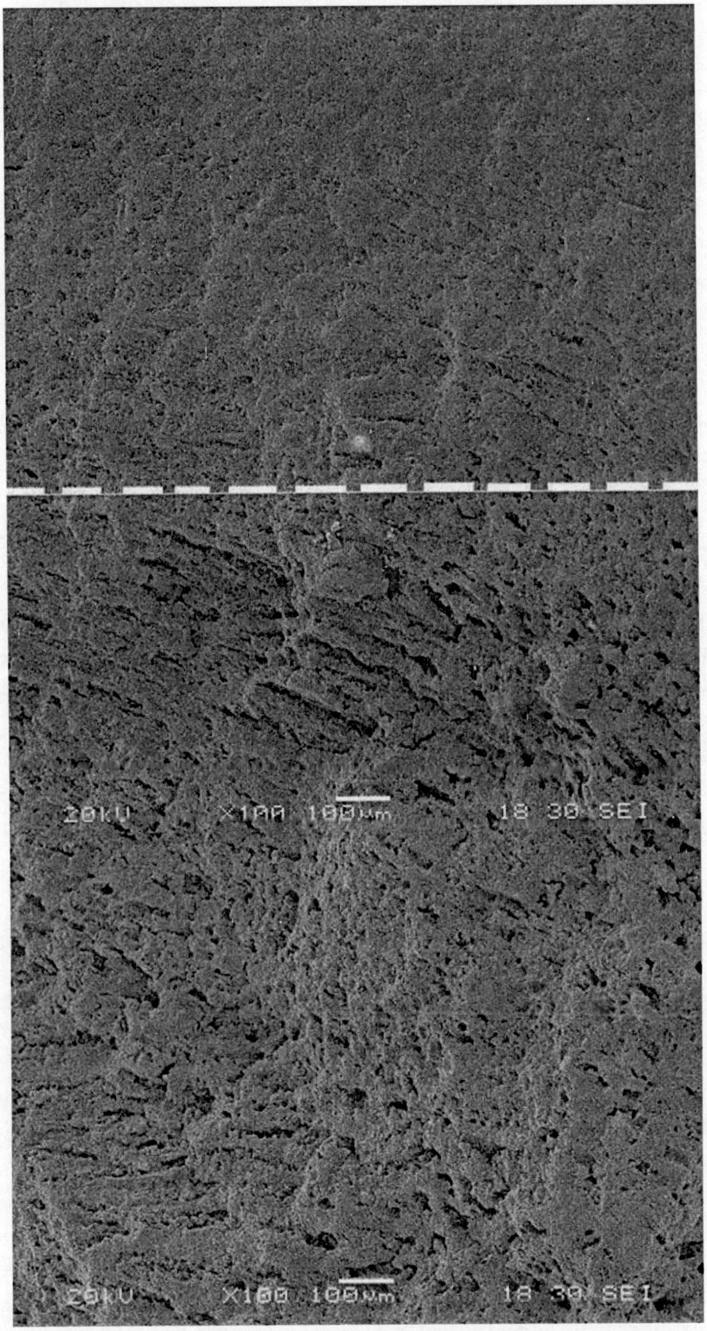

Figure 7.19 The gradient porous alumina ceramics fabricated by the freeze-gel-casting technique under the freezing temperature gradient of 5℃/mm

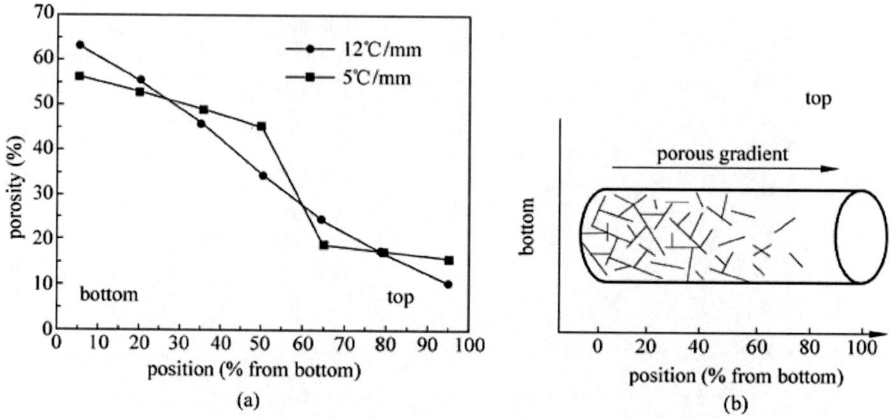

Figure 7.20 The distribution of porosity over thickness in the sintered gradient porous alumina, with the bottom as the starting position (a) top temperature = 60℃, bottom temperature = –10℃; (b) top temperature = 30℃, bottom temperature = 0°C

determined by controlling the temperature gradient. Furthermore, method (b) produced weaker green bodies than method (a); no obvious polymerization occurred at the dense end, because in the freezing step under this gradient temperature, the top end was neither "hot" enough to initiate the polymerization of AM, nor "cold" enough to avoid evaporation of TBA solvent. Hence, the gelation of AM/TBA solution on the top side was restricted. This results in low green body strength before firing. By setting the dense side as the tension face, the bending strength of the gradient porous alumina fabricated under these two different temperature gradients were measured, and the results were shown in Table 7.3.

Table 7.3 The bending strength of porous alumina with different gradients

fabrication method	green strength (MPa)	strength after sintering at 1500℃ (MPa)
method (a)	11 ± 1	260 ± 20
method (b)	2 ± 0.5	226 ± 15

The effect of AM polymerization on the gradient porous structure was also investigated. The same experiment was carried out under the temperature condition given in Fig. 7.13(a), but without AM in TBA, and the microstructure was shown in Fig. 7.21. By comparing the microstructure given in Fig. 7.18 with that presented in Fig. 7.21, it can be seen that the pore distribution in Fig. 7.18 is more uniform and homogeneous than in Fig. 7.21. The main reason for this difference is that when the polymerization of AM occurred, a rapid rise of temperature on the hot side takes place, indicating that the temperature gradient through the body is enhanced.

Figure 7.21 The pores in alumina ceramics fabricated by TBA freezing without AM addition

7.2.4 The Mechanical Properties and Applications of Alumina Ceramics with Ultra Low Density

The preparation of ceramics with ultra low density is similar to that of porous ceramics by freeze-gel-casting. From the paper "Ceramics with ultra low density fabricated by gel-casting: an unconventional view" by Chen Ruifeng et al., in 2007, many meaningful conclusions could be obtained (Chen Ruifeng, et al., 2007).

The main properties of dry green bodies with low relative density (from 5 to 15% of initial solid loading) are shown in Table 7.4. As a result of polymerization, the strength of the green body samples was greatly improved; they were much stronger than ceramic green bodies with ultra high porosity fabricated by the other methods. Even when the relative green body density was as low as 5%, the compression strength was over 1 MPa, which was strong enough to allow the materials to be handled in the subsequent procedures.

Table 7.4 The properties of green bodies with ultra low density (from 5 to 15% of initial solid loading)

solid loading of slurry (vol%)	compression strength (MPa)	linear shrinkage after drying (%)	relative density (%)
5	1.1 ± 0.2	0.42 – 0.48	5.05 – 5.1
10	2.1 ± 0.2	0.25 – 0.35	10.1 – 10.2
15	2.9 ± 0.2	0.14 – 0.22	15.1 – 15.2

The density and mechanical properties of the sintered samples are summarized in Table 7.5. The results show that after being sintered at a high temperature, the samples acquired from 5 and 10 vol% slurries were so light that the bulk densities were even lower than that of water (1 g/cm^3). As the initial solid loading was increased, the compression strength increased from 31 to 84 MPa.

Table 7.5 The properties of alumina with different initial densities after being sintered at 1500℃ for 2 h

slurry solid loading (vol%)	bulk density after sintering (g/cm^3)	compression strength (MPa)
5	0.48 – 0.52	31.5 ± 3.1
10	0.99 – 1.02	56.4 ± 3.9
15	1.55 – 1.60	83.9 ± 4.3

The effect of sintering temperature on the mechanical properties was also investigated, and the results are shown in Table 7.6.

Table 7.6 The properties of alumina (made from the slurry containing 5 vol% of solid loading) after being sintered at different temperatures

sintering temperature (℃)	bulk density after sintering (g/cm^3)	compression strength (MPa)
1400	0.31 – 0.33	11.1 ± 1.1
*1500	0.49 – 0.51	31.5 ± 1.5
1600	0.79 – 0.82	50.6 ± 2.1

Note: symbol "*" refers to the examples mentioned in the section of "Application".

Table 7.7 lists some other important properties of alumina ceramics with ultra low density fabricated by this gel-casting method, in which the open porosity varies from 60 to 92%, the pore size varies from 0.1 to 2.2 μm, and the specific surface area varies from 4 to 15 m^2/g. It can be concluded that in these light-weight alumina ceramics, the increase of porosity is often related to larger pore sizes, decreased strength and greater specific areas.

Table 7.7 The open porosity, pore size and specific surface area of low density alumina acquired by TBA-based gel-casting

slurry solid loading (vol%)	sintering temperature (℃)	open porosity (%)	mean pore size (μm)	specific surface area (m^2/g)
5	1400	92	0.1, 1.0	14.82
	1500	88	1.2	10.69
	1600	80	2.2	6.52
10	1400	85	0.1, 0.8	8.75
	1500	75	1.0	4.97
15	1400	77	0.1, 0.6	5.20
	1500	64	0.8	4.00

With controllable microstructures and properties, the gel-casting technique for the fabrication of ceramics with ultra low density can be expanded from the laboratory to actual industrial applications, as illustrated in Table 7.8. For each application, ceramics with the required properties can be produced, as shown in Tables 7.5, 7.6 and 7.7. For example, alumina acquired from 5 vol% slurry and sintered at 1500℃ (designated with an asterisk in Tables 7.6 and 7.7) had an open porosity of 88%, a pore size of 1.2 μm and a compression strength over 30 MPa, which makes it suitable for infiltrating bacteria in biomedicine.

In this section, a new freeze-gel-casting technique, which combines the merits of freeze-casting and gel-casting has been introduced. For this method, alumina/TBA/AM slurry was prepared, with TBA as the freezing vehicle and AM as the gelation agent. Freezing occurred rapidly at a relatively high temperature (0 – 5℃),

Table 7.8 The application of alumina with low density acquired by TBA-based gel-casting (Michael Scheffler, et al., 2005)

application	required properties
filter	open porosity > 80%, pore size 1 μm
supports for catalysts	specific surface area > 10 m^2/g, compression strength >10 MPa
artificial bones	bulk density < 1 g/cm^3, compression strength > 50 MPa
lightweight parts for high-temperature manufacturing	bulk density < 0.5 g/cm^3, compression strength > 30 MPa

and generated frozen TBA prisms with diameters of several tens of micrometers as templates of pore channels. By controlling the freezing direction, unidirectional porous ceramics with a long range were successfully fabricated, with almost all the pore channels being open pores. The temperature gradient has a great influence on the porosity and the pore size distribution in the ceramic body over the thickness of the sample. By choosing appropriate temperature gradient in the freezing step, a uniform linear pore gradient can be achieved, with high porosity (⩾65%) at the bottom and low porosity (⩽ less than 10%) at the top after sintering. The polymerization of AM in TBA was realized by subsequent heat treatment, providing a way to acquire green porous ceramic bodies with high strength, i.e., compression strength over 10 MPa of unidirectional porous alumina, thus, more complex manufacturing can be handled. The porous ceramics have good mechanical properties, i.e., the unidirectional porous alumina reaches a compression strength of 150 MPa after being sintered at 1500℃, because the pore channels formed by the TBA template are surrounded by almost complete dense walls without any noticeable defects. These results show the prospect of the freeze-gel-casting technique for fabricating porous ceramics with possible applications such as infiltration and catalytic support. In addition, this technique can be applied in many other ceramic systems.

7.3 The Solidification of Concentrated Si_3N_4 Suspensions for Gelcasting by Ultrasonic Effects

7.3.1 Gelcasting by Ultrasonic Effects

Gel-casting is a near-net-shaped forming method for the production of advanced ceramic materials. The green body prepared by gel-casting has a similar homogeneous microstructure like the precursor suspensions and shrinks more uniformly during drying and sintering than those prepared using conventional methods. As a result, the structure homogeneity and reliability of ceramics are improved. Hence, ceramics scientists have paid great attention to the gel-casting technique since its initial development. However, the controllability of the gel-casting process is not good, and the amounts of initiator and catalyst added to ceramic suspensions are restricted within a narrow range. Ceramic suspensions solidify too fast to cast with slightly more initiator and catalyst, whereas slightly less initiator and catalyst result in an incomplete solidification of the suspensions. Therefore, in most cases, initiator and catalyst were added to cooled ceramic suspensions to prolong the processing time, and then the polymerization of the monomers was initiated to solidify the suspensions by increasing the temperature of the mold. The temperature induced polymerization of monomers results in another problem, namely the asynchronous solidification of the suspension, which can cause cracks or potential cracks between the parts solidifying at different times, owing to the thermal transmission from the outside to the center of the suspension. So, it is necessary to develop another mechanism to solidify the suspensions homogeneously. Yang Jinlong, et al., (2002) reported that a pressure-induced solidification of ceramic suspensions could avoid the effect of the temperature gradient on gel-casting process, but this method increases the complexity of the equipment—microwaves can heat the suspension uniformly, but it is not very easy to control and it usually overheats the suspension and forms pores in it. It is not applicable for metal mold, which is necessary for the formation of a good surface of the green body. On the other hand, to reduce cracks caused by excessive shrinkage during drying of the green body, suspensions for gel-casting must contain the solid powders not less than 45 vol%. Unfortunately, most suspensions prepared directly with raw Si_3N_4 powders could not reach this lower limit. In this section, acid-washing and preoxidation were used to increase the solid loading of Si_3N_4 suspensions from 40 to 50 vol%, and ultrasonic waves were used to accelerate the decomposition of initiator (AP) and solidification of Si_3N_4 suspensions for gel-casting homogeneously. The obtained results showed that the flexibility of gel-casting could be improved with the ultrasonic effect.

The Si_3N_4 powder, M11 from Stark Corporation, Germany, was used as raw material. 15 wt% AM premix solution, which contains some MBAM (2 wt% to AM), was used as disperse media, and 10 wt% TMAH solution was used as dispersant. The groups on the surface of Si_3N_4 powders were analyzed with infrared Fourier transformation spectroscopy (Model FTIR-750, Nicolet Corp., USA); the chemical state of Si on the surface of Si_3N_4 powders was observed by XPS (Model PHI-5300ESCA, USA). The absorption spectra of the solutions were recorded by UNICAM500 spectral photometer (Unicam Corp., USA). The rheological curves of Si_3N_4 suspension were measured by a MCR300 rheometer (Physica Corp., Germany). After the addition of AP (8 wt% solution) and TEMED, in the volume ratio of 2:1 to the deaired suspension, the temperature of the suspension during solidification was recorded by a computer linked with the temperature detector. The strength of the samples was measured by three point flexural tests. The frequency and capacity of the ultrasonic wave were 28 kHz and 0.3 W/cm^2 respectively.

7.3.2 The Preparation of Concentrated Si_3N_4 Suspensions

It's well known that it is difficult to prepare concentrated suspensions more than 40 vol% solid loading for gel-casting with raw M11 Si_3N_4 powders, because of the presence of hydrophobic groups such as amine structures (Si_2NH and $SiNH_2$) on the powders surface. Generally, various surface treatments were used to modify the surface groups of Si_3N_4 powders to improve its dispersibility in aqueous suspensions. In this work, preoxidation after acid-washing was used to eliminate the hydrophobic groups and to introduce a hydrophilic group such as silanol (Si—OH) on the surface of Si_3N_4 powders. First, Si_3N_4 powders were mixed with 5% HCl solution in the ratio of 100 g:200 mL, and the mixture was ball milled for 24 h to ensure the effect of acid-washing; then, the slurry was filtered and washed three times with deionized water during filtration, and the obtained mud was dried at 110℃ for 24 h; finally, the dried mud was smashed and calcined at 600℃ for 6 h for the preoxidation of Si_3N_4 powders. The preoxided Si_3N_4 powders were analyzed with diffusion reflection infrared Fourier transformation (DRIFT) spectrum (see Fig. 7.22(a)) and XPS spectrum (see Fig. 7.22(b)). Figure 7.22(b) shows that after preoxidation the hydrophobic amine groups (corresponding to the absorption peak at about 3400 cm^{-1}) in untreated Si_3N_4 powders were eliminated, and the hydrophilic Si—OH groups (corresponding to the absorption peak at about 3750 cm^{-1}) were introduced on the powder surface. In addition, the peak of binding energy of Si_{2p} obviously shifted to larger values after preoxidation (see Fig. 7.22(a)). The shifted peak could be confirmed to consist of the Si_{2p} peak of Si_3N_4 (101.8 eV) and Si_2N_2O (102.3 eV) after

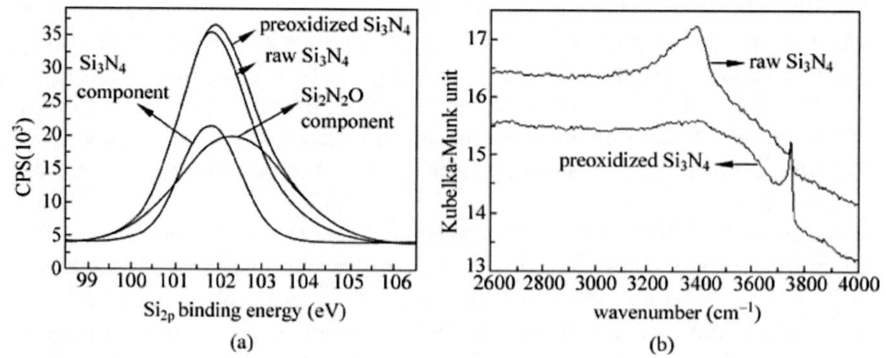

Figure 7.22 The DRIFT spectrum (a) and XPS spectrum (b) of raw and preoxided Si_3N_4 powders

data-processing, so it could be considered that Si_2N_2O appeared on the surface of the Si_3N_4 powder. Because of the formation of Si_2N_2O, the amount of oxygen on the Si_3N_4 surface was increased, and the raised oxygen amount could increase the zeta potential in the Si_3N_4 surface (Nagel A, 1988). As a result of the combination of the above mentioned effects, the solid loading of the Si_3N_4 suspensions was increased from 40 vol% (untreated powders) to 50 vol% (preoxided powders) with preoxidation treatment, and the rheological behavior was also improved (see Fig. 7.23).

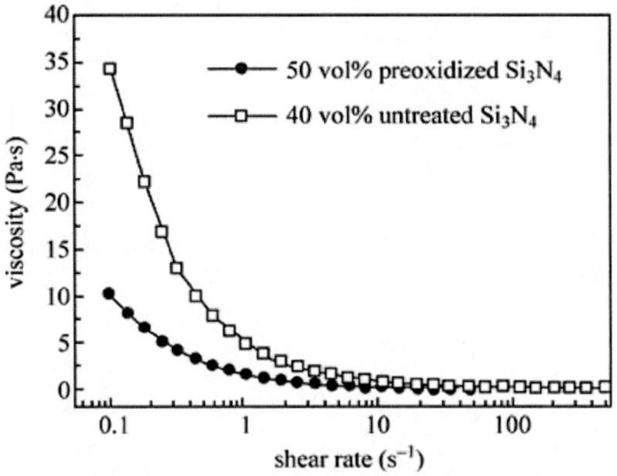

Figure 7.23 The rheological curves of suspensions prepared with raw and preoxided Si_3N_4 powders

7.3.3 The Ultrasonic Accelerated Solidification

The temperature changes of Si_3N_4 suspensions without and with AP (3 μL/mL suspension) during ultrasonic treatment are shown in Fig. 7.24. During the process of ultrasonic solidification, the temperature increase of the suspension may be owing to two reasons: one is the ultrasonic heating effect; the other is the polymerization of monomers.

In Fig. 7.24, curve A shows the ultrasonic heating effect, and curve B is the overall temperature rise caused by monomer polymerization and ultrasonic heating effect. If the ultrasonic heating effect (curve A) is deducted from the overall temperature rise in the suspension (curve B) during ultrasonic treatment, the temperature increment from the polymerization of monomers in Si_3N_4 suspension can be obtained, which is shown as curve C in Fig. 7.25. The temperature increment from polymerization induced by ultrasonic waves (curve C) and heating at 30℃ in a water bath without ultrasonic wave (curve D) are compared in Fig. 7.25. The acceleration of the polymerization of monomer by ultrasonic waves can be predicated by the comparison, because the Si_3N_4 suspension had a much shorter induced time under ultrasonic waves.

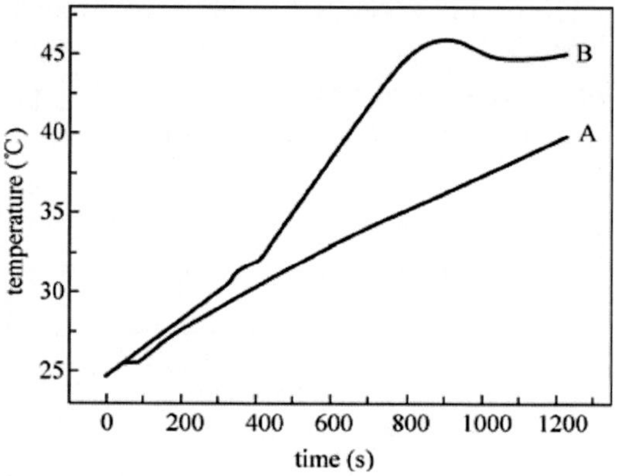

Figure 7.24 The temperature rise of 50 vol% Si_3N_4 suspensions under ultrasonic condition without initiator (A) and with initiator (B)

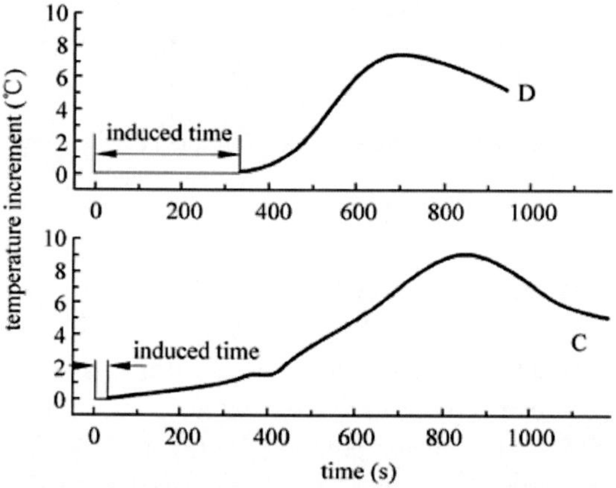

Figure 7.25 The temperature increment caused by solidification of 50 vol% Si_3N_4 suspensions under ultrasonic treatment (C) and at 30℃ water bath (D)

The polymerization of AM monomers was initiated by free radicals. The free radicals from AP-TEMED system were found to be superoxide anion (Xiao Huashan, et al., 1991), and the superoxide anion could react with the hydroxylamine solution according to the formula:

$$2O_2^{\cdot-} + NH_2OH + H^+ = NO_2^- + H_2O_2 + H_2O \qquad (7.1)$$

The resultant nitrite ions, after coloration reaction with sulfanilic acid and α-naphthylamine, could be quantified by the specific absorption peak at a wavelength of 530 nm (Wang Aiguo, et al., 1990; Xiao Huashan, et al., 1999). The linear relationship between the concentration of nitrite ions and the adsorption coefficient of the solution at 530 nm was calibrated with a $NaNO_2$ solution, as shown in Fig. 7.26. The original mole concentration of the superoxide anions could be determined by doubling that of nitrite ions.

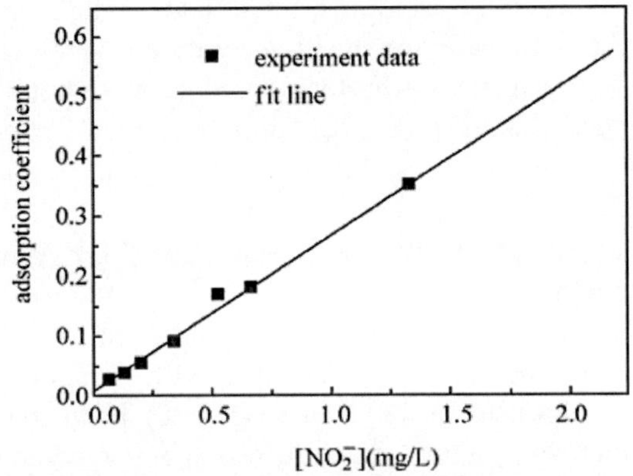

Figure 7.26 The dependence of adsorption coefficient at a wavelength of 530 nm on the concentration of nitrite ions in the solution

According to the above principle, the concentration of superoxide anions in AP-TEMED aqueous solution (6 μL, 8 wt% AP and 3 μL TEMED in 1 mL deionized water) under various conditions was determined (see Fig. 7.27). From this figure it can be seen that the concentration of superoxide anions at 40℃ is much larger than that at 25℃, but less than half of that treated by ultrasonic wave, although the solution temperature during ultrasonic treatment never reached 40℃. Figure 7.27 shows that the concentration of superoxide anions changed little with

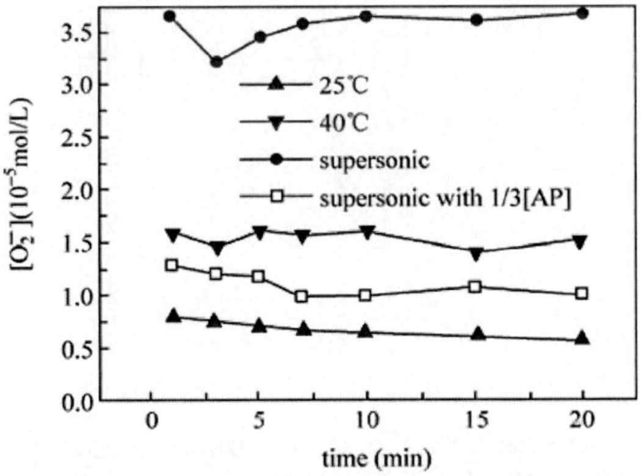

Figure 7.27 The effects of ultrasonic wave and temperature on the concentration of superoxide anions in AP-TEMED solutions

time, indicating that the decomposition of AP reaches equilibrium in a shorter time under a specified condition. Hence, the solidification of the suspension is difficult to control at room temperature. However, the cage effect, which restricts the diffusion and promotes the recombination of the free radicals in the solution, was weakened under ultrasonic treatment or at raised temperature, and this effect resulted in higher concentrations of superoxide free radicals (see Fig. 7.27). Figure 7.27 shows that the ultrasonic treatment had a much larger effect on the decomposition of AP in the solution than temperature. Therefore, the accelerated solidification of Si_3N_4 suspensions by the ultrasonic wave could be attributed to the increased decomposition degree of the initiator (AP).

7.3.4 The Comparison between Thermal and Ultrasonic Activated Solidifications

Thermal activated solidification was accomplished by dipping the glass beaker containing the Si_3N_4 suspension at 25℃ (see Fig. 7.28) into a 40℃ water bath in a thermostat, while ultrasonic activated solidification was accomplished by dipping the beaker containing a suspension at 25℃ into 25℃ water bath in the ultrasonic generator. In both the cases, the beaker was fixed in a polyurethane shoe to avoid the influence of axial thermal transfer from the bottom of the beaker. The water surface out of the beaker was about 10 mm higher than the suspension surface in the beaker. The temperature changes at points A, B, and C (see Fig. 7.28) were recorded after the beaker was dipped into the thermostat or after the ultrasonic generator was turned on.

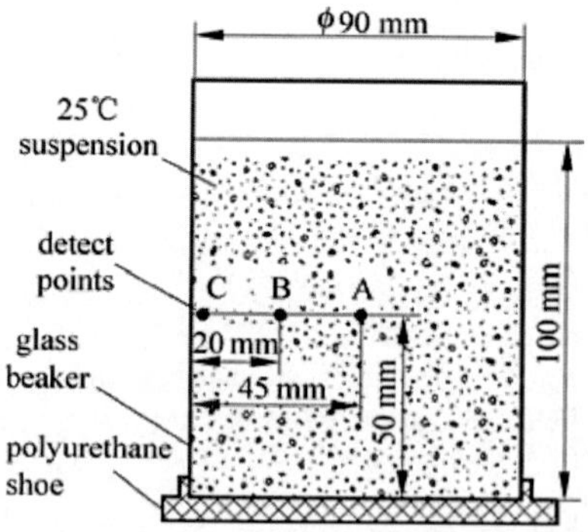

Figure 7.28 The schematic representation of the apparatus to study the asynchronous solidification of Si_3N_4 suspensions

The temperature increase from the polymerization at different points in Si_3N_4 suspensions during solidification activated by ultrasonic wave and temperature is compared in Fig. 7.29. It should be noted that the effects of ultrasonic heating and at 40℃ in a water bath were deducted according to the same method as mentioned above. In the case of solidification by heating at 40℃ in a water bath, an annular temperature gradient appeared in the suspension after the beaker was dipped into the water, so the solidification proceeded radically from the beaker wall to the center of the suspension (see Fig. 7.29(b)). This makes the solidification of the suspension near the center constrained by the outer solidified ring. As a result, the center part of the green body endured radial and circumferential tensile stresses, which may result in cracks in green body during drying and demoulding. The asynchronous solidification caused by the temperature gradient could be suppressed using ultrasonic waves. Figure 7.29(a) shows that the Si_3N_4 suspension from the edge to the center of the beaker began solidification at almost the same time. The temperature difference at the points A, B, C during solidification may owing to the thermal transfer from the solidifying suspension to the water out of the beaker.

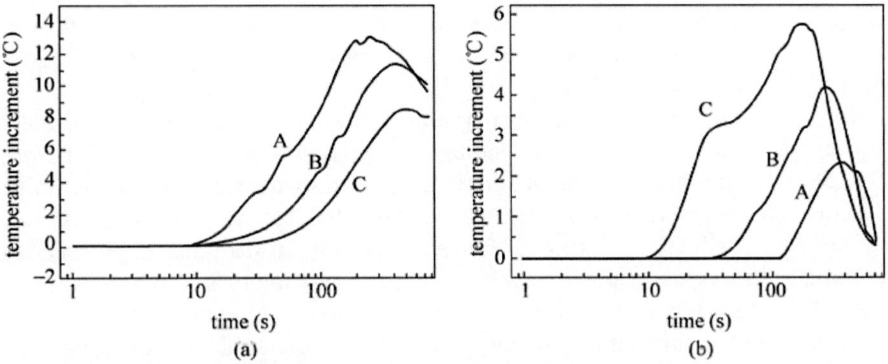

Figure 7.29 The temperature increments at different positions in Si_3N_4 suspensions during solidification activated by ultrasonic wave (a) and temperature (b) (in a water bath at 40℃; A,B,C represent different places in Fig. 7.28)

The strength of Si_3N_4 green bodies prepared by gel-casting of 50 vol% suspension is shown in Table 7.9. 3 μL and 1 μL AP solution were added to a 1 mL suspension in the cases Ⅰ and Ⅱ respectively. Table 7.9 indicates that the thermal solidified green body has a higher strength than that solidified with ultrasonic wave in the case Ⅰ. The reason might be owing to the fact that ultrasonic treatment yielded too many superoxide anions in the suspension (see Fig. 7.27), thus reducing the length and the crosslinkage of the polyacrylamide, which bound the Si_3N_4 powders and determined the strength of the green body. For the same reason, the suspension ultrasonically solidified with less amount of AP solution (case Ⅱ) resulted in a higher strength of the green body. Contrastingly, the suspension with less amount of AP solution could not be solidified by thermal activation (40℃). This shows that in the case of less initiator in the suspension, thermal activation could not yield enough free radicals to initiate the solidification of the suspension, while the ultrasonic wave could. Hence, ultrasonic treatment could be used to improve the controllability of the solidification of Si_3N_4 suspensions thus improving the flexibility of the gel-casting process.

Table 7.9 The strength (MPa) of Si_3N_4 green bodies prepared by gel-casting of 50 vol% suspension

case	I	II
ultrasonic activated	23.6	43.6
thermal activated (40℃ water bath)	30.5	incomplete solidification

Notes: I : 3 μL AP in 1 mL solution;
II : 1 μL AP in 1 mL solution.

In addition, Si_3N_4 suspensions with less amount of initiator were solidified, yielding high strength green body (43.6 MPa). This suggests that Si_3N_4 suspensions with less amount of monomer could be solidified by ultrasonic waves and yielded enough strength to maintain the configuration of the green body, which was in favor of the debonding of the green body. However, this suggestion should be confirmed in the future studies.

The hydrophobic amine groups were removed from the surface of Si_3N_4 powders by preoxidation, hence the solid loading of Si_3N_4 suspensions was raised from 40 to 50 vol% and the rheological behavior of Si_3N_4 suspensions was improved.

Both temperature and ultrasonic waves promoted the decomposition of AP, thus accelerating the solidification of concentrated Si_3N_4 suspensions for gel-casting. However, ultrasonic treatment resulted in apparently more superoxide anionic free radicals than temperature increase. Therefore the Si_3N_4 suspensions could be solidified with less amount of initiator, and hence, the controllability of suspension solidification and flexibility of gel-casting process could be improved. At the same time, asynchronous solidification of the suspension, caused by the temperature gradient, which led to crack potential during the drying and debonding of the green bodies, could be suppressed by ultrasonic activation solidification. Additionally, Si_3N_4 suspensions with less amount of initiator yielded a green body with high strength after ultrasonic activated solidification. This implied that Si_3N_4 suspensions with less amount of monomer could be solidified by ultrasonic waves.

7.4 Novel Laser Machining Technology for Al_2O_3 Green Ceramic

7.4.1 Laser Machining Technology

The machining of ceramics to their final dimensions by conventional methods is extremely laborious and time-consuming. Laser machining is a non-contacting, abrasionless technique, which eliminates tool wear, machine-tool deflections, vibrations and cutting forces, reduces limitations to shape formation and inflicts less sub-surface damage. So, as an advanced manufacturing method, it is used in different areas such as electronics, aerospace, and material processing, etc. (Chryssolouris G., et al., 1997; Chryssolouris G., 1991). Structural ceramics exhibit good high temperature performances, abrasion resistance, anti-corrosion and other unique properties, and have been widely used in the fields of machinery, chemicals, petroleum, textiles, aviation, aerospace, etc. However, the current marketing characteristic of structural ceramics is multi-purpose and limited. So if the manufacturing costs are high, the promotion and application of structural ceramics are limited. A new laser machining method has been proposed to produce ceramics that satisfies the requirements mentioned above. This technology combines laser machining and gel-casting technique to machine complex-shaped ceramic parts. This method builds parts by selectively removing the materials, layer by layer, as specified by a computer program to generate geometrical objects directly from a three-dimensional computer image without part-specific tooling or human intervention.

The laser machining process of ceramics in the sintered state has some shortcomings owing to the presence of spatter and microcracks. Hence, many researches are related with the ceramics in their green states. At present, this study has attempted to discuss the laser machining mechanism by means of laser-material interaction parameters, and to realize the laser-machinability of Al_2O_3 green bodies by changing the laser parameters. Laser machining of ceramics is also used extensively in the microelectronics industry for scribing and via hole drilling (Lumpp J. K., et al., 1997).

7.4.2 Practical Application of Laser Machining Tcchnology

7.4.2.1 The Spatter-Free Laser Drilling Processing

Laser drilled holes are inherently associated with spatter deposition due to the incomplete expulsion of ejected material from the drilling site, which subsequently resolidifies and adheres onto the material surface around the hole periphery. In addition, high hardness and brittleness may lead to fracture (microcracks) of the ceramic material during laser machining. Hence, to prevent spatter deposition and microcracks during laser machining, many techniques based on either chemical or physical mechanisms have been developed (Orita N., 1988; Sharp, C. M., et al., 1997; Low D. K. Y., et al., 2001; Murray A. J., et al., 2000), but the results were not encouraging.

In this study, an anti-spatter and anti-crack laser drilling method that use Nd: YAG laser to drill holes directly on a gelcast green ceramic body is reported. Owing to the difference between the microstructure of the green body and that of sintered ceramics, formation of spatter and microcracks can be successfully prevented in the green body.

The gel-casting process is similar to that described in the previous studies. First, the ceramic powder was added to the premix solution (containing 20 wt% AM) and thoroughly ball milled. Then the slurry was deaired by a vacuum pump. After adding initiator and catalyst the slurry was casted into a nonporous mold without additional pressure. After polymerization and drying, a green body with high mechanical strength was obtained, and then laser drilling was carried out. The detailed flowchart of the gel-casting and drilling process is shown in Fig. 7.30. The microstructure of the sample was observed by SEM. The hole size was measured by optical microscopy. The mechanical strength of the samples was determined by the three-point flexure test. A pulsed Nd: YAG laser, typical of many lasers used in industry for drilling and fine cutting applications, was selected for processing. The laser beam had a focal spot size of about 100 μm and a wavelength of 1.064 μm. Drilling was operated at a pulse duration of 0.2 ms and pulse repetition rate of 2 Hz without any assist gases.

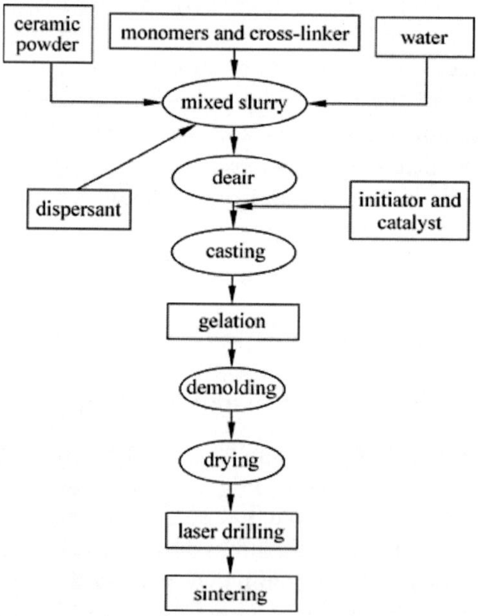

Figure 7.30 The detailed flowchart of the gel-casting and laser drilling process

7.4.2.2 The Basic Mechanism of Laser Drilling and the Comparison of Hole Morphologies

Figure 7.31 shows the SEM photos of the holes drilled on the green alumina samples obtained from the slurry with 50 vol% solid loading. Holes on sintered ceramics drilled under the same conditions are also illustrated for comparison. The basic mechanism of laser drilling is based on the thermal action of the light. When high intensity light strikes the material surface, the photons of the beam are absorbed and converted into thermal energy. The temperature rises locally and a combination of melting and vaporization occurs. Under the continual effect of the laser beam, the molten material is ejected by the material vapor and a hole is thus formed. When the laser pulse comes to an end, the decrease of the flux density, the blurring of the light spot edges, as well as the power decrease at the pulse trailing edge, contributes to an increase of resolidified liquid residue in the hole, resulting in recast layers and a spatter deposited surface (Metev S. M., et al., 1996). As can be seen in Fig. 7.31(a), the holes drilled on sintered ceramics are encircled with large areas of irregular spatter deposits. In contrast, it is clearly evident from Fig. 7.31(b) that very little spatter is present around the holes drilled on the gelcast green body. Figure 7.31(b) also exhibits better surface quality and a smooth hole periphery. Figure 7.31(c) – (f) show further detail of the surface quality of the holes presented in Fig. 7.31(a), (b), respectively. From Fig. 7.31(c), (e) it can be seen that the spatter deposits at the periphery of the hole on the sintered alumina show characteristics of a resolidified molten phase, and some microcracks can be seen, which may be owing to laser induced thermal shock and stress. On the other hand, the periphery of the hole on the green body shown in Fig. 7.31(d), (f) show a structure composed of particles, without any cracks.

7.4.2.3 The Different Character of Holes on the Different Green and Sintered Body

Figure 7.32 compares the cross section features of the two samples, which also indicates that the spatter on the sintered ceramics does not appear on the green body.

From the fracture microstructures of the two samples shown in Fig. 7.33, it can be seen that the sintered alumina has a polycrystalline character, i.e., it is composed of grains of 7 – 8 μm in size. On the other hand, the green body is composed of particles of about 2 μm in size.

As a result of the irregular spatter deposition, the consistency/repeatability of the geometry of the holes on the sintered body is inferior to that of the holes on the green body. This can be clearly seen by comparing the standard deviation of the entrance hole diameter and hole depth values obtained by the two methods (see Table 7.10).

The relatively large standard deviation of the hole depth are owing to the difficulty in getting precisely fractured hole sections for measurement. More regular and uniform hole shapes can be obtained on gelcast green samples than on sintered ceramics. In addition, the holes drilled on the green body show larger

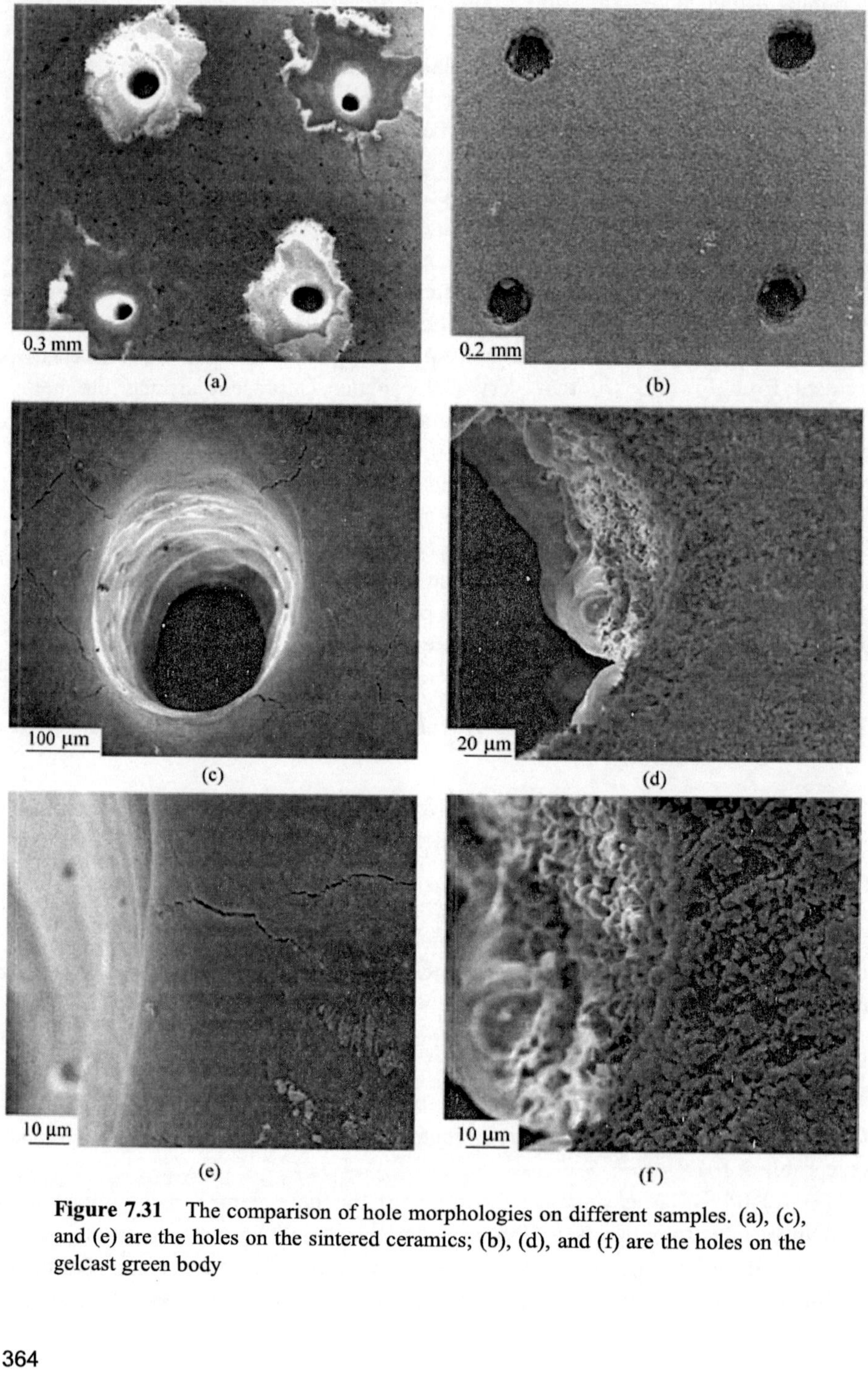

Figure 7.31 The comparison of hole morphologies on different samples. (a), (c), and (e) are the holes on the sintered ceramics; (b), (d), and (f) are the holes on the gelcast green body

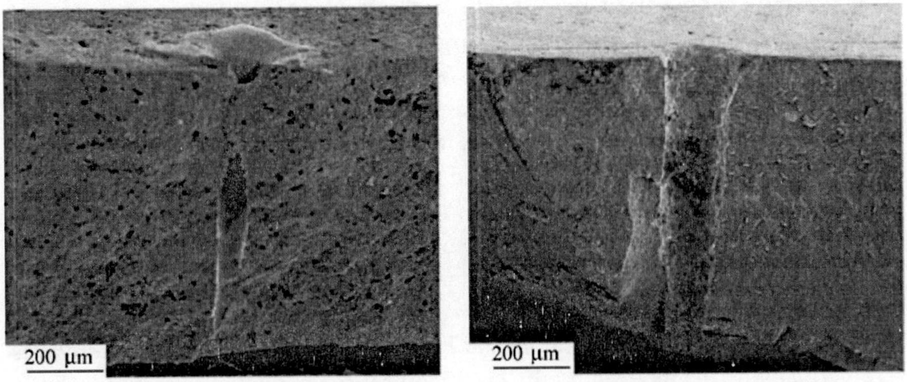

(a) (b)

Figure 7.32 The illustration of the cross section features of the holes (a) sample of sintered ceramics; (b) sample of gelcast green body

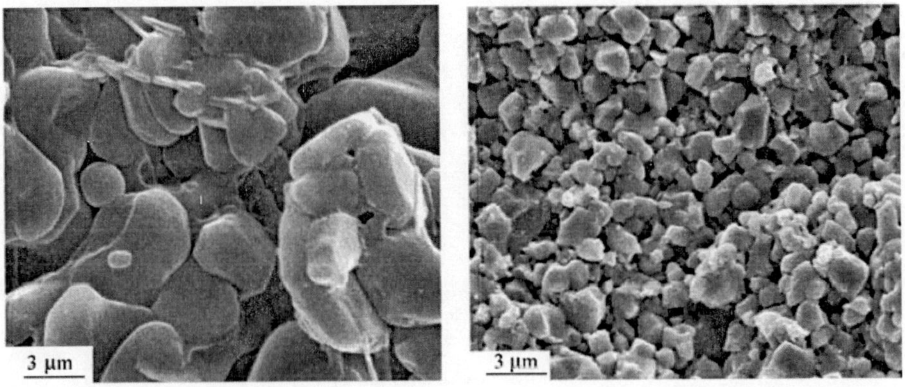

(a) (b)

Figure 7.33 The SEM micrographs showing the fracture microstructures of the two samples (a) and (b) are the fracture graphs of the sintered ceramics and gelcast green body

Table 7.10 The comparison of the entrance hole diameter and hole depth by two different drilling methods (6 pulses)

sample	average hole diameter [a] (mm)	standard deviation (mm)	average hole depth (mm)	standard deviation (mm)
gelcast green body	0.21	0.016	1.1	0.18
sintered ceramic body	0.19	0.033	0.8	0.25

[a] Calculated from 10 holes.

diameter and depth values than those on the ceramics. A lower total heat of evaporation and melting will result in larger hole depth and hole diameter values under the same pulse energy (Metev S. M., et al., 1996). This indicates a lower heat of evaporation for the green bodies. Thus, from the thermal nature of the radiation and from the examination of the microstructure, it is reasonable to conclude that the absence of spatter and microcracks in the gelcast green body (which contains about 4 wt% of polymer material) may be attributed to the lower heat of evaporation and the relatively loose microstructure of the green body, when compared with the sintered ceramics.

The conventional laser drilling of green alumina bodies containing PVA or PVB as binders is not a suitable method because the propagation of shock waves and thermal stresses induced by the laser may cause damage to the alumina body (Metev S. M., et al., 1996). The influence of AM content in the premix solution on the green body strength derived from 50 vol% alumina slurries is shown in Fig. 7.34.

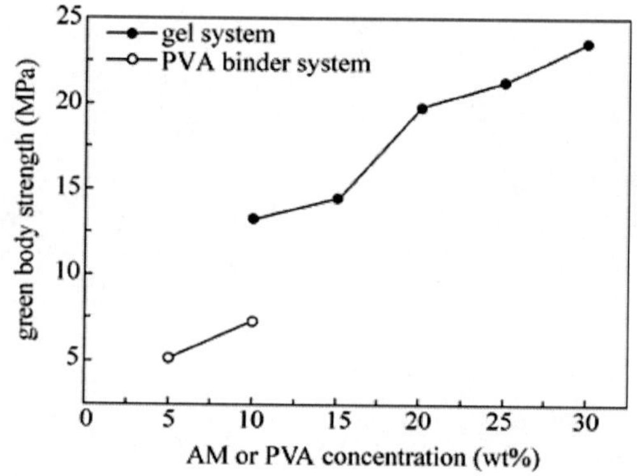

Figure 7.34 The influence of AM or PVA concentrations on the green body strength

For the PVA system, the samples were die pressed at a pressure of 75 MPa.

It can be seen that an AM concentration of 20 wt% can provide a strength of about 20 MPa. In comparison, die pressed samples prepared using even a 10 wt% of PVA solution as the binder show a much lower strength of about 7.3 MPa. Consequently, the holes are damaged under the radiation of the laser beams. This can be clearly seen from the SEM microphotos shown in Fig. 7.35. A conical shape in the upper part of the hole was formed during laser drilling. The hole also exhibits a conical shape in the lower part near the bottom of the material. This implies that the relatively high quality holes obtained on the gelcast green body can be attributed to its excellent mechanical properties.

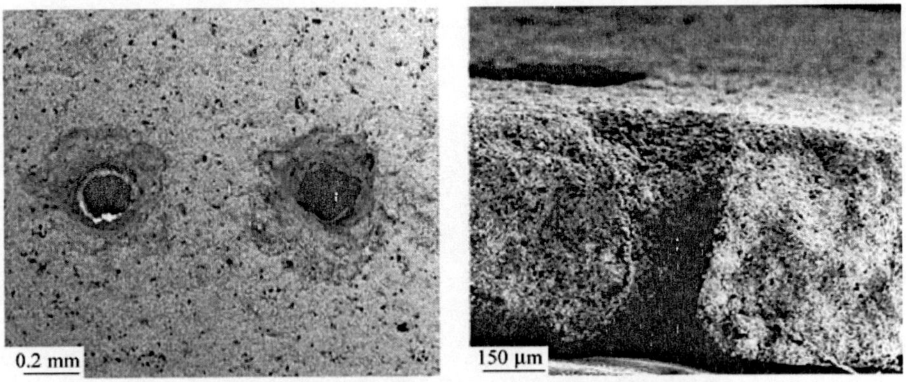

(a) (b)

Figure 7.35 The SEM micrographs of the holes drilled on the conventional die pressed green alumina body

In addition, the absence of spatter and microcracks in the holes produced by the method described earlier is independent of the Nd:YAG laser and special drilling parameters, although these conditions can affect the hole quality to a large extent. Figure 7.36 shows the compact hole arrays obtained after varying the pulse energy and pulse repetition rate, which exhibits reasonably encouraging results. Furthermore, when this laser drilling method was tested on Si_3N_4 samples, similar results were obtained (see Fig. 7.37). It should be pointed out that gel-casting is a highly versatile fabrication process. It is not limited to the usage with any particular ceramic powder. It can be quickly adapted for use with new materials and new applications, and it works as well for metal powders as for ceramic

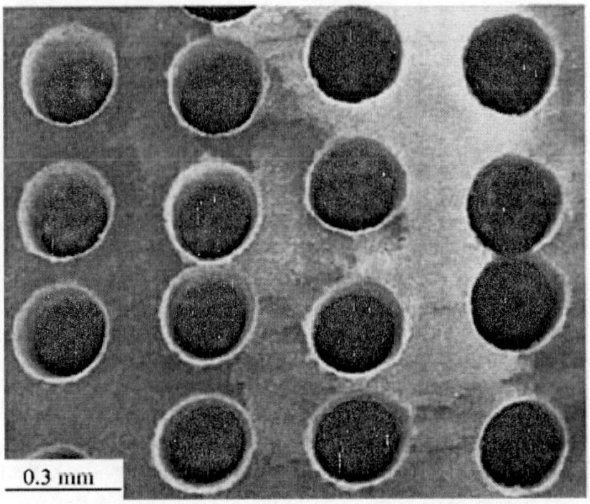

Figure 7.36 The SEM photo of compact hole arrays drilled on the gelcast alumina green body

powders. Therefore laser processing based on the gel-casting technique may be applied to a variety of materials.

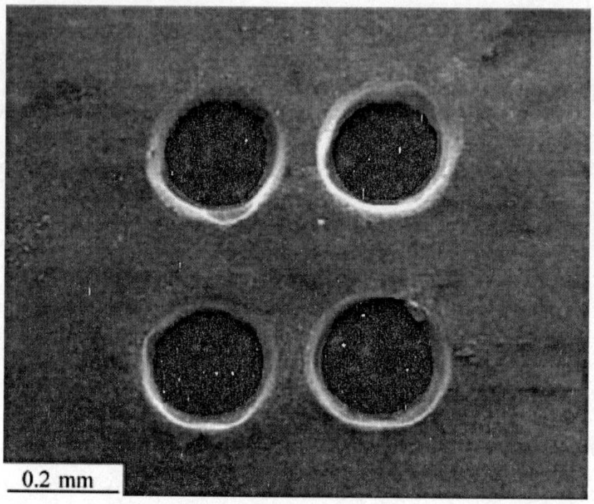

Figure 7.37 The SEM photo of compact hole arrays drilled on the gelcast Si_3N_4 green body

A novel anti-spatter and anti-crack laser drilling method based on the gel-casting technique was developed. Laser drilling of a gelcast alumina green body proved to be highly effective in preventing spatter and microcracks during laser percussion drilling of the material. Micro-holes with regular and uniform shapes, but without cracks, were successfully obtained by this technique. From the thermal nature of the drilling processing and from the examination of the microstructure, it was deduced that the absence of spatter and cracks in the gelcast green body might be attributed to the lower heats of evaporation and melting, as well as the relatively loose microstructure of the green body, when compared with the sintered ceramics.

References

Agarwals M. K., Weeren R., 1996. Am. Ceram. Soc. Bull., 75: 60.

Araki K., Halloran J. W., 2005. Porous ceramic bodies with interconnected pore channels by a novel freeze casting technique. J. Am. Ceram. Soc., 88(5): 1108 – 1114.

Barnes H. A., Hutton J. F., Walers K., 1989. An Introduction to Rheology. Oxford: Elsevier Press.

Chang A. S., Shih T. S., 1999. Permeability of coating in the lost foam casting process. Int. J. Cast. Metal. Res., 12: 263 – 275.

Chen Ruifeng, Huang Yong, Wang Chang'an, et al., 2007. Ceramics with ultra low density fabricated by gel-casting: an unconventional view. Journal of the American Ceramic

Society, 1 – 31.
Chen Y. L., Xie Z. P., Yang J. L., et al., 1999.Alumina casting based on gelation of gelatine. J. Eur. Ceram. Soc., 19: 271 – 275.
Chryssolouris G., 1991. Laser Machining: Theory and Practice. New York: Springer-Verlag.
Chryssolouris G., Anifantis N., Karagiannis S., 1997. Laser assisted machining: an overview. J. Manufact. Eng.—Trans. ASME, 119(4B): 766 – 769.
Danforth S. C., 1995. Mat. Tech., 144: 10.
Ferreira J. M. F., Diz H. M. M., 1992. Effect of slurry structure on the slip casting of silicon carbide powders. J. Eur. Ceram. Soc., 10: 59 – 64.
Fukasawa T., Ando M., Ohji T., et al., 2001. Synthesis of porous ceramics with complex pore structure by freeze-dry processing. J. Am. Ceram. Soc., 84(1): 230 – 232.
Griffin C., Daufenbach J., Mcmillim S., 1994. Am.Ceram. Soc. Bull., 73:109.
Griffin M. L., Halloran J. W., 1996. J. Am. Ceram. Soc., 79: 2601.
Guha S., Ray B., Mandal B. M., 2001. Anomalous solubility of poly acrylamide prepared by dispersion (precipitation) polymerization in aqueous tert-butyl alcohol. J. Polym. Sci.: Part A: Poly. Chem., 39: 3434 – 3442.
Hinczewski C., Corbel S., Chartier T., 1998. J. Euro. Ceram. Soc., 583: 18.
Huzzard R. J., Blackburn S., 1997. A water-based system for ceramic injection molding. J. Eur. Ceram. Soc., 17: 211 – 216.
Koh Y. H., Song J. H., Lee E. J., et al., 2006. Freezing dilute ceramic/camphene slurry for ultra-high porosity ceramics with completely interconnected pore networks. J. Am. Ceram. Soc., 89(10): 3089 – 3093.
Low D. K. Y., Li L., Corfe A. G., et al., 2001. Spatter-free laser percussion drilling of closely spaced arrayholes Int. J. Mach. Tool Manu., 41: 361 – 377.
Lumpp J. K., Allen S. D., 1997. Excimer laser machining and metallization of vias in aluminum nitride. IEEE Trans. Compon. Pack, B20: 241 – 246.
Mcnulty T. F., Shanefield D. J., Danforth S. C., et al., 1999. Dispersion of PZT for fused deposition of ceramics. J. Am. Ceram. Soc., 82: 1757 – 1760.
Meng G. Y., Wang H. T., Zheng W. J., et al., 2000. Preparation of porous ceramics by gel-casting approach. Mater. Lett., 45: 224 – 227.
Menon M., Decourcelle S., Ramousse S., et al., 2006. Stabilization of ethanol-based suspension. J. Am. Ceram. Soc., 89(2): 457 – 464.
Metev S. M., Veiko V. P., 1996. Laser Assisted Microtechnology. Berlin: Springer-Verlag.
Michael Scheffler, Paolo Colombo, 2005. Cellular ceramics: structure, manufacturing, properties and applications. Weinheim: Wiley-VCH, Chichester: John Wiley [distributor].
Murray A. J., Tyrer, J. R., 2000. New numerical evaluation techniques for laser processed ceramic substrates. J. Laser Appl., 12: 251 – 260.
Nagel A., 1988. Slip casting of sinterable silicon nitride (Ph.D. Thesis). University of Stuttgart.
Omateta O. O., Janney M. A., Strehlow R. A., 1991. Gel-casting—a new ceramic forming process. Ceram. Bull., 70(10): 1641 – 1649.
Orita N., 1988. Laser cutting method for high chromium steel and a device to carry out that method. US Patent 4774392.
Pham D. T., Gault R. S., 1998. A comparison of rapid prototyping technologies. Int. J. Mach. Tool Manu., 38: 1257 – 1287.

Shanti N., Araki K., Halloran J. W., 2006. Particle redistribution during dendritic solidification of particle suspensions. J. Am. Ceram. Soc., 89(8): 2444 – 2447.
Sharp C. M., Mueller M. E., Murthy J., et al., 1997. A novel antispatter technique for laser drilling: applications to surface texturing. In: Proceedings of the 6th International Congress on Applications of Lasers and Electro-Optics, San Diego, 41 – 50.
Studart A. R., Gonzenbach U. T., Tervoort E., et al., 2006. Processing routes to macroporous ceramics: a review. J. Am. Ceram. Soc., 89(6): 1771 – 1789.
Tari G., 2003. Gel-casting ceramics: a review. Am. Ceram. Soc. Bul., 82(4): 42 – 46.
Wang Aiguo, Luo Guanghua, 1990. Plant Physiology Communication, 6: 55.
Xiao Huashan, Fu Wenqing, Zhao Baolu, et al., 1991. Acta Biophysica Sinica, 8: 334.
Xiao Huashan, He Wenjin, Fu Wenqing, et al., 1999. Prog. Biochem. Biophys., 26: 180.
Xie Z. P., Huang Y., 2001. New progress of gel-casting technology application in ceramic processes. J. Ceram., 22(3): 142 – 146.
Yang Jinlong, Su Liang, Ma Liguo, et al., 2002. Key Eng. Mater., 224 – 226: 667.
Young A. C., Omatete O. O., Janney M. A., et al., 1991. Gel-casting of alumina. J. Am. Ceram. Soc., 74(3): 612 – 618.
Zhang Fazhi, Takeaki Kato, Masayoshi Fuji, et al., 2006. Gel-casting fabrication of porous ceramics using a continuous process, J. Eur. Ceram. Soc., 26: 667 – 671.

Appendix 1 The Testing and Analyzing Methods Used in Authors' Research

1 The flexural strength was measured by standard three-point bending method for ceramics on samples of size 4 mm × 3 mm × 40 mm with a loading speed of 0.5 mm/min. Weibull modulus was derived from 20 such tests on each material.
Fracture toughness K_{IC} was measured by four-point bending test with a single edge notch beam (SENB), on an universal test machine (Model AG-2000G, Shimazu, Japan) with a sample size of 4 mm × 6 mm × 30 mm and a loading speed of 0.1 mm / min. Each set of data involves 4 – 5 samples.

2 Zeta potentials of ceramic powders were determined by a Zetaplus Analyzer (Brookhaven Instrument Corp., USA) using an electrophoretic light-scattering method at 25℃. 1 mmol/L NaCl was used as the solvent and the powder concentration was maintained at 0.05 vol.%. The pH adjustment was carried out by using HCl and NaOH solutions.

3 The particle size distributions were measured by an X-ray centrifugal sedimentation technique (Instrument type: BI-XDC, made in Brookhaven Instrument Co., New York, USA).

4 Rheological behaviors of the slurry were tested using a rheometer (Model MCR300, Paar Physica USA, Glen Allen, Pa.).

5 The surface tension and wetting angel of the slurry were determined (Mode Processor Tensionmeter K12, Kruss USA, Charlotte, N.C).

6 SEM (Model S-450, Hitachi Corp., Tokyo) was used to characterize the microstructures.

7 Thermogravimetric analysis was carried out with the thermogravimetric analyser (TGA92-18, SETARAM Corp., France).

8 Density of specimens was determined by the Hg immersion method or by the Archimedes principle.

Notes: The above testing and analyzing methods are not specially indicated in the book. In order to avoid the iterative introductions in the every part, they are list altogether here. The other testing and analyzing methods used were marked in every chapter.

Appendix 2 The Raw Materials Used in Authors' Research

1 Alumina, Henan Xinyuan Alumina Industry Co. Ltd., >99.7 wt% purity, a mean particle size 2.9 μm, and specific surface area 0.434 m^2/g.

2 Monomer: acrylamide (AM), Mitsui Toatsu Chemical Inc., Japan, chemical purity.

3 Crosslinker: *N*, *N'*-methylenebisacrylamide(MBAM), Beijing Hongxing Chemical Factory, chemical purity.

4 Catalyst: *N*, *N*, *N'*, *N'*-tetramethylethylenediamine(TEMED), Xingfu Fine Chemical Institute, chemical purity.

5 Initiator: ammonium persulfate, $(NH_4)_2S_2O_8$, Beijing Chemical Reagent Company, chemical purity.

6 Dispersant: ammonium citrate(AC), the Beijing Chemical Reagents Company, chemical purity.

7 Dispersant: tri-ammonium citrate (TAC), the Beijing Chemical Reagents Company, chemical purity.

Notes: The above raw materials (alumina, monomer, crosslinker, catalyst and initiator) are not specially indicated in the book. In order to avoid the iterative introductions about raw materials in the every part, they are list altogether here. The other raw materials used were marked in every chapter.

Index of Terms

T

U

W

X

Y

Z

Index of Scholars

Postscript

When this book will be published we want to express the appreciation to our graduate students. This book "Novel Colloidal Forming of Ceramics" is also derived from their diligent work in our group. Their names and degrees are listed as below.

(1) Ph D students and their research directions

No.	Ph D Student	Periods	Research direction
1	Junhui Xiang	1997 – 2002	Aqueous gel tape casting
2	Longjie Zhou	1997 – 2002	Ceramic colloidal forming
3	Xiaolin Liu	1997 – 2002	Ceramic colloidal forming
4	Jianqing Dai	1997 – 2002	Ceramic colloidal forming
5	Liguo Ma	1998 – 2002	Ceramic colloidal forming
6	Zhiyong Yu	1999 – 2003	Ceramic freeform fabrication and computer-assisted forming technique
7	Jingtao Ma	1999 – 2003	Ceramic colloidal forming
8	Kai Cai	1999 – 2003	Colloidal forming techqniue for ceramics microbeads
9	Ma Tian	2000 – 2004	Ceramic colloidal forming
10	Yang Song	2001 – 2005	Forming technique for superconducting wires and tapes
11	Liming Zhang	2001 – 2005	Ceramic colloidal forming
12	Xuemin Cui	2002 – 2004	Ceramic freeform fabrication process
13	Ruifeng Chen	2002 – 2007	New processing for porous super-light and high-strenghth ceramics

(2) Master students and their research directions

No.	Master Student	Periods	Research direction
14	Yali Chen	1996 – 1998	Application of natural macro-molecule gel in ceramic *in-situ* solidification
15	Yu Jia	1997 – 2000	Application of sodium alginate *in-situ* solidification of ceramic suspension
16	Chunlei Ma	1999 – 2002	Aqueous gel tape casting of alumina
17	Lei Zhao	2000 – 2003	Ceramic colloidal forming
18	Liang Su	2000 – 2003	Pressure-induced colloidal injection molding for ceramics
19	Chunlei Dai	2001 – 2004	Key techniques in gelcasting process

(Continued)

No.	Master Student	Periods	Research direction
20	Tiechao Wang	2002 – 2005	New RTV rutile composites for ozone generator
21	Jiesheng Luo	2002 – 2005	Injection molding of super-fine zirconia
22	Yan Gao	2003 – 2005	Injection molding of zirconia and SiC and debindering process through water extraction
23	Xiaojun Liu	2003 – 2006	Injection molding of alumina
24	Xinyue Zhang	2004 – 2006	Properties and characterization of ceramic microbeads
25	Ming Yue	2004 – 2006	Wear behavior of the ceramic microbeads prepared by gel-beads forming
26	Xiaoqing Xi	2004 – 2007	Key techniques and equipments for industrialized production of high performance ceramic microbeads
27	Huang Lin	2005 – 2007	Particle-stabilized porous ceramics by gelcasting
28	Yu Ding	2006 – 2008	Influences of forming techniques on ceramic reliability
29	KeZeng	2006 – 2008	New process of laser machining of ceramic green bodies
30	Yuanyuan Cui	2007 – 2010	New process of laser machining of ceramic green bodies

Authors
Jun., 2010